Artificial Intelligence, Internet of Things (IoT) and Smart Materials for Energy Applications

This reference text offers the reader a comprehensive insight into recent research breakthroughs in blockchain, the Internet of Things (IoT), artificial intelligence and material structure and hybrid technologies in their integrated platform, while also emphasizing their sustainability aspects.

The text begins by discussing recent advances in energy materials, energy conversion materials using machine learning and recent advances in optoelectronic materials for solar energy applications. It covers important topics including advancements in electrolyte materials for solid oxide fuel cells, advancements in composite materials for Li-ion batteries, progression of materials for supercapacitor applications, and materials progression for thermochemical storage of low-temperature solar thermal energy systems.

This book:

- Discusses advances in blockchain, Internet of Things, artificial intelligence, material structure and hybrid technologies
- Covers intelligent techniques in materials progression for sensor development and energy materials characterization using signal processing
- Examines the integration of phase change materials in construction for thermal energy regulation in new buildings
- Explores the current happenings in technology in conjunction with basic laws and mathematical models

Connecting advances in engineering materials with the use of smart techniques including artificial intelligence, machine learning and Internet of Things (IoT) in a single volume, this text will be especially useful for graduate students, academic researchers and professionals in the fields of electrical engineering, electronics engineering, materials science, mechanical engineering and computer science.

Smart Engineering Systems: Design and Applications

Series Editor:
Suman Lata Tripathi

Internet of Things
Robotic and Drone Technology
Edited by Nitin Goyal, Sharad Sharma, Arun Kumar Rana, and Suman Lata Tripathi

Smart Electrical Grid System
Design Principle, Modernization, and Techniques
Edited by Krishan Arora, Suman Lata Tripathi, and Sanjeevikumar Padmanaban

Artificial *Intelligence*, Internet of Things (IoT) and Smart Materials for Energy Applications
Edited by Mohan Lal Kolhe, Kailash J. Karande, and Sampat G. Deshmukh

For more information about this series, please visit: https://www.routledge.com/

Artificial Intelligence, Internet of Things (IoT) and Smart Materials for Energy Applications

Edited by
Mohan Lal Kolhe, Kailash J. Karande and
Sampat G. Deshmukh

CRC Press
Taylor & Francis Group
Boca Raton London New York

CRC Press is an imprint of the
Taylor & Francis Group, an **informa** business

First edition published 2023
by CRC Press
6000 Broken Sound Parkway NW, Suite 300, Boca Raton, FL 33487-2742

and by CRC Press
4 Park Square, Milton Park, Abingdon, Oxon, OX14 4RN

CRC Press is an imprint of Taylor & Francis Group, LLC

ISBN: 9781032115023 (hbk)
ISBN: 9781032115030 (pbk)
ISBN: 9781003220176 (ebk)

DOI: 10.1201/9781003220176

Typeset in Times
by codeMantra

Contents

Preface

In the past decade, significant developments have taken place in computing techniques and materials development. There is a global trend to use artificial intelligence techniques for processing and analysing complex data for specific applications in decision-making. Also, Internet of Things (IoT) technology is widely used in developing integrated cyber-physical systems that have machine-to-machine networking for automation and system monitoring. Applications of computing techniques and perceptible advancements in material developments have emerged over the past few years. Recent technological advancements in materials significantly contribute to the development of sustainable technologies.

This monograph presents applications of artificial intelligence techniques in the context of opportunities for applications in selected cases using collected complex data for a meaningful purpose. Specific examples of IoT technologies are discussed to collect and process complex data to perform meaningful operations and take appropriate actions. Also, chapters on energy materials cover some aspects of fundamental and applied research on design/developments of materials for specific applications spanning metallurgy and energy conversion. The authors have attempted to present individual chapters as research manuscripts. However, the authors of respective chapters remain accountable for all enduring inaccuracies.

Finally, the editors wish to express their sincere gratitude to the authors and peer reviewers who have significantly contributed to the preparation of this book. We are also very grateful to our publisher CRC Press of Taylor and Francis Group, for providing an opportunity to publish a monograph on emerging topics of artificial intelligence, the Internet of Things and smart materials for energy applications.

Editors

Prof. (Dr.) Mohan Lal Kolhe is a full-time professor at the University of Agder, Norway. He is a leading renewable energy technologist with three decades of academic experience at the international level and previously held academic positions at the world's prestigious universities such as University College London (UK/Australia), University of Dundee (UK), University of Jyvaskyla (Finland) and Hydrogen Research Institute, QC (Canada). In addition, he was a member of the Government of South Australia's first Renewable Energy Board (2009–2011) and worked on developing renewable energy policies. He received the offer of a full professorship in smart grid from the Norwegian University of Science and Technology (NTNU). He won competitive research funding from the prestigious research councils (e.g. Norwegian Research Council, EU, EPSRC, BBSRC, and NRP) for his work on sustainable energy systems. He has published extensively in the Energy Systems Engineering. He has been invited by many international organizations for delivering expert lectures/keynote addresses. His research works in Energy Systems have been recognized within the top 2% of scientists globally by Stanford University's 2020 and 2021 matrices. He is working as editor for various journals published by Springer, Elsevier, AIP Publishing, and MDPI. He is an internationally recognized pioneer in his field, whose top 10 published works have an average of over 175 citations each.

Prof. (Dr.) Kailash J. Karande is the Director and Principal of SKN Sinhgad College of Engineering, Pandharpur, India. His research interest encompasses a wide range of activities with a focus on face recognition, and he has developed a comparative approach to ICA analysis. A strong electronics background combined with FRT has enabled him and his team to produce PCA-ICA modules for specific applications. He has published ten books and 80+ research papers (h-index 10, i-10 index 10) in various national and international journals. He is working as PI on a research project funded by Punyashlok Ahilyadevi Holkar Solapur University, Solapur, India. He received a European fellowship in Erasmus Mundus Program at Aalborg University, Denmark. He was honoured with Albert Nelson Lifetime Achievement Award. Also, he was awarded as the Best Principal from Punyashlok Ahilyadevi Holkar Solapur University, Solapur and many more for his contribution in the field of Education. He is acting as editor for various journals/books published by Lambert Publications, Springer, CRC Press/Taylor & Francis Group, IOP Publishing and AIP Publishing.

Dr. Sampat G. Deshmukh is working as Dean at SKN Sinhgad College of Engineering, Pandharpur. He has completed his Ph.D. in Sardar Vallabhbhai National Institute of Technology, an Institute of National Importance, Surat, India, giving his societal contribution in the field of Nanoscience and Nanotechnology. He has 24 years of experience in research, teaching and industrial settings. Dr. Sampat and his team have continuously worked in and contributed to the field of Material Science. He has published 15 research articles in journals, 15 papers in various national and international conferences and 2 book chapters (h-index 7, i-10 index 7). He is the Editorial Board Member of Electronics Science Technology and Application (ESTA) and *Frontiers of Mechatronical Engineering*. He received the Best Paper award in ICEM-2014 at Chennai and a consolation award in ICC-2015 at Bikaner, India. He is working as PI on a research project funded by Punyashlok Ahilyadevi Holkar Solapur University, Solapur. He is also acting as editor for conference proceedings/books published by IOP Publishing, AIP Publishing, CRC Press/Taylor & Francis Group, etc.

Contributors

Dheeraj Agarwal
Department of ECE
Maulana Azad National Institute of Technology
Bhopal, India

D. Sharmila Banu
Department of Electronics and Communication Engineering
Ultra College of Engineering and Technology
Madurai, India

A. K. Bhosale
Department of Physics
Raje Ramrao Mahavidyalaya, Jath
Kolhapur, India

Snehal A. Bhosale
E&TC Department
RMD Sinhgad School of Engg.
Pune, India

D. Janith Kavindu Dassanayake
Department of Electrical and Information Engineering
University of Ruhuna
Matara, Sri Lanka

Rohan S. Deshmukh
Department of Mechanical Engineering
SKN Sinhgad College of Engineering
Pandharpur, India

Sampat G. Deshmukh
Department of Engineering Physics
SKN Sinhgad College of Engineering
Pandharpur, India

Nilesh Dhobale
Department of Mechanical Engineering
RMD Sinhgad School of Engineering
Pune, India

M.H.M.R.S. Dilhani
Department of Interdisciplinary Studies
University of Ruhuna
Matara, Sri Lanka

Vaibhav.V. Dixit
RMD Sinhgad School of Engineering
Pune, India

Mayuresh B. Gulame
G. H. Raisoni College of Engineering and Management,
Pune, India

Makarand M. Jadhav
NBN Sinhgad School of Engineering,
Pune, India

Shubham Jain
Department of Mechanical Engineering
Maulana Azad National Institute of Technology
Bhopal, India

Shital B. Kale
Centre for Interdisciplinary Research
D.Y. Patil Education Society (Deemed to be University)
Kolhapur (M.S.), India

K. Kanagaraj
Department of Computer Applications
MEPCO Schlenk Engg. College
Sivakasi, India

Kailash J. Karande
Department of Electronics and Telecommunication Engineering
SKN Sinhgad College of Engineering, Pandharpur, India

K. Kavitha
Department of Electronics and Communication Engineering
Vellammal College of Engineering and Technology,
Viraganur, India

Vipul Kheraj
Department of Applied Physics
S. V. National Institute of Technology
Surat, India

Mohan Lal Kolhe
Faculty of Engineering and Science
University of Agder
Kristiansand, Norway

Konara Mudiyanselage Sandun Y. Konara
Department of Interdisciplinary Studies
University of Ruhuna
Matara, Sri Lanka
and
Faculty of Engineering and Science
University of Agder
Kristiansand, Norway

Nandkumar Kulkarni
SKNCOE
Pune, India

Sanjay S. Latthe
Department of Physics
Raje Ramrao Mahavidyalaya, Jath
Kolhapur, India

Chandrakant D. Lokhande
Centre for Interdisciplinary Research
D.Y. Patil Education Society (Deemed to be University)
Kolhapur (M.S.), India

Dhanaji B. Malavekar
Centre for Interdisciplinary Research
D.Y. Patil Education Society (Deemed to be University)
Kolhapur (M.S.), India

Dnyaneshwar S. Mantri
SIT Lonavala
Pune, India

Pushyamitra Mishra
Department of Mechanical Engg.
Maulana Azad National Institute of Technology
Bhopal, India

A. M. More
Department of Physics
K.N. Bhise Arts, Commerce and Vinayakrao Patil Science College
Kurduwadi, India

Altaf O. Mulani
SKN Sinhgad College of Engineering, Pandharpur, India

Sharad Mulik
Department of Mechanical Engineering
RMD Sinhgad School of Engineering
Pune, India

Mahmadasraf A. Mulla
Department of Electrical Engineering
Sardar Vallabhbhai National Institute of Technology
Surat, India

K. Nalini
Department of Microbiology
Ayya Nadar Janaki Ammal College
Sivakasi, India

Koki Ogura
Department of Electrical Engineering
Kyushu Sangyo University
Fukuoka, Japan

Ashish K. Panchal
Department of Electrical Engineering
Sardar Vallabhbhai National Institute of Technology
Surat, India

Vishal Parashar
Department of Mechanical Engineering
Maulana Azad National Institute of Technology
Bhopal, India

Pranav M. Pawar
BITS
Pillani, Dubai

Ramjee Prasad
Arhus University
Aarhus, Denmark

Kanchan Pujari
Department of Electronics and Telecommunication Engineering
Smt. Kashibai Navale College of Engineering,
Pune, India

O. Seifunnisha
Department of Physics
Avinashilingam Institute for Home Science and Higher Education for Women
Coimbatore, India

Mahesh Seth
TECH-CITY Research and Consulting (OPC) Pvt. Ltd.
Pune, India

J. Shanthi
Department of Physics
Avinashilingam Institute for Home Science and Higher Education for Women
Coimbatore, India

Raj Kumar Singh
Department of Mechanical Engineering
Rewa Engineering College, University Road
Rewa, India

S. S. Sonavane
Mechatronics Department
Symbiosis Skills and Professional University
Pimpri Chinchwad, India

Sanjay Soni
Department of Mechanical Engineering
Maulana Azad National Institute of Technology
Bhopal, India

Rajaram S. Sutar
Department of Physics
Raje Ramrao Mahavidyalaya, Jath
Kolhapur, India

R. Swathi
Department of Physics
Avinashilingam Institute for Home Science and Higher Education for Women
Coimbatore, India

Sanjay N. Talbar
Department of Electronics and Telecommunication Engineering
Shri Guru Gobind Singhji Institute of Engineering and Technology
Nanded, India

Praveen Kumar Tyagi
Department of ECE
Maulana Azad National Institute of Technology
Bhopal, India

Shweta Tyagi
Department of Electronics and Telecommunication Engineering
Shri Guru Gobind Singhji Institute of Engineering and Technology
Nanded, India

1 A Review of Automated Sleep Apnea Detection Using Deep Neural Network

Praveen Kumar Tyagi, Dheeraj Agarwal and Pushyamitra Mishra
Maulana Azad National Institute of Technology

CONTENTS

DOI: 10.1201/9781003220176-1

1.1 INTRODUCTION

Sleep apnea (SLA) is defined by the American Academy of Sleep Medicine (AASM) [1] as a sleep-related condition caused by respiratory difficulties while sleeping. It involves apnea and hypopnea, and has been caused by recurring episodes of reduced or absent cardiovascular flow of air caused by upper airway disintegration or even other respiratory failures [2]. The Apnea–Hypopnea Index (AHI) is known as the much more relevant measurement system for diagnosing the presence and seriousness of the condition, representing the number of apnea occurrences per hour of sleep. This abnormality is extremely common, with a worldwide estimated population of 0.2 billion people [3]. SLA can increase the likelihood of developing cardiovascular disease, hypertension, chronic kidney disease, diabetes, anxiety, and cognitive impairment [4]. This effect is responsible for the observed physiological approaches such as respiratory depression and enhanced sympathetic regulation, which in the longer-term impacts the rhythm of the heart. In certain situations, road accidents can sometimes occur as a result of drowsiness caused due to lack of sleep [5]. Obstructive sleep apnea (OSA) is a frequently reported sleep condition interspersed by frequent pharyngeal breakdown involving a partial or total interruption in the upward airway that prevents air from reaching the lung, disrupting airflow during sleep [6]. OSA is prevalent in heart failure patients, and it's directly connected to high blood pressure, coronary artery disorder, arrhythmia, and heart disease. Central sleep apnea (CEN) is a form of SLA that is less frequent than OSA. When there has been a significant loss of breathing but also no cardiovascular activities, these are called CEN episodes [7]. People with CEN have such a core nervous control disorder. This indicates either that the human brain of the respiratory system failed to stimulate respiratory or the impulse to breathe is not correctly distributed. Occurrences of cessation of respiratory, sudden awakenings during sleep, difficulties remaining asleep (Insomnia), sleepiness during daytime (hypersomnia), and snoring are also major symptoms of CEN. Mixed apnea (MIX) is a combination of the OSA and CEN apnea forms, characterized by a decrease in respiratory effort that results in an upward respiratory obstruction.

A full-night PSG report in a specialized sleep center lab is used to diagnose SLA [8]. Various physiological parameters, including respiration, oxygen levels, cardiopulmonary activity, and sleeping condition, are collected throughout that PSG. Following that, a professional research technician analyzes the information of the overnight data and examines every aspect of the waveform for the existence of sleep disorders [8]. Besides that, since PSG is an unpleasant, time-consuming, and costly standardized test, several current researchers have worked on the implementation of a compact and far less cost-effective OSA clinical diagnosis incorporating fewer physiological data, such as blood oxygen saturation, echocardiogram, heart and abdomen breathing signals, respiratory motions, and the cumulative signals [9]. Data indicates that many of the techniques focused on a singular sensor, single-lead electrocardiographic monitoring systems have the largest global detection by analyzing patient ratios [9].

To resolve these concerns, a variety of approaches have been developed in the research, the majority of which require two steps: handcrafting a collection of specific features and developing a sufficient classification model to provide an automated diagnosis. Some existing research has concentrated on feature processing, which usually requires the use of a specialized feature extraction process to obtain electrocardiogram (ECG) characteristics through ECG signal, Heart Rate Variability (HRV), R to R intervals, and ECG-derived respiration (EDR) signals [10]. Several researches used wavelet transforms to derive characteristics of ECG output waveform [11,12], as well as recurrence quantitative analysis of HRV statistical data to record the complex changes within the cardiopulmonary cycle during OSA [13]. Varon et al. [14] used orthonormal feature space projections to obtain two characteristics resulting from improvements within morphological characteristics of QRS sets, as well as heart rhythm and EDR. Changyue et al. [15] developed a hidden Markov model–based feature extracting process and a feature selection technique that used leave-one-out cross-validation to exclude irrelevant features. Neural network (NN) [11,13], k-nearest neighbor (kNN) [16], hidden Markov model [15], support vector machine (SVM) [3,14,16], linear discriminant analysis (LDA) [15], least-square SVM [16,17], and fuzzy logic [18] are some of the classifiers used in these approaches. These approaches have two issues: first, the unlimited number of features that may be selected, which is amplified either by the assumption that merging two or more distinct characteristics, selected as the highest, may not ensure a superior feature set, and second, the requirement for extensive knowledge within a particular area in order to build significant features. Deep-learning models, which automatically produce features by identifying correlation patterns from sensor reference signals, could also solve such two main problems. Even though existing studies in the research area of SLA diagnosis have also been conducted, including such testing devices for residential diagnosis of SLA [19], diagnostic models focused on cardiovascular and oximetry sensors [9], and various identification approaches [20]. However, there has never been a comprehensive study of the existing state of deep-learning–based approaches for identifying SLA. Furthermore, recently published studies indicate that deep networks outperform shallow networks mostly in order to increase accuracy.

The remainder of the chapter is structured as follows. Section 1.2 provides a review method, Section 1.3 briefly describes signal and dataset, the data preprocessing is discussed in Section 1.4, and data performance metric and classifiers are discussed in Sections 1.5 and 1.6, respectively. Discussion and conclusion of the chapter are presented in Sections 1.7 and 1.8, respectively.

1.2 MATERIALS AND METHODS

The study was performed to analyze research presented over the last decade, taking into consideration the period of time. Web of Science, IEEE Explorer, SENSORS, PubMed, MPDI, Science Direct, ELSEVIER, and arXiv were all used to perform a comprehensive search. Due to the various words and phrases of the word apnea, the search words were sleep apnea OR sleep apnea, along with the AND operation, semi-supervised function learning, unsupervised function learning; 'DBF', deep belief network; 'DNN', deep NN; 1D and 2D Convolution Neural Network (CNN), one-dimensional and two-dimensional convolution NN; autoencoder; 'RNN', recurrent NN; deep learning. Each chapter's title and abstract were evaluated, and most of the selected chapters were found to be important to the subject. The key terms apnea and deep network were evaluated as part of the inclusion criteria. Works that were not created specifically for the diagnosis of sleep disorders but could be modified for such a reason also were exempted. Primarily due to the significance, several articles were included while not appearing in the search, while some chapters were excluded despite appearing throughout the search. After not featuring with in search results, a related article has been included after reviewing the references of previously selected publications.

1.3 SIGNAL AND DATASET

All physiological inputs used in the systems were obtained either by researchers or their collaborators or have been previously collected and recovered from datasets (Table 1.1). The existence of apnea can also be detected using a combination of sensors and detectors. As a result, an evaluation of the most frequently implemented sensors and detectors was conducted, and the dataset used for the study was identified, providing prospective scholars with a summary of existing resources.

1.3.1 Based on Pulse Oxygen Saturation Signal

The level of oxygen in the bloodstream is measured by the pulse oxygen saturation level (SpO_2). It is defined as follows: SpO_2 increases and decreases in proportion to how effectively a patient breathes and how efficiently blood is distributed throughout the body. Due to the extreme repeated episodes of apnea, that are often followed by oxygen desaturations, substantial changes could be seen in patients with OSA. The Physionet Apnea ECG Database (AED) [21] and the University College Dublin Sleep Apnea Database (UCD Dataset) [22] were both obtained from the Physionet web page that is free and accessible. Just eight of the 70 recordings in the AED have SpO_2 signals [23]. These records, which ranged in length from 7 to 10 hours and include minute-by-minute annotation [23], were used. This dataset is sampled at a rate of 50 Hz. Pathinarupothi et al. [24], Almazaydeh et al. [25], and Mostafa et al. [26] have used it from Apnea ECG Dataset. Twenty-one men and four women were among the 25 referred (identified SLA) cases at UCD [22]. Hypopnea (HYP) or obstructive (OA), central (CA), and mixed (MA) apnea are all chronically annotated throughout this database. The SpO_2 signal was sampled every 8 seconds and was used by Almazaydeh et al. [25], Mostafa et al. [26], and Cen et al. [27].

Ravelo-García et al. [28] used data from Dr. Negrin's Sleep Unit and compiled a sample from Gran Canaria University Hospital of 70 subjects, which would be referred to as the HuGCDN2008 dataset. The total amount of sleep duration was at least 3 hours, and the SpO_2 signal was sampled at a rate of 50 Hz. VIASYS Healthcare, Inc.'s (Wilmington, MA, USA) computerized system was used to collect data. Biswal et al. [29] collected two types of data: the Massachusetts General Hospital (MGH) sleeping lab, which collected data from six channels of 10,000 objects, and the Sleep Heart Health Study (SHHS) database [30]. While the MGH database contains five sensors, only messages from four of them (pulse-ox, chest, abdomen, and AF) were used across

TABLE 1.1
Overview of Dataset Information

Paper	Year	Number of Patients	Sampling Frequency (Hz)	Window Size (Sec)	Signals Type	Database
[34]	2008	70	100	60	ECG-HRV signal	AED
[25]	2012	8	50	-	SpO_2	AED
[45]	2013	8	100	60	Respiratory signals (Chest and abdominal) and RI	AED
[28]	2015	70	50	60	SpO_2	HuGCDN2008 dataset
[46]	2016	8	100	-	Nasal Airflow signal	AED
[24]	2017	35 8	10,050	60	IHR-ECG SpO_2	AED
[26]	2017	8 25	508	60	SpO_2	AED UCD dataset
[35]	2017	35	100	60	ECG-IHR signal	AED
[48]	2017	100	32	30	Nasal airflow signal	MESA dataset
[27]	2018	23	8	1	Nasal AF, SpO_2, ribcage and abdomen motion	UCD dataset
[29]	2018	10,005,804	200	1	Chest belt, abdomen belt, SpO_2 and airflow	MGH dataset SHHS dataset
[52]	2020	1,507	32	60	Respiratory signals	MESA dataset

both datasets. In Mostafa et al. [31], data was obtained in the University Hospital of Gran Canaria Dr. Negrin using VIASYS Healthcare, Inc. in (Wilmington, USA) system. There are 70 patients in the dataset, ranging in age between 18 and 82 years, including 51 men and 19 women. This dataset was named HuGCDN2008 [32]. Only the SpO_2 signal, with a sampling rate of 50 Hz, has been used in the study. Choi et al. [33] obtained PSG sensor information among 129 subjects over the age of 20 at Seoul National University Hospital's Center for Sleep and Chronobiology. The values were recorded using the NEUVO method (Software Project Ltd., Victoria, Australia).

1.3.2 Based on Electrocardiogram (ECG)

ECG analyses the heart's electrical activity by connecting electrodes to the body and detecting minor electrical variations caused by ventricular depolarization and repolarization of ventricles of cardiac muscle. One of the most widely used datasets for ECG research is the Apnea ECG Database [21]. This dataset contains a distributed set of data of 35 records and a suppressed set of data of 35 records, all of which were digitized at a sampled rate of 100 Hz [23]. All 35 observations are labeled down to the minute to differentiate between apnea and nonapnea. Experts inspect and mark each section of a minute. There has been a total of 16,988-minute episodes. The apnea segment has 6,496 minutes, while the nonapnea segment has 10,492 minutes. This dataset was used by Novak et al. [34], Pathinarupothi et al. [35], Wang et al. [36], Li et al. [37], DeFalco [38], and Chang et al. [39].

In Banluesombatkul et al. [40] dataset collected from the recruitment for the osteoporotic fractures in men study (MrOS) sleep study (Visit 1) database. The data were obtained from 2,911 people aged 65 and above in six health centers through a baseline test. The ECG signals within that dataset were collected using Ag/AgCl patched sensors at a sample rate of 512 Hz. There were 545 patients, 364 of whom were normal and 181 of whom had extreme OSA.

Urtnasan et al. [41] used the normal night-time PSG observations of 86 patients to identify with OSA who were examined for this analysis. PSG data were collected at the Laboratory of Samsung Medical Clinic (Seoul, Korea) [41] SCSMC86. During night-time PSG, the datasets were collected using a solitary-lead transducer only at lead II and processed at a sample rate of 200 Hz. In the same laboratory, night-time PSG records were obtained by 82 patients (63 M and 19 F) [42], and 92 patients (74 M and 18 F) [43] provided the databases SCSMC82 and SCSMC92, respectively. Erdenebayar et al. [44] used PSG records from 86 patients at SCSMC. During the PSG processing, ECG data were captured at 200 Hz for around 6 hours. This research procedure was approved by the Samsung Medical Center's institutional review panel (IRB:2012-01063).

1.3.3 Based on Airflow (AF)

Thommandram et al. [45] used Apnea ECG Dataset [23] in the analysis of their study. The archive comprises 70 recordings, which were about 8 hours long and sampled at 100 Hz. However, only eight records (age: 43.3 ± 8.3 years, 7 Males and 1 Female) provide respiratory signs such as nasal breathing, abdominal pressure, and oxygen saturation. Every minute of the dataset has an annotation identifying the present state of apnea [23]. These apnea AF datasets were used by Minu et al. [46]. Choi et al. [33] have used PTAF-2 (Pro-Tech, Woodinville, USA) for assessing respiratory signal that used a pressure transducer to measure the respiratory signals of 129 patients. Biswal et al. [29] examined AF signals data from the sleep lab at MGH, which had 10,000 patients, as well as sets of data from the SHHS [30], which had 5,804 patients.

Cen et al. [27] evaluated oronasal AF from 23 subjects of UCD [22] dataset records. Haidar et al. [47,48] and McCloskey et al. [49] used respiratory AF signals collected at a sampling frequency of 32 Hz from the Multi-Ethnic Analysis of Atherosclerosis (MESA) database [50]. ElMoaqet et al. [51] used PSG data of 17 subjects collected at the Charité-Universitätsmedizin Berlin's Interdisciplinary Unit of Sleep Medicine The signal from the nasal airflow transducer was sampled at a rate of 256 Hz. Professional clinicians of sleep medicine annotated and classified SLA cases in the data collection. Haidar et al. [52] have used MESA dataset, which contains PSG records for 1,507 subjects. Every subject has a complete night recording along with a minimum 7 hours of information collected from respiratory signals [52]. Cardiopulmonary tracheal motions were used to detect respiratory episodes and determine the severity of SLA [50].

1.3.4 Based on Sound

A respiratory activity generates distinct sounds that could be selected to identify abnormalities. When such a subject is asleep, the microphone is the more widely used detector to capture respiratory sounds. Kim et al. [53] obtained complete PSG records of 120 subjects at the Seoul National University Bundang Hospital (SNUBH) sleep center. The respiratory sound was captured as part of PSG that used a PSG-embedded microphone mounted on the rooftop just above the subject's bed, at a height of 1.7 m. The recording's sampling rate was 8 kHz. Choi et al. [33] used recorded snoring sounds from 129 subjects in the Seoul national university hospital (SNUH) [33] dataset to use a microphone. Apnea is the result of rapid breathing and has been detected by a detector dependent on ribcage and abdominal movements. Biswal et al. [29] evaluated the chest and abdomen motions of 5,804 patients, in the MGH sleeping center of 10,000 observations, which are accessible in the SHHS dataset [30].

1.4 DATA PREPROCESSING

1.4.1 Raw Signal

Biswal et al. [29] used unprocessed respiratory signals: AF, pulse oximetry, chest, and abdominal belts as inputs for a recurrent CNN (RCNN). Almazaydeh et al. [25] used pulse oximetry SpO_2 signals as inputs to a NN. Cen et al. [27] used a number of physiological signals, including SpO_2, oronasal AF, and ribcage and abdominal motion. The number of signal samples in every instance is denoted by

$$N_S = w \times f_s \times N_{ch} \tag{1.1}$$

where w represents the window length of 5, f_s is the sampling frequency of 16 Hz, and the number of channels is denoted as N_{ch}. This sample length $N_S = 240$ was reconstructed into a matrix of 16×15 dimensions and it's zero-padded to produce a 16×16 size matrix as CNN input.

The raw respiratory signal can be used directly as a classification algorithm input, as proposed by Steenkiste et al. [54] by respiratory belts in the abdominal and thorax surrounding it, and Mostafa et al. [26] used signal from two distinct databases, to evaluate both database frequency resample to 1 Hz.

1.4.2 Filtered Signal

Noise reduction methods are effective in every SLA detection technique and are commonly used to collect respiratory relevant data. In order to minimize interference, source ECG signal was filtered using a 60 Hz notch filter, and then a 2-order bandpass Butterworth filter with a cut-off frequency of 5 and 35 Hz was used [40]. To minimize high-frequency noise and baseline drift from the raw signal, input signal filtered with first HBF (0.01 Hz) and the consecutively LPF (3 Hz) with fifth-order IIR Butterworth filter in [33], and finite impulse response (FIR) bandpass filter with a cut-off frequency of 0.5 and 30 Hz [44] were used.

Signals were altered from muscle movement during long periods of respiratory recording. To remove unwanted noise from the raw ECG data, a bandpass filter with a cut-off frequency of 5 and 11 Hz was applied [34]. All respiratory signals have the same sampling rate, the nasal pressure signal (NPRE) has been down-sampled to 32 Hz [51]. Preprocessing was performed on signals using an LP-FIR filter (0.5 Hz).

A third-order Butterworth bandpass filter of 0.1 and 25 Hz was used to filter tracheal motions of signals related to respiratory [50].

1.4.3 Signal Normalization

Choi et al.'s [33] adaptive normalization technique [55] was used to get that segment at which the magnitude of respiration was low due to a long period of asleep posture. For each second, the area $a(s)$ and standard deviation (SD) $\sigma(s)$ of the filtered signal were calculated [33]. Each sample was normalized dependent on the mean (μ) value and SD (σ) value of the common episodes for every subject to balance the differences in signal (nasal, thoracic, and abdominal) [47]. The normalized signal defined as,

$$X_{s,n} = \frac{x_{s,n} - \mu_{x_{m,s,n}}}{\sigma_{x_{m,s,n}}} \tag{1.2}$$

where $x_{s,n}$ is raw signal for subject s and type n (which is either thoracic, nasal, or abdomen) and m is the total number of normal subject's sample. McCloskey et al. [49] used only AF signal for all segments between the start of sleep and the end of sleep event, and also used normalized equation $X_{s,n}$ as (2) of type n =1 signal.

1.4.4 Spectrogram

Biswal et al. [29] have segmented each 30-second epoch into the subepochs of 2-second length with 1-second overlapping using the spectrogram characterization of EEG and EMG records. Thomson's multitaper technique has been used to approximate the power spectral density (PSD) for each 2-second sub-epoch. Erdenebayar et al. [44] used the short-time Fourier transformation to transform ECG into 2D spectrogram images in order to obtain 2D input as

$$y[k,f] = \sum_{l=0}^{n-1} w[l] \cdot y[k+l] \cdot e^{-jl(2\pi f/K)} \tag{1.3}$$

where k and f represent time and frequency when the signal was received, respectively, and $w[l]$ is window function with a 128-point window length and a 127-point overlap.

1.4.5 Feature Analyses

Almazaydeh et al. [25] used SpO_2 signal-based delta index and oxygen destruction of 3% as two oximetric features and one nonlinear metric signal. Ravelo-Garca et al. [28] used a combination of SpO_2 and RR interval features. Both time and frequency domain characteristics were derived from the RR sequence for the oxygen saturation signal. Variables dependent on symbolic parameters have also been used for power ratios in various frequency ranges. Novak et al. [34] also obtained time domain and frequency domain features parameters from ECG-based HRV data. DeFalco [38] used 12 standard parameters from HRV signal of ECG data, relating to time domain, frequency domain, and nonlinear domain parameter, developed by Kubios [38].

1.5 PERFORMANCE METRICS

A number of metrics are used to test the performance of SLA prediction algorithms. Table 1.2 presents the parameters and calculations used to determine the metrics. The true positive (TP), true negative (TN), false positive (FP), and false negative (FN) values have been used to measure some most similar parameters defined across most studies. The receiver operating characteristic (ROC) has been used to assess apnea detection efficiency of various classification thresholds, and the area under the ROC curve (AUC) is measured to assess overall performance [10]. Positive-predictive value and sensitivity are used to calculate the F-measure, also defined as the F1 Score. A weighted proportion, w_i, is added into the F1 to determine $F1_w$, where class index is i, N represents the total number and, n_i is the number of i class.

1.6 CLASSIFIERS

In general, the researchers used three types of deep networks (Table 1.3): CNN, RNN, and deep vanilla neural network (DVNN).

1.6.1 CNN

CNNs are a form of DNN that is commonly used in image recognition, speech recognition, and signal analysis. In comparison to handcrafted features, such networks

TABLE 1.2
Parameters for Evaluating Performance Metrics

Parameters	Calculation
Accuracy (Acc)	$\frac{(TP+TN)}{(TP+FP+TN+FN)}$
Specificity (Spe)	$\frac{TN}{(TN+FP)}$
Sensitivity (Sen)	$\frac{TP}{(TP+FN)}$
Positive-predictive value (PPV)	$\frac{TP}{(TP+FP)}$
Negative-predictive value (NPV)	$\frac{TN}{(TN+FN)}$
F_1	$2\frac{PPV*Sen}{PPV+Sen}$
F_{1w}	$\sum_i 2\cdot w_i \frac{PPV_I*Sen_i}{PPV_I+Sen_i}$ where $w_i = n_i/N$

TABLE 1.3
Different Classification Model Performance Analysis

Study	Analysis Model	Classifier[a] Type	Performance Metric (%)						
			Acc	Spe	Sen	PPV	NPV	AUC	Others
[31]	D1CNN (AED)	AH/N	88.49	93.80	73.64	-	-	-	-
		G	95.71	-	-	-	-	-	-
	D1CNN	AH/N	95.14	97.08	92.36	-	-	-	-
	(HuGCDN2008)	G	100	-	-	-	-	-	-
[32]	D1CNN (AED)	A/N	94.24	96.61	9..04	-	-	-	-
	D1CNN (UCD)	85.79	93.90	67.35	-	-	-	-	
	D1CNN (HuGCDN 2008)	89.32	94.60	74.75	-	-	-	-	
[33]	D1CNN	AH/N	96.6	98.5	81.1	87	97.7	-	-
	G (AHI ≥ 5)	96.2	84.6	100.0	95.1	100	99.0	F1 0.98	
	G (AHI ≥1.5)	92.3	86.5	98.1	87.9	97.8	99.0	F1 0.93	
	G(AHI ≥ 30)	96.2	96.2	96.2	89.3	98.7	100	F1 0.93	
[36]	D1CNN	A/N	90.97	83.04	95.50	-	-	88.0	-
	Residual network	94.39	93.04	94.95	-	-	-	-	
[39]	D1CNN	A/N	87.9	92.0	81.1	-	-	94.0	-
[41]	D1CNN	OA/H/N	90.8	87.0	87.0	87.0	-	-	$F1_w$ 87.0
[42]	D1CNN	OA/N	96.0	96.0	96.0	-	-	-	$F1_w$ 96.0
[44]	D1CNN	A/N	98.5	99.0	99.0	-	-	-	-
	D2CNN	95.9	96.0	96.0	-	-	-	-	
	LSTU	98.0	98.0	98.0	-	-	-	-	
	GRU	99.0	99.0	99.0	-	-	-	-	
[47]	D1CNN	OA/H/N	83.5	-	83.4	83.4	-	-	F1 83.4
[48]	D1CNN	OA/N	74.70	-	74.70	74.50	-	-	F1 75.0
[49]	D1CNN	OA/H/N	77.6	-	77.6	77.4	-	-	F1 77.5
	D2CNN		79.8		79.7	79.8	-	-	F1 79.7
[52]	D1CNN	A/N	80.78	-	81.73	80.78	-	-	F1 80.63
[27]	D2CNN	OA/H/N	79.61	-	-	-	-	-	-
[24]	LSTM	OA/N(SpO2)	95.5	-	92.9	99.2	-	98.0	-
	OA/N(IHR)	89.0	-	99.4	82.4	-	99.0	-	
	OA/N (SpO_2 +IHR)	92.1	-	84.7	99.5	-	99.0	-	
[34]	LSTM	A/N	82.1	80.1	85.5				
[35]	LSTM	G (IHR)	100	-	-	-	-	-	F1 100
[43]	LSTM	A/H/N	98.5	98.0	98.0	-	-	-	$F1_w$ 98.0
	GRU		99.0	99.0	99.0	-	-	-	$F1_w$ 99.0
[51]	LSTM	A/N(NPRE)	85.1	83.8	90.0	58.9	97.0	91.7	F1 71.2
	BiLSTM		85.0	83.7	90.3	58.8	97.1	92.4	F1 71.2
[25]	MHLNN	OA/N	93.3	100	87.5	-	-	-	-
[38]	MHLNN	OA/N	68.37	-	-	-	-	-	-
[53]	MHLNN	G	75.0	-	-	-	-	-	-

(Continued)

TABLE 1.3 (*Continued*)
Different Classification Model Performance Analysis

Study	Analysis Model	Classifier[a] Type	Performance Metric (%)						
			Acc	Spe	Sen	PPV	NPV	AUC	Others
[37]	SSAE	OA/N	G	84.7	82.1	88.9	-	-	86.9
			100	100	100	-	-	-	-
[26]	DBN (AED)	A/N	97.64	95.89	78.75	-	-	-	-
	DBN (UCD)	85.26	91.71	60.36	-	-	-	-	
[29]	RCNN (MGH)	G	88.2	-	-	-	-	-	-
	RCNN (SHHS)	G	80.2	-	-	-	-	-	-
[40]	CNN-LSTM-DNN	G	79.45	80.10	77.60	-	-	-	F1 79.07
[50]	CNN-LSTM	G (AHI 15)	84.0	87.0	81.0	-	-	-	F1 88.0

[a] A, apnea; H, hypopnea; N, normal; G, global or OSA severity; O, obstructive.

will effectively remove hierarchy trends in information using scaled learnable filters or kernels, requiring relatively less data preprocessing. A standard CNN is made up of many convolution operations, a normalization layer, dropout layer, nonlinear activation function, pooling layer commonly used as max and average, and a fully connected (FC) layer. For ECG classification, two different types of CNN are frequently used: one-dimensional CNN (DICNN) and two-dimensional CNN (D2CNN).

1.6.1.1 D1CNN

Mostafa et al. [31] used the greed-based optimization algorithm to optimize a CNN model effectiveness in detecting apnea events from a 1D SpO_2 signal. Two datasets of HuGCDN2008 and AED are used, which were split into 1-, 3-, and 5-minute windows with a sliding window of one-minute range. For apnea, the weighted-topology transfer with rough estimation was found to be the most accurate, with an accuracy of 88.49% for the HuGCDN2008 dataset [31] and 95.14% for the AED [23]. Mostafa et al. [32] used the non-dominated sorting genetic algorithm-II (NSGA-II) to develop a CNN for abnormal respiration based on signal segments varying in 1, 3, and 5 minutes on three different datasets. The model parameters of CNN were optimized by using NSGA-II, a multiobjective evolution model, which was used as an optimization [56]. The CNN used structure CONV layers, nonlinear layers (ReLU), FC layer with two outputs with softmax layer. Three distinct input widths and datasets were evaluated, with the effective one achieving an average accuracy of 94%. Choi et al. [33] used CNN and a single-channel NPRE for AHI identification. The input signals were adaptively normalized before being segmented at 1-second intervals by sliding a 10-second window. A CNN structure of three convolutions (CONV) layers, two max-pooling (MPL) layers, and two FC layers were used; the first CONV layer involves 15 filters, resulting in 15 filtered data signals. The first MPL layer then subsamples the filtered signal. This process was repeated also for second CONV-third CONV-second MPL layers, resulting in a total of 30 output signals. Such signals were further attached to the FC layers of 50 units and obtained an accuracy of 96.6%.

Wang et al. [36] used a residual and CNN network with RR intervals for the detection of apnea segments that used 30 subjects (training model) and five subjects (validation model). Dynamic autoregressive representation, a method of representing RR intervals through convolutional layers, was also used. The CNN has seven CONV layers and two dense layers, while the residual network has 33 CONV layers in 31 residual blocks and one dense layer. The accuracy of CNN and the residual network was 90.9% and 94.4%, respectively.

Chang et al. [39] used an SLA prediction model based on a deep D1CNN using 1D ECG data signals. CNN architecture is composed of ten equivalent CNN-based layers, a flattened layer, and four equivalent classification layers primarily made up of FC networks. The method obtained the highest accuracy of 87.9% for per-minute apnea diagnosis and 97.1% for per-recording classification. Urtnasan et al. [41] used a multiclass CNN structure to diagnose SAH cases in 10-second ECG segments from 86 individual SAHS subjects. The CNN model's signal was a 1D time-series data obtained from a single ECG signal. 1D-CONV, MPL (size of 1×2) layer, dropout ($p = 0.25$), and FC layers with softmax function were used in the CNN structure. The six-layer CNN obtained a mean F1-score of 87% and a mean accuracy of 90.8% for all classes. Urtnasan et al. [42] proposed an automatic system for detecting OSA from such a single-lead ECG dataset of 82 patients across 10-second ECG segments. CNN model consists of various CONV layers, an activation function (ReLU), MPL (size of 1×2), dropout (rate of 0.25) and FC layer by softmax function. The CNN with six layers obtained the best accuracy of 96% while using the F1 measure as a classification variable. Erdenebayar et al. [44] analyzed the utility of six deep-learning models, including D1CNN, in order to detect apnea events. The accuracy of 98.5% was achieved using a six-layer CONV with three kernels (sizes of 50×1, 30×1, and 10×1) and 1D pooling (Size 1×2), followed by the activation feature (ReLu), dropout layer (0.25), with FC layer.

Three 1D signals (oronasal AF, abdomen, and thoracic plethysmography) segmented into 30-second epochs were used to feed a D1CNN with three channel inputs for OSA detection by Haider et al. [47]. The CNN network consists of six CONV layers (32 filters), 3 MPL (1 × 2 sizes), and 1 FC softmax output layer. Both individual and paired combinations of channels were significantly outperformed by the collective usage of the three channels, which achieved an accuracy of 83.5%. Using D1CNN, Haidar et al. [48] proposed a method for detecting apnea–hypopnea episodes from a raw signal of nasal AF data, segmented in a 30-second sample, consisting of three 1-D CONV layers (30 filters, kernel size 5) followed by an MPL and an FC layer with a softmax activation feature. To optimize the descriptive cross-entropy optimal solution, the CNN was trained by using backpropagation and Adam optimizer [57]. This resulted in an average accuracy and F-score of 75% and 72%, respectively.

McCloskey et al. [49] analyzed the nasal AF directly normalized with 30 seconds episodes of 1,507 subject's dataset and detected OSA events using a CNN and wavelets. The normalized signal of each epoch was fed into the D1CNN. Each 30-seconds epoch had 960 attributes. Each CONV layer of kernel size (3 × 1) is preceded by MPL. The obtained accuracy of the CNN1D was 77.6%. The CNN model was used by Haidar et al. [52]. Data from the past 1 minute have been used in the analysis of data from respiratory signals over 30 second intervals to detect SLA instances. The number of CONV layers was assigned to 32 along with ReLU activation variable, and a pooling layer size of two (CONV-MPL) was used in a three-cascading structure. The predictive CNN model achieved the highest outcomes, with an accuracy of 80.78% as well as an F1 measure of 80.63%.

1.6.1.2 D2CNN

Cen et al. [27] developed a method to detect the events based on 1-second annotation using a mixture of SpO_2, oronasal airflow, and ribcage and abdominal motions on two layers of convolution and subsampling. The first convolution operation has six feature vectors, while the next CONV layer increases the feature vectors to twelve. Every convolutional layer is followed by two subsampling levels with scale sizes of 2. The output layer, which has three nodes, is attached to an FC layer, and with the UCD Database's leave-one-out cross-validation obtained an accuracy rate of 79.6%.

Erdenebayar et al. [44] analyzed multiple deep-learning models, including D2CNN, in order to detect apnea events from the single-lead ECG data signal. Seven-layer CONV, MPL (size 2×2), followed by the activation feature and dropout layer (p = 0.25), with FC layer, achieved an accuracy of 95.9%. McCloskey et al. [49] proposed a D2CNN approach for detecting three types of incidents (OSA, hypopnea, and normal) using wavelet transform spectrogram images of nasal AF as an input. CNN was comprised of two CONV layers of 56 filters of (10×10) kernel size followed by activation function layers, one maximum 2-D pooling layer (size of 2×2), an FC layer, and three nodes softmax layer. It obtained a 79.8% accuracy rate.

1.6.2 RNN

RNN is the optimal learning method for learning sequential data inputs and time-series data processing since its feedback and current value are feeding back across that network as well as the output includes the addition of variables in memory. For each stage in the process, the RNN collects information, updates its hidden layer, and makes a prediction. The analyzed works use two forms of RNN: long short-term memory (LSTM) and gated recurrent unit (GRU).

1.6.2.1 LSTM

The SLSTM-RNN technique was used to detect and rate OSA on a minute basis using only SpO_2 and IHR data in the work by Pathinarupothi et al. [24]. The structure is composed of three layers: the input layer containing 30 neurons (or 60 within a case for SpO_2), the hidden layer consisting of 32 system memory with one neuron each, and the output layer including two neurons containing two classes. Minute-to-minute IHR data had an accuracy of 89.0%, and SpO_2 had an even higher accuracy of 95.5%. Pathinarupothi et al. [35] used a two-layer SLSTM-RNN with double memory blocks for every layer of HRV in form of IHR with only a fixed value of beats. The existing algorithms used a single hidden layer of different LSTM blocks of between 2 and 32, a learn value of 0.1, for 150 epochs. This method achieved perfect precision and an F1 calculate value of 1. Novak et al. [34] used a three-layered LSTM of feature input data to measure apnea episodes using HRV signal with features as feedback. The network's hidden layers comprise five blocks, which are composed of seven memory cells, with an accuracy of 82.1%. Urtnasan et al. [43] used a single-lead ECG data segmented to 10-seconds incidents to diagnose apnea occurrences using a deep LSTM-RNN model, which consisted of six-layered recurrent layers. This approach achieved 98.5% accuracy and a weighted F1 measure of 98%, respectively. Erdenebayar et al. [44] analyzed multiple deep-learning models, including LSTM, in order to detect apnea events. The LSTM model, which consists of three layers of RNNs, each with 60, 80, or 120 memory cells, was proceeded by output function maps that performed batch normalization and dropout to achieve an accuracy of 98%.

ElMoaqet et al. [51] used an LSTM and bidirectional LSTM (BiLSTM) deep RNN network for extracting features and diagnosis of apnea episodes from single respiration channel inputs of oronasal AF signal, NPRE, and abdominal RIP signal (ABD). The numbers of memory blocks for the first and second LSTM layers were placed as 100 and 40, respectively, after testing and optimizing the training data sample. Furthermore, the first and second BiLSTM layers' memory cell counts were set at 100 and 40, respectively. In comparison to the LSTM, the BiLSTM-based model performed better with oronasal thermal AF and ABD signals. For NPRE signal accuracy, the LSTM and BiLSTM obtained 85.1% and 85%, respectively.

1.6.2.2 GRU

Urtnasan et al. [43] used a GRU-RNN to evaluate night-time ECG records from 92 subjects (74 for training and 18 for testing) to develop a method for automatically identifying sleep-disordered breathing (SDB) episodes. All ECG segments used to have the same length of 10 seconds and were formed as 2,000 × 1. The structure is made up of six layers of RNNs, each with distinct numbers of memory cells. GRU-RNN method obtained accuracy and the weighted F1 measure was both 99.0%. Erdenebayar et al. [44] analyzed multiple deep-learning models, including GRU, in order to detect apnea events. The GRU model, which consists of three layers of RNNs, each with 60, 80, or 120 memory cells, was proceeded by output function maps that performed batch normalization and dropout to achieve an accuracy of 99%.

1.6.3 Deep Vanilla Neural Network (DVNN)

The researchers of the reviewed papers used three forms of DVNN: multihidden layers NNs (MHLNN), stacked sparse autoencoders (SSAE), and DBF.

1.6.3.1 MHLNN

Almazaydeh et al. [25] used a three-layered feed-forward NN as a classifier for OSA detection using SpO_2 signal. It feeds an oxygen desaturation index (ODI), and delta index into a NN, which creates an output space separated into two regions: OSA positive and OSA negative and achieved epoch-based accuracy of 93.3% on Apnea database. DeFalco et al. [38] used evolutionary equations (EAs) in combination with a dataset subsampling method to minimize simulation time in order to select the optimal MHLNN hyperparameters. The optimal DNN structure obtained consists of two hidden layers of 23 and 24 units, respectively, and rectified linear unit as an activation function to achieve an accuracy of 68.37%. Kim et al. [53] used an MHLNN with two hidden layers and two dropout layers for four classes to analyze respiratory sounds during the night. Using ten-fold cross-validation, four window sizes of 2.5, 5, 7.5, and 10 seconds were measured, with 5 seconds outperforming the others. Obtained an accuracy of 88.3% in the four-category classification and 92.5% in the binary classification.

1.6.3.2 SSAE

A stacked autoencoder is an NN made up of many layers of sparse autoencoders, each with its output linked to the input of the next hidden layer. Li et al. [37] used a sparse autoencoder and a hidden Markov model to diagnose OSA. The R-peaks were detected using the Pan-Tompkins algorithm [58], and the physiologically irrelevant and redundant points were removed using the median filter [59]. For primary extraction of features, a hidden layer SAE unsupervised training was used first, followed by fine-tuning with a logistic regression layer. The highest accuracy of 84.7% was achieved by analyzing two deep network architectures.

1.6.3.3 DBN

Mostafa et al. [26] developed a deep-learning method for performing OSA classification by providing the raw SpO_2 input to a DBN. They proceed by calculating the preliminary weights, using an unsupervised learning technique. The weights are then standardized using supervised fine-tuning. The learning model has three layers: the first two are for the Boltzmann machine and the final is a softmax layer. The accuracy obtained from the UCD dataset [22] is 85.26%, while the accuracy obtained from the Apnea ECG Dataset [23] is 97.64%.

1.6.4 Combined DNN Approach

Biswal et al. [29] used a hybrid network of deep RCNN for sleeping AHI value based on spectrogram representation of PSG data of nasal AF, SpO_2, and abdominal signal.

The combined effect of CNN and RNN allows one to derive features from the input data using CNN (two filter dimensions of 100 and 200 sizes) and model long-term temporal correlations in the dataset with RNN. This hybrid approach obtained 80.2% and 88.2% accuracy with RCNN classifiers, respectively, using the SHHS and MGH databases [25]. Banluesombatkul et al. [40] used 15-second raw ECG records of AHI incidents with a sequence of D1CNN for automated extraction of features, RNN with LSTM for temporal data retrieval, and FC-DNN for function encoding out of a wide set of features. A stack of D1CNNs with 64, 128, and 256 units, and that each CNN layer is accompanied by batch normalization, the activation function (ReLU), and the MPL (size of 2) in order to find only essential features from the CNN layer. an LSTM structure with the same CNN units, with the intermittent dropout set to 0.4, and then stacked DNNs with five layers, and four hidden nodes for feature encoding, proceeded by the softmax layer. For OSA severity classification, an accuracy of 79.45% was obtained.

To diagnose physiological occurrences from respiratory-related motion inputs and calculate the AHI, a deep-learning method based on a combination of CNN and LSTM has been used in the work by Hafezi et al. [50]. A kernel scale of four with the strides of 1 is used for each CNN layer to obtain 64 features followed by batch normalization and the ReLU activation function, as well as dropout (probability of 0.5) on the second and fourth CNN layer outputs. The LSTM model has hidden units of

128 and produced the same no of hidden states and output states for each time cycle, with an FC layer with a sigmoid activation feature, achieving an accuracy of 84% to detect SLA (Table 1.3).

1.7 DISCUSSION

The single source signal used in apnea detection, SpO_2 [24–33], ECG [34–44], respiration [45–49,51,52], and sound [33,53] signals have all been analyzed as input variables. The ECG data signal has been the most widely used signal which could also be validated by Li et al [37]. The most widely used feature of the RR series is that over a source signal, ECG data has given the maximum global accuracy classification. Among all the reviewed sensor signals, pulse oximetry that measures oxygen saturation shows promising results for convenient and effective SLA identification. Therefore, the increased precision of ECG data signals could've been attributed to the use of publicly available data which are less likely to be affected by interference. Pathinarupothi et al. [24] obtained the highest performance by using SpO_2 signal in comparison to IHR from ECG for studies related to a single signal source. However, even with the use of various algorithms and datasets, a comparison between distinct input signal efficiency metrics is also not realistic for such an analysis. As stated in Refs [29,47,51], the use of more than one signal data from the source signal increases the accurate predictions of all model types. Furthermore, the major study aim of the majority of the experiments is to obtain a decent outcome with minimal detectors and sensors.

CNN has been the most widely applied classification method, that was based on both D1CNN and D2CNN. Generally, CNN was developed for 2D images with multiple channels as input, but that could be used for signals with only a single channel [39,42,44]. Several researchers used SpO_2 signal [32,39], nasal AF [33,42,49] or a mixture of SpO_2, oronasal AF, as well as ribcage and abdomen motions [47,52] and transformed these 1D signals into a 2D input for apnea detection and applying the CNN2D [27,49] directly. However, McCloskey et al. [49] evaluated both two and observed that 2D images spectrogram with the nasal AF outperformed raw 1-D data signal for CNN. Biswal et al. [29] reported a significant result, with RCNN and with spectrogram presentation obtained better accuracy. Wang et al. [36] observed that a residual network performed significantly better than a CNN with a reduced number of input samples. Urtnasan et al. used D1CNN [41,42] and RNN [43] information on the same data from the research lab, which was able to determine that RNN performed better than CNN. Evaluating the studies that have used LSTM [34,35] and GRU [43] can proceed to a similar kind of conclusion. ElMoaqet et al. [51] evaluated three expiratory signals using automatic function extraction of temporal features and classification of apneic episodes and obtained that the NPRE signal outperformed with deep BiLSTM-based method.

Optimization of hyperparameters is also an issue in deep network implementation. Several studies [41,42] found that simply boosting the number of neurons or layers throughout the hidden layers in the network did not improve results. Others attempted to figure out the solution by using a predefined search space [26]. DeFalco et al. [38] proposed an alternative approach in which the hypermeters have been chosen using evolutionary algorithms.

1.8 CONCLUSION

The study summarized the results of an analysis of several techniques for identifying SLA detection. The analysis method's purpose was to assess some most effective deep-learning approaches for detecting SLA from a different form of the input signal. It is found that a significant number of works have been presented throughout the last 4 years, highlighting the potential interest in this topic among the academic community. The comparative analysis of deep NNs as well as the parameter selection of deep-learning models is also an area of active study and a key discussion. In addition to discovering new deep learning techniques, other detection methods such as automatic SLA detection techniques are promising. As epoch-based diagnosis is difficult, an automatic and effective SLA detection solution would be required. When developing a new automatic SLA detection method, considering several data signals (ECG, oximetry, and respiratory) would result in an increased detection rate.

REFERENCES

1. Sateia, M. J. (2014). International classification of sleep disorders. *Chest*, *146*(5), 1387–1394.
2. Olson, E. J., Moore, W. R., Morgenthaler, T. I., Gay, P. C., & Staats, B. A. (2003, December). Obstructive sleep apnea-hypopnea syndrome. In *Mayo Clinic Proceedings* (Vol. 78, No. 12, pp. 1545–1552). Elsevier.
3. Zhang, J., Zhang, Q., Wang, Y., & Qiu, C. (2013, April). A real-time auto-adjustable smart pillow system for sleep apnea detection and treatment. In *2013 ACM/IEEE International Conference on Information Processing in Sensor Networks (IPSN)*, Philadelphia, PA, USA (pp. 179–190). IEEE.
4. Peppard, P. E., Szklo-Coxe, M., Hla, K. M., & Young, T. (2006). Longitudinal association of sleep-related breathing disorder and depression. *Archives of Internal Medicine*, *166*(16), 1709–1715.
5. Ancoli-Israel, S., DuHamel, E. R., Stepnowsky, C., Engler, R., Cohen-Zion, M., & Marler, M. (2003). The relationship between congestive heart failure, sleep apnea, and mortality in older men. *Chest*, *124*(4), 1400–1405.
6. Mannarino, M. R., Di Filippo, F., & Pirro, M. (2012). Obstructive sleep apnea syndrome. *European Journal of Internal Medicine*, *23*(7), 586–593.
7. Sezgin, N., & Tagluk, M. E. (2009). Energy based feature extraction for classification of sleep apnea syndrome. *Computers in Biology and Medicine*, *39*(11), 1043–1050.
8. Berry, R. B., Brooks, R., Gamaldo, C. E., Harding, S. M., Marcus, C., & Vaughn, B. V. (2012). The AASM manual for the scoring of sleep and associated events. *Rules, Terminology and Technical Specifications, Darien, Illinois, American Academy of Sleep Medicine*, *176*, 2012.
9. Mendonca, F., Mostafa, S. S., Ravelo-Garcia, A. G., Morgado-Dias, F., & Penzel, T. (2018). A review of obstructive sleep apnea detection approaches. *IEEE Journal of Biomedical and Health Informatics*, *23*(2), 825–837.
10. Mostafa, S. S., Mendonça, F., G Ravelo-García, A., & Morgado-Dias, F. (2019). A systematic review of detecting sleep apnea using deep learning. *Sensors*, *19*(22), 4934.
11. Lin, R., Lee, R. G., Tseng, C. L., Zhou, H. K., Chao, C. F., & Jiang, J. A. (2006). A new approach for identifying sleep apnea syndrome using wavelet transform and neural networks. *Biomedical Engineering: Applications, Basis and Communications*, *18*(03), 138–143.

12. Hassan, A. R., & Haque, M. A. (2017). An expert system for automated identification of obstructive sleep apnea from single-lead ECG using random under sampling boosting. *Neurocomputing*, *235*, 122–130.
13. Nguyen, H. D., Wilkins, B. A., Cheng, Q., & Benjamin, B. A. (2013). An online sleep apnea detection method based on recurrence quantification analysis. *IEEE Journal of Biomedical and Health Informatics*, *18*(4), 1285–1293.
14. Varon, C., Caicedo, A., Testelmans, D., Buyse, B., & Van Huffel, S. (2015). A novel algorithm for the automatic detection of sleep apnea from single-lead ECG. *IEEE Transactions on Biomedical Engineering*, *62*(9), 2269–2278.
15. Song, C., Liu, K., Zhang, X., Chen, L., & Xian, X. (2015). An obstructive sleep apnea detection approach using a discriminative hidden Markov model from ECG signals. *IEEE Transactions on Biomedical Engineering*, *63*(7), 1532–1542.
16. Sharma, H., & Sharma, K. K. (2016). An algorithm for sleep apnea detection from single-lead ECG using Hermite basis functions. *Computers in Biology and Medicine*, *77*, 116–124.
17. Atri, R., & Mohebbi, M. (2015). Obstructive sleep apnea detection using spectrum and bispectrum analysis of single-lead ECG signal. *Physiological Measurement*, *36*(9), 1963.
18. Álvarez-Estévez, D., & Moret-Bonillo, V. (2009). Fuzzy reasoning used to detect apneic events in the sleep apnea-hypopnea syndrome. *Expert Systems with Applications*, *36*(4), 7778–7785.
19. Mendonça, F., Mostafa, S. S., Ravelo-García, A. G., Morgado-Dias, F., & Penzel, T. (2018). Devices for home detection of obstructive sleep apnea: A review. *Sleep Medicine Reviews*, *41*, 149–160.
20. Mendonca, F., Mostafa, S. S., Ravelo-Garcia, A. G., Morgado-Dias, F., & Penzel, T. (2018). A review of obstructive sleep apnea detection approaches. *IEEE Journal of Biomedical and Health Informatics*, *23*(2), 825–837.
21. PhysioNet. Available online: www.physionet.org.
22. *St. Vincent's University Hospital/University College Dublin Sleep Apnea Database*. Available online: https://physionet.org/pn3/ucddb/.
23. Penzel, T., Moody, G. B., Mark, R. G., Goldberger, A. L., & Peter, J. H. (2000). The apnea-ECG database. In *Computers in Cardiology 2000*. Vol. 27 (Cat. 00CH37163), Cambridge, Massachusetts, USA (pp. 255–258). IEEE.
24. Pathinarupothi, R. K., Rangan, E. S., Gopalakrishnan, E. A., Vinaykumar, R., & Soman, K. P. (2017, August). Single sensor techniques for sleep apnea diagnosis using deep learning. In *2017 IEEE International Conference on Healthcare Informatics (ICHI)*, Park City, Utah, USA (pp. 524–529). IEEE.
25. Almazaydeh, L., Faezipour, M., & Elleithy, K. (2012). A neural network system for detection of obstructive sleep apnea through SpO2 signal. *Editorial Preface*, *3* (5), 7–11.
26. Mostafa, S. S., Mendonça, F., Morgado-Dias, F., & Ravelo-García, A. (2017, October). SpO2 based sleep apnea detection using deep learning. In *2017 IEEE 21st International Conference on Intelligent Engineering Systems (INES)*, Larnaca, Cyprus (pp. 000091–000096). IEEE.
27. Cen, L., Yu, Z. L., Kluge, T., & Ser, W. (2018, July). Automatic system for obstructive sleep apnea events detection using convolutional neural network. In *2018 40th Annual International Conference of the IEEE Engineering in Medicine and Biology Society (EMBC)*, Honolulu, Hawaii, USA (pp. 3975–3978). IEEE.
28. Ravelo-García, A. G., Kraemer, J. F., Navarro-Mesa, J. L., Hernández-Pérez, E., Navarro-Esteva, J., Juliá-Serdá, G., … Wessel, N. (2015). Oxygen saturation and RR intervals feature selection for sleep apnea detection. *Entropy*, *17*(5), 2932–2957.

29. Biswal, S., Sun, H., Goparaju, B., Westover, M. B., Sun, J., & Bianchi, M. T. (2018). Expert-level sleep scoring with deep neural networks. *Journal of the American Medical Informatics Association, 25*(12), 1643–1650.
30. Sleep Heart Health Study. Available online: https://sleepdata.org/datasets/shhs.
31. Mostafa, S. S., Baptista, D., Ravelo-García, A. G., Juliá-Serdá, G., & Morgado-Dias, F. (2020). Greedy based convolutional neural network optimization for detecting apnea. *Computer Methods and Programs in Biomedicine, 197*, 105640.
32. Mostafa, S. S., Mendonça, F., Ravelo-Garcia, A. G., Juliá-Serdá, G. G., & Morgado-Dias, F. (2020). Multi-objective hyperparameter optimization of convolutional neural network for obstructive sleep apnea detection. *IEEE Access, 8*, 129586–129599.
33. Choi, S. H., Yoon, H., Kim, H. S., Kim, H. B., Kwon, H. B., Oh, S. M., ... Park, K. S. (2018). Real-time apnea-hypopnea event detection during sleep by convolutional neural networks. *Computers in Biology and Medicine, 100*, 123–131.
34. Novák, D., Mucha, K., & Al-Ani, T. (2008, August). Long short-term memory for apnea detection based on heart rate variability. In *2008 30th Annual International Conference of the IEEE Engineering in Medicine and Biology Society*, Vancouver, Canada (pp. 5234–5237). IEEE.
35. Pathinarupothi, R. K., Vinaykumar, R., Rangan, E., Gopalakrishnan, E., & Soman, K. P. (2017, February). Instantaneous heart rate as a robust feature for sleep apnea severity detection using deep learning. In *2017 IEEE EMBS International Conference on Biomedical & Health Informatics (BHI)*, Orlando, Florida, USA (pp. 293–296). IEEE.
36. Wang, L., Lin, Y., & Wang, J. (2019). A RR interval based automated apnea detection approach using residual network. *Computer Methods and Programs in Biomedicine, 176*, 93–104.
37. Li, K., Pan, W., Li, Y., Jiang, Q., & Liu, G. (2018). A method to detect sleep apnea based on deep neural network and hidden markov model using single-lead ECG signal. *Neurocomputing, 294*, 94–101.
38. De Falco, I., De Pietro, G., Sannino, G., Scafuri, U., Tarantino, E., Della Cioppa, A., & Trunfio, G. A. (2018, June). Deep neural network hyper-parameter setting for classification of obstructive sleep apnea episodes. In *2018 IEEE Symposium on Computers and Communications (ISCC)*, Natal, Brazil (pp. 01187–01192). IEEE.
39. Chang, H. Y., Yeh, C. Y., Lee, C. T., & Lin, C. C. (2020). A sleep apnea detection system based on a one-dimensional deep convolution neural network model using single-lead electrocardiogram. *Sensors, 20*(15), 4157.
40. Banluesombatkul, N., Rakthanmanon, T., & Wilaiprasitporn, T. (2018, October). Single channel ECG for obstructive sleep apnea severity detection using a deep learning approach. In *TENCON 2018-2018 IEEE Region 10 Conference*, Jeju, South Korea (pp. 2011–2016). IEEE.
41. Urtnasan, E., Park, J. U., & Lee, K. J. (2018). Multiclass classification of obstructive sleep apnea/hypopnea based on a convolutional neural network from a single-lead electrocardiogram. *Physiological Measurement, 39*(6), 065003.
42. Urtnasan, E., Park, J. U., Joo, E. Y., & Lee, K. J. (2018). Automated detection of obstructive sleep apnea events from a single-lead electrocardiogram using a convolutional neural network. *Journal of Medical Systems, 42*(6), 1–8.
43. Urtnasan, E., Park, J. U., & Lee, K. J. (2020). Automatic detection of sleep-disordered breathing events using recurrent neural networks from an electrocardiogram signal. *Neural Computing and Applications, 32*(9), 4733–4742.
44. Erdenebayar, U., Kim, Y. J., Park, J. U., Joo, E. Y., & Lee, K. J. (2019). Deep learning approaches for automatic detection of sleep apnea events from an electrocardiogram. *Computer Methods and Programs in Biomedicine, 180*, 105001.

45. Thommandram, A., Eklund, J. M., & McGregor, C. (2013, July). Detection of apnoea from respiratory time series data using clinically recognizable features and kNN classification. In *2013 35th Annual International Conference of the IEEE Engineering in Medicine and Biology Society (EMBC)*, Osaka, Japan (pp. 5013–5016). IEEE.
46. Minu, P., & Amithab, M. (2016). SAHS detection based on ANFIS using single channel airflow signal. *International Journal of Innovative Research in Science, Engineering and Technology, 5*(7), 13053–13061.
47. Haidar, R., McCloskey, S., Koprinska, I., & Jeffries, B. (2018, July). Convolutional neural networks on multiple respiratory channels to detect hypopnea and obstructive apnea events. In *2018 International Joint Conference on Neural Networks (IJCNN)*, Rio de Janeiro, Brazil (pp. 1–7). IEEE.
48. Haidar, R., Koprinska, I., & Jeffries, B. (2017, November). Sleep apnea event detection from nasal airflow using convolutional neural networks. In *International Conference on Neural Information Processing* (pp. 819–827). Springer, Cham.
49. McCloskey, S., Haidar, R., Koprinska, I., & Jeffries, B. (2018, June). Detecting hypopnea and obstructive apnea events using convolutional neural networks on wavelet spectrograms of nasal airflow. In *Pacific-Asia Conference on Knowledge Discovery and Data Mining* (pp. 361–372). Springer, Cham.
50. Hafezi, M., Montazeri, N., Saha, S., Zhu, K., Gavrilovic, B., Yadollahi, A., & Taati, B. (2020). Sleep apnea severity estimation from tracheal movements using a deep learning model. *IEEE Access, 8*, 22641–22649.
51. ElMoaqet, H., Eid, M., Glos, M., Ryalat, M., & Penzel, T. (2020). Deep recurrent neural networks for automatic detection of sleep apnea from single channel respiration signals. *Sensors, 20*(18), 5037.
52. Haidar, R., Koprinska, I., & Jeffries, B. (2020, July). Sleep apnea event prediction using convolutional neural networks and Markov chains. In *2020 International Joint Conference on Neural Networks (IJCNN)*, Glasgow, United Kingdom (pp. 1–8). IEEE.
53. Kim, T., Kim, J. W., & Lee, K. (2018). Detection of sleep disordered breathing severity using acoustic biomarker and machine learning techniques. *Biomedical Engineering Online, 17*(1), 1–19.
54. Van Steenkiste, T., Groenendaal, W., Ruyssinck, J., Dreesen, P., Klerkx, S., Smeets C. and Dhaene, T., (2018). Systematic comparison of respiratory signals for the automated detection of sleep apnea. In *2018 40th Annual International Conference of the IEEE Engineering in Medicine and Biology Society (EMBC)*, 449–452.
55. Tian, J. Y., & Liu, J. Q. (2006, January). Apnea detection based on time delay neural network. In *2005 IEEE Engineering in Medicine and Biology 27th Annual Conference*, Shanghai, China (pp. 2571–2574). IEEE.
56. Deb, K., Pratap, A., Agarwal, S., & Meyarivan, T. A. M. T. (2002). A fast and elitist multiobjective genetic algorithm: NSGA-II. *IEEE Transactions on Evolutionary Computation, 6*(2), 182–197.
57. Kingma, D. P., & Ba, J. (2014). Adam: A method for stochastic optimization. *arXiv preprint arXiv:1412.6980.*
58. Pan, J., & Tompkins, W. J. (1985). A real-time QRS detection algorithm. *IEEE Transactions on Biomedical Engineering*, BME-32 (3), 230–236.
59. Chen, L., Zhang, X., & Song, C. (2014). An automatic screening approach for obstructive sleep apnea diagnosis based on single-lead electrocardiogram. *IEEE Transactions on Automation Science and Engineering, 12*(1), 106–115.

2 Optimization of Tool Wear Rate Using Artificial Intelligence–Based TLBO and Cuckoo Search Approach

Vishal Parashar, Shubham Jain and P. M. Mishra
Maulana Azad National Institute of Technology

CONTENTS

2.1 INTRODUCTION

The major aim of Artificial Intelligence (AI) is to improve a computer system that can solve critical and complex problems. Machine learning aims to develop a machine that can gain knowledge from data and produce accurate results. Therefore, it is clear that the scope of AI is very large, while the scope of machine learning is limited [1]. Optimization is a technique by which one can maximize desired results while minimizing harmful consequences. In the present work, cuckoo search algorithm (CSA) and teaching–learning-based optimization (TLBO) techniques were used to decrease the tool wear rate (TWR). The CSA is a metaheuristic method that uses the concept of the cuckoo's egg-laying strategy to find optimal results [2], whereas TLBO is a populace-based metaheuristic optimization algorithm. To find optimal results, the TLBO imitates the learning behavior of teacher and student. TLBO is becoming a very popular optimization technique, as it has no tuning parameters [3].

DOI: 10.1201/9781003220176-2

With continuous research for the development of new materials, metal matrix composites (MMCs) are becoming very popular as some suitable fabrication techniques give them some beneficial properties. MMCs are typically cast by dispersing some reinforcement material into the matrix material. MMCs are known for their elevated melting point, great thermal conductivity, extreme electrical conductivity, higher strength, high hardness, excellent stiffness, great specific modulus, decent wear resistance, small heat expansion coefficient, etc. MMCs have a large range of usages, for example in the aerospace industry, automobile sector, and biomedical sector. In this work, Al-TiB_2 MMC is cast by selecting aluminum as the matrix material and TiB_2 as the reinforcing material. Due to TiB_2's high strength and lightweight aluminum, Al-TiB_2 is highly used in industrial applications [4]. Palanisamy et al. [5] performed EDM on aluminum-MMC (Al-MMC) with input process parameters such as pulse off time, discharge current, and pulse on time. The responses were chosen as TWR, surface roughness (SR), and material removal rate (MRR). The optimization technique adopted was Taguchi and Gray Relational Analysis (GRA). It is found that discharge current was the greatest significant factor for SR and MRR. The future scope was identified as the influence of other process parameters to improve responses. Kamal Kishore et al. [6] have studied EDM on hybrid MMCs, which consider process parameters such as pulse on time, varying weight % of SiC and graphite reinforcement, and current. The output responses were MRR, TWR, and SR. Kamal Kishore found that current is the most important factor for MRR and SR, while the pulse on time is the most important factor for TWR. Selvarajan et al. studied spark EDM on Si_3N_4-TiN ceramic composite by considering dielectric pressure, pulse off time, current, spark gap voltage, and pulse on time as input process parameters. SR, TWR, MRR, and bottom radical overcut were chosen as output responses. GRA was used as an optimization technique. Research shows that spark gap voltage, pulse on time, and the current were the very important parameters [7]. Bhaskar et al reviewed EDM on Al-MMC. They determined that MRR increases with an increasing pulse on time and peak current [8]. Amina et al investigated the WEDM process on TiB_2 nanocomposite ceramics by selecting MRR and SR as the output response. The responses were improved by an artificial neural network (ANN) and genetic algorithm (GA). Amini concluded that the predicted outcomes were in consent with the actual results [9]. Kumar et al investigated EDM on aluminum (6,351) based MMC (5% silicon carbide and 5% boron carbide reinforced). Duty factor, current, pulse on time, and voltage were chosen as input process parameters, while power consumption, TWR, and SR were chosen as responses. Research shows that increasing the pulse current raises the TWR. SR rises with increasing current and pulse on time. This SR increased because of the matter of fact that by increasing the current and pulse on time, the crater size and waviness increase [10]. Some researchers chose the functionally graded aluminum matrix composite as a workpiece material to study EDM. In addition, the pulse on time, current, and zone position in the brake disc were chosen as input process parameters. The GRA technique was used as an optimization technique to reduce overcut, power consumption, SR, and electrode wear ratio. It is found that the pulse current is a very important process parameter that impacts the EDM process on MMC. However, the machinability of the composite is very difficult in the reinforcement particle-rich region [11].

2.2 ARTIFICIAL INTELLIGENCE

AI began with the development of the first electronic computer by Charles Babbage [12]. Subsequently, primitive robots were developed, which could perform some basic tasks automatically [13]. Then the Turing test came into existence, and it was successful in the field of AI. Newell and Simon discovered a 'heuristic search' for a combinatorial problem by some logistic algorithm [14]. Sefridge developed computer vision for face and pattern recognition [15]. The work began in the 1960s for natural language processing [16]. A wheeled robot named 'shakey' was introduced to machine learning by SRI internationally [17]. Rosenblutt formed the ANN to train data, extract knowledge from data, and model fitness [18]. Feigenbaum developed an expert system in the field of AI. This expert system only performs specific tasks such as medicine and chemistry diagnostics [19]. The Turing test can estimate the level of AI. Alan Turing developed this test in the 1950s. This test determines how well a machine can simulate human conversation. In this test, computer systems are placed in front of human subjective knowledge and experience. The Turing test is also known as the imitating test [20]. Thus, this Turing test has become a pivot in the field of AI.

It can be concluded that by the end of 2040, there will be robots as professors in colleges, as doctors in hospitals, and as drivers in cars. According to Bostrom, that period will be the era of transhumanism where both humans and machines create more powerful and efficient cyber logs, which will be superior to both humans and machines [21]. In general AI, a computer is capable of reading and analyzing problems like a human. It is becoming very difficult for scientists to develop machines of human intelligence type. This technology will be fully achieved by 2040 [22].

2.3 ELECTRIC DISCHARGE MACHINING (EDM)

Electric Discharge Machining (EDM) can machine electrically conductive materials without considering the hardness or toughness of the material. Since the tool (or electrode) does not contact the workpiece, no cutting force is generated on the workpiece. Thus, no mechanical stress is generated during the material removal process. Therefore, very delicate parts such as webs or feathers can be easily made without any distortion. Hardened workpieces can also be machined using EDM. It can produce irregularly shaped cavities, complex die sections, and molds, which are fast with great precision, accuracy, and very low cost. Uneven erosion in the surface created by EDM results in many small craters, which can help with oil holding and improved lubrication, particularly in sections in which lubrication is a hindrance. The high level of automation, workpiece changer, and tool changer allows the machine to operate seamlessly overnight or during weekends.

2.4 ANALYSIS OF VARIANCE (ANOVA)

The design of experiment approach is a well-known technique for process improvement. It is mainly divided into three main phases: planning, execution, and the analysis phase. The basic step is to fit a model to the response. The predicted model is then tested. Then validate the predicted model. A three-level three-factor

Box–Behnken design (BBD) type response surface methodology (RSM) was applied to perform and plan the experimentations. RSM develops a correlation between input process parameters and output response. RSM uses both mathematical and statistical techniques to analyze and model data. In this research, each process parameter has three levels—upper point (+1), center point (0), and lower point (−1). The RSM approach empowers us to inspect the dependence of more input factors on one response [23].

The RSM predicts the model of the output response by minimizing the variance error. In this research, a BBD was employed to estimate the TWR in terms of process parameters, namely, gap voltage, pulse on time, and peak current. In this work, there are 12 factorial runs and 5 center runs. Therefore, 17 runs are required to perform. The response y can be expressed in terms of process parameters by Taylor's expansion series. If the response is modeled as a first-order equation, it can be represented as Eq. (2.1):

$$y = \beta_0 + \beta_1 x_1 + \beta_2 x_2 + \beta_3 x_3 \cdots\cdots + \varepsilon \tag{2.1}$$

If there is a curvature effect in response modeling, it can be presented as Eq. (2.2):

$$y = \beta_0 + \sum_{j=1}^{n} \beta_j x_j + \sum_{j=1}^{n} \beta_{jj} x_j^2 + \sum_{j<k} \sum \beta_{jk} x_j x_k \tag{2.2}$$

where β is the coefficient of estimation to reduce the sum of squared error. x_1, x_2, x_3,... are process parameters [24].

2.5 OPTIMIZATION

Modern metaheuristic algorithms are being widely used to solve optimization problems. These metaheuristic algorithms mimic the best features of nature and biological systems that have evolved over millions of years. The two most important features of these algorithms are exploitation and exploration. During the exploitation phase, the algorithm seeks a solution near the current best solution, though in the exploration phase, the algorithm explores the solution randomly. Therefore, a combination of these characteristics makes the algorithm more perfect. In this work, CSA and TLBO have been used to reduce TWR.

2.5.1 Cuckoo Search Algorithm

Yang and Deb established a CSA and this algorithm outperforms other existing algorithms, for example, particle swarm optimization. In the CSA, the cuckoo's breeding manner is used to optimize responses.

The cuckoo is recognized not only for its attractive voice but also for its generative approach. Some species of cuckoo lay their eggs in different birds' nests and eliminate the eggs of another bird to raise the hatching possibility of their eggs. The basic brood parasitisms used by the cuckoo are:

- Intraspecific
- Cooperative
- Nest takeover

In this, when the host bird identifies the cuckoo's eggs, they remove those eggs or leave the nest and make another nest. A variety of other cuckoo species such as Tapera have a strategy such that they choose a host nest such that their eggs are similar to host eggs in color and pattern. This reduces the chances of cuckoo eggs being removed. While the timing of laying some cuckoo eggs is also wonderful, these cuckoo selects birds that are just laying eggs as the host nests. Therefore, the hatching probability of cuckoo eggs increases.

Once the cuckoo chicks come out of the egg, the mother cuckoo removes the eggs of the host bird from its nest. Therefore, the cuckoo chick's chances of getting food are increased. These cuckoo chicks also mimic the host chicks to increase their chances of getting food from the host bird.

Levy flight is the flight of birds or animals in search of food. It is random and is used mathematically in solving an optimization problem [2].

The following rules are applied in the cuckoo search optimization algorithm:

- All cuckoos lay an egg and put it in a random host nidus.
- The top nests and first-class eggs are chosen to be accepted in the following generation.
- The host eggs' number is set, so by some probability, the alien eggs are removed or the host bird builds a new nest.
- The host cuckoo uses levy flight to make a new nidus.

The pseudocode of the cuckoo search optimization algorithm is shown in Figure 2.1.

2.5.2 Teaching–Learning-Based Optimization

Teaching Learning Optimization (TLBO) has been developed by R.V. Rao. This algorithm mainly depends on the effect of learning between teachers and students. It is a populace-based technique and leads to achieving global solutions. The TLBO is split into two portions: the teacher phase and the learner phase. In this algorithm, first, the population is generated at random. Subsequently, the best solution is to select a teacher from a randomly generated population. Then compute the mean of each process parameter. When the teacher is chosen as the best solution, all random solutions move to the teacher as shown in Eq. 2.3.

$$Z_{\text{new}\,j} = Z_{\text{old}\,j} + r \times \left(Z_{\text{Teacher}} - (TF) \times \text{Mean}\right) \quad (2.3)$$

Cuckoo Search Algorithm

Start

Define objective function as $y = f(x_1, x_2........ x_d)$;

Initiate the host nest population and its size (n);

While (iter < maximum generation)

A cuckoo (say j) selects by levy flight & evaluate its fitness;

Select a host nest (say k) randomly;

If ($Fitness_j > Fitness_k$)

Replace cuckoo k by cuckoo j (due to maximization approach);

End

The bad nest gets abandon by some fraction (P_a);

The best solution keeps;

All solution gets ranked & current best solution finds;

End

The global solution found;

End

FIGURE 2.1 Pseudocode of Cuckoo Search Algorithm (CSA).

where $Z_{new\,j}$ is the new position of random population j, $Z_{old\,j}$ is the old position of random population j, $Z_{Teacher}$ is the best position of the teacher, TF is the teaching factor, mean is the mean of each process parameter, and r is the random number.

If it is obtained that a better solution exists than the existing solution from Eq. (2.3), it will be updated; otherwise, it will remain the same. This is called the teacher phase. In the learner's stage, the two particular solutions (called learners) compare and update in such a way that one learner moves to another learner if he or she has more knowledge because it is quite clear that the learning or understanding of two learners can be developed between, by group discussion, formal communication and presentation [3].

Therefore, if there are two learners (say j and k) such that $j \neq k$. If their knowledge as a fitness value is $f(Z_j)$ and $f(Z_k)$, then their position can be updated by Eqs. (2.4) and (2.5):

If $f(Z_j)$ is better than $f(Z_k)$,

$$Z_{new\,j} = Z_{old\,j} + r_j \times (Z_j - Z_k) \tag{2.4}$$

Else

$$Z_{new\,j} = Z_{old\,j} + r_j \times (Z_k - Z_j) \tag{2.5}$$

This process is repeated until the termination criterion is met and the best solution is, therefore, obtained. The pseudocode of TLBO is shown in Figure 2.2.

Teaching Learning Based Optimization

```
Start
Define objective function, process parameters and their range;
Initialize population in a random manner;
Select best solution as teacher;
    For iteration = 1: maximum iteration
        Teacher Phase
            For j = 1: number of populations
            Calculate the mean of population;
            Explore the position of population by equation 3;
                If (f (Z new j) is better than f (Z old j))
                Update Z old j as Z new j
                Else
                Not update Z old j
                End
            End
        Learner Phase
            For j = 1: number of populations
                For k = 1: number of populations
                    If (f (Z j) is better than f (Z k))
                    Update Z j by equation 4;
                    Else
                    Update Z j by equation 5;
                    End
                End
            End
    Update The best solution as a Teacher
    End
The global solution found
End
```

FIGURE 2.2 Pseudocode of Teaching–Learning-Based Optimization (TLBO).

2.6 EXPERIMENTAL DETAILS AND RESULTS

In this work, the material is prepared using a stir casting setup available at Nano-composite Material Laboratory MANIT Bhopal. Al-TiB_2 is manufactured with stir-casting, as stirring ensures the uniform chemical composition of the fabricated composites, as well as it is not a very expensive technique. Various salts such as potassium hexafluoro titanate (K_2TiF_6) and potassium tetrafluoroborate (KBF_4) are also mixed with aluminum alloy to improve the properties of Al-TiB_2. The stir casting machine mainly consists of various arrangements such as preheater, electric furnace, and motor-controlled stirrer, etc. The furnace is maintained at 900°C temperature before melting the metal. The die is preheated in a heating furnace up to 250°C to ensure proper solidification for the growth of desirable grains and ease of flow of molten liquid metal. Aluminum bars are placed inside a clay crucible, and the crucible is placed in a furnace to melt aluminum bars. After some time, all the solid metals were converted into molten liquid form. The stir casting motor is switched on, and by adjusting the slider rod, the stirrer is joined with a crucible. The stirring rod is designed to rotate at 600–700 rpm. When titanium-di-boride (TiB_2) particles begin to disperse into the liquid molten metal, stirring continues until the entire powder of TiB_2 is mixed inside the molten metal. Simultaneously, potassium hexafluoride salts are added to the particles to improve stability. Slag was also removed during continuous stirring. After ensuring the proper mixing of all the particles, the molten metal is poured directly into the preheated dies at the optimum speed. After proper solidification at room temperature, a cast composite of $116 \times 116 \times 14\,mm^3$ dimensions is obtained. In this study, copper is selected as an EDM electrode tool material. The workpiece after machining is shown in Figure 2.3. To investigate the rate of tool wear during spark EDM on the Al-TiB_2 composite, the BBD is used. Three levels of process parameters are presented in Table 2.1. The analysis of variance (ANOVA) for the TWR is shown in Table 2.2. From the ANOVA, it can be inferred that a linear model with logarithmic transformation is best suited to the TWR due to its high F value. The R^2 value is 0.9331. Therefore, it suggests the good fitness of the model. Adequate precision is 23.538, which is much higher than 4. Hence, the model can navigate the design space correctly. By ANOVA, it is found that there is nonnormality in the data. Therefore, it is necessary to change the response by Box–Cox transformation.

The regression equation for the TWR is shown in Eq. (2.6);

$$\text{Ln}(\text{TWR}) = -7.26946 + 0.073293 \times \text{Peak Current} + 0.017278 \times \text{Pulse On Time} + 0.006104 \times \text{Gap Voltage}. \quad (2.6)$$

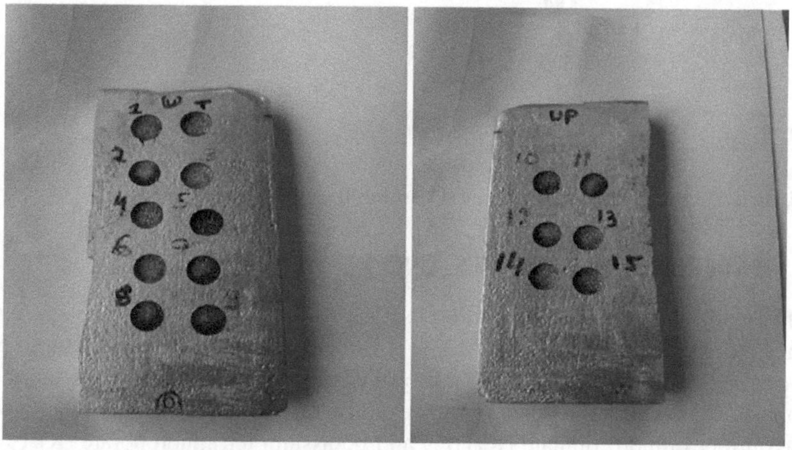

FIGURE 2.3 Al-TiB_2 composite after machining.

TABLE 2.1
Input Parameters and Their Levels for EDM

Input Parameter	Symbol	Unit	Level −1	Level 0	Level +1
Peak current (*A*)	IP	Ampere (A)	10	20	30
Pulse on time (*B*)	TON	Microsecond (μs)	20	30	40
Gap voltage (*C*)	V	Volts (V)	40	50	60

TABLE 2.2
Analysis of Variance for Tool Wear Rate (TWR)

Source	Sum of Squares	DF	Mean Square	*F* Value	Prob > *F*	
Model	4.57	3	1.52	60.47	< 0.0001	Significant
A	4.30	1	4.30	170.74	< 0.0001	
B	0.24	1	0.24	9.49	0.0088	
C	0.030	1	0.030	1.18	0.2962	
Residual	0.33	13	0.025			
Lack of fit	0.22	9	0.025	0.97	0.5558	Not significant
Pure error	0.10	4	0.026			
Cor total	4.89	16				

The perturbation plot, as shown in Figure 2.4, clearly indicates that by increasing the current, the TWR increases rapidly. After fitting the linear model to the TWR, it is optimized by two metaheuristic algorithms, namely the CSA and (TLBO) TLBO. The tuning parameters for CSA include the abandon probability (0.25), beta levy flight

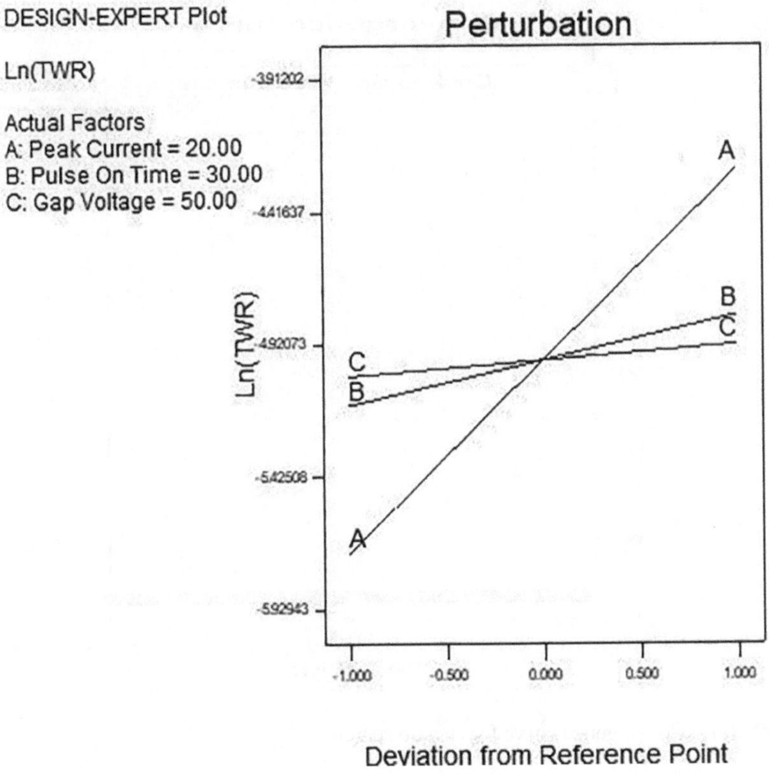

FIGURE 2.4 Perturbation plot for TWR.

index (3/2), and swarm size (20). For TLBO, the tuning parameters include teaching factor (1–2) and swarm size (20). For both the techniques, the maximum numbers of iterations chosen are one thousand. The optimal outcomes are presented in Table 2.3. However, tool wear optimization is a linear model optimization. Therefore, both algorithms give the same result, but from the convergence curve, it is clear that the TLBO achieves the best solution for this problem at a very early stage. The convergence curves of both algorithms are shown in Figures 2.5 and 2.6, respectively.

TABLE 2.3
Optimization Results by Cuckoo Search and TLBO

Type of Optimization	Optimum Position	Optimization Results
Cuckoo search optimization	10 A, 20 μs, 40 V.	TWR = 0.0026 mm³/min
Teaching–learning-based optimization	10 A, 20 μs, 40 V.	TWR = 0.0026 mm³/min

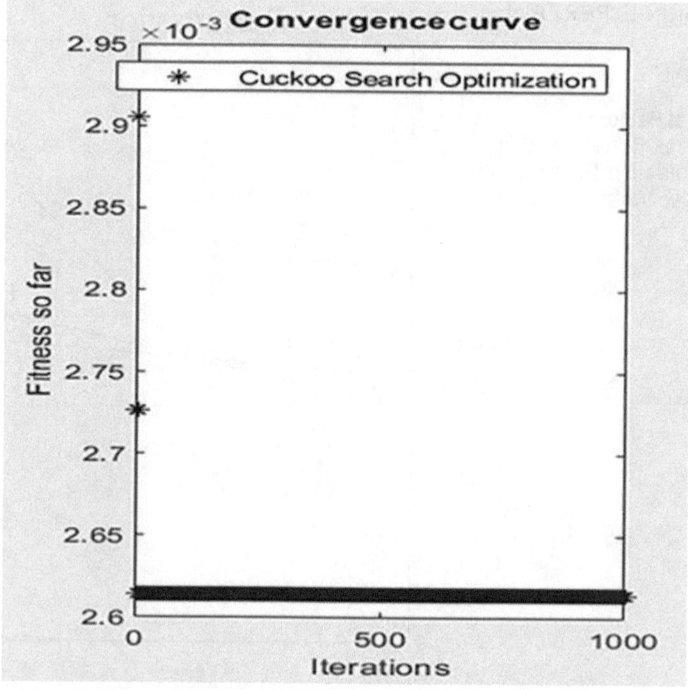

FIGURE 2.5 Convergence curve for cuckoo search.

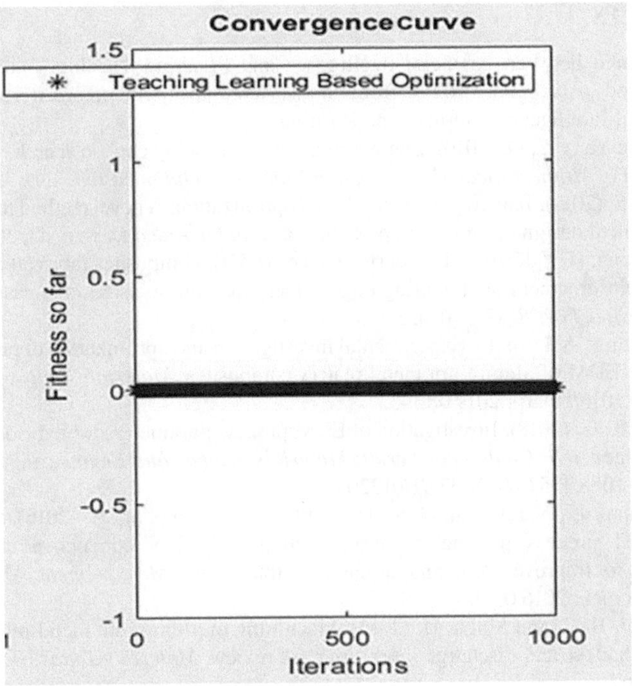

FIGURE 2.6 Convergence curve for TLBO.

2.7 CONCLUSION

In the present study, the TiB_2-reinforced aluminum-MMC is prepared using the most efficient technique of stir casting. The addition of some salts like K_2TiF_6 and potassium tetrafluoroborate improved the properties of the fabricated composite material. The selection of copper as a tool electrode provided good conductivity for the spark erosion process of machining. Experimental results show that a linear model with logarithmic transformation is appropriate to the TWR due to its high F value. The R^2 value is 0.9331. So, it suggests the good fitness of the model. It has been found that peak current is the most significant factor influencing the TWR. The peak current value should be kept low for better tool life. As the peak current increases, the TWR also increases. Other factors like gap voltage and pulse on time showed a relatively lesser impact on the TWR. TWR during EDM was optimized by cuckoo search and TLBO techniques. Both the optimization results showed the optimum value for TWR is 0.0026 mm³/min when the same values of input parameters are considered. The TLBO achieves the best solution for this problem at a very early stage, hence, found to be the best approach for the present study. However, the future scope of this work may be to consider more process parameters that affect TWR.

REFERENCES

1. Difference between Artificial intelligence and Machine learning. (n.d.). Javatpoint. Retrieved April 05, 2021, from https://www.javatpoint.com/difference-between-artificial-intelligence-and-machine-learning.
2. Xin-She Yang, S. D. (2010). Engineering optimisation by cuckoo search. *International Journal of Mathematical Modelling and Numerical Optimisation*, 1(4), 330–343.
3. Rao, V. S. (2011). Teaching–learning-based optimization: A novel method for constrained mechanical design optimization problems. *Computer-Aided Design*, 43, 303–315.
4. Xixi Dong, H. Y. (2019). High performance Al/TiB_2 composites fabricated by nanoparticle reinforcement and cutting-edge super vacuum assisted die casting process. *Composites Part B*, 177, 107453.
5. Palanisamy, A. D. (n.d.). Experimental investigation and optimization of process parameters in EDM of aluminium metal matrix composites. *Materials Today: Proceedings*. Doi: 10.1016/j.matpr.2019.08.145.
6. Ingole, K. G. (2018). Investigation of EDM process parameters for hybrid metal matrix composites. *IOP Conference Series: Materials Science and Engineering*, 377, 012204. Doi: 10.1088/1757-899X/377/1/012204.
7. Selvarajan, L., Narayanan, C. S., Jeyapaul, R. and Manohar, M. (2016). Optimization of EDM process parameters in machining Si_3N_4-TiN conductive ceramic composites to improve form and orientation tolerances. *Measurement*. Doi: 10.1016/j.measurement.2016.05.018.
8. Kandpal, B. C. and Singh, H. (2015). Machining of aluminium metal matrix composites with electrical discharge machining - A review. *Materials Today, Proceedings*, 2, 1665–1671.
9. Amini, H., Soleymani Yazdi, M. R. and Dehghan, G. H. (2011). Optimization of process parameters in wire electrical discharge machining of TiB_2 nanocomposite ceramic. *Proceedings of the Institution of Mechanical Engineers, Part B: Journal of Engineering Manufacture*, 225, 2220–2227. Doi: 10.1177/0954405411412249.
10. Kumar, S. S., Uthayakumar, M., Kumaran, S. T. and Parameswaran, P. (2014). Electrical discharge machining of Al(6351)–SiC–B_4C hybrid composite. *Materials and Manufacturing Processes*, 29(11–12), 1395–1400. Doi: 10.1080/10426914.2014.952024.
11. Uthayakumar, M., Babu, K. V., Kumaran, S. T., Kumar, S. S., Jappes, J. W. and Rajan, T. P. D. (2017). A study on the machining of Al–SiC functionally graded metal matrix composite using die–sinking EDM. *Particulate Science and Technology*. Doi: 10.1080/02726351.2017.1346020.
12. Shurkin, J. N. (1996). *Engines of the Mind: The Evolution of the Computer from Mainframes to Microprocessors*. New York: W. W. Norton & Company.
13. Walter, W. G. (1950). An electromechanical animal. *Dialectica*, 4, 42–49.
14. Allen Newell, J. C. (1959, June 15–20). *Report on a General Problem-Solving Program*. Paris: Unesco/Oldenbourg/Butterworths.
15. Selfridge, O. G. (1959). Pandemonium: A paradigm for learning. *Proceedings of the Symposium on Mechanization of Thought Processes* (pp. 511–531). London.
16. Reddy, D. R. (1976). Speech recognition by machine: A review. *Proceedings of the IEEE*, 64(4), 501–531.
17. Samuel, A. (1959). Some studies in machine learning using the game of checkers. *IBM Journal of Research and Development*, 3(3), 210–229.
18. Rosenblatt, F. (1957). *The Perceptron—A Perceiving and Recognizing Automaton*. New York: Buffalo: Cornell Aeronautical Laboratory.
19. Buchanan, E. A. (1993). DENDRAL and meta-DENDRAL: Roots of knowledge systems and expert system applications. *Artificial Intelligence*, 59(1–2), 233–240.

20. Fulmer, R. (2019). Artificial intelligence and counseling: Four levels of implementation. *Theory & Psychology*, 00(0), 1–13. Doi: 10.1177/0959354319853045.
21. Gyanendra Singh, A. M. (2013). An overview of artificial intelligence. *SBIT Journal of Sciences and Technology*, 2(1), 1–4.
22. Krishna, N. and Patel, S. R. (2020, February). Artificial intelligence and its models. *Journal of Applied Science and Computations*, VII(II), 95–97.
23. Thirumalpathy Padmanabhan, V. K. (2011). Experimental investigation on the operating variables of a near-field electrospinning process via response surface methodology. *Journal of Manufacturing Processes*, 13(2), 104–112.
24. Sergio Luiz Moni Ribeiro Filho, C. H. (2016). Influence cutting parameters on the surface quality and corrosion behavior of Ti–6Al–4V alloy in synthetic body environment (SBF) using response surface method. *Measurement*, 88, 223–237. Doi: 10.1016/j.measurement.2016.03.047.

3 Lung Tumor Segmentation Using a 3D Densely Connected Convolutional Neural Network

Shweta Tyagi and Sanjay N. Talbar
Shri Guru Gobind Singhji Institute of Engineering and Technology

CONTENTS

DOI: 10.1201/9781003220176-3

3.1 INTRODUCTION

Cancer poses a big threat to mankind and is one of the deadliest diseases worldwide. Lung cancer has the second highest incidence rate among all other types of cancer, that is, 11.4%, and it caused the highest number of cancer deaths worldwide with a mortality rate of 18%, according to GLOBOCAN report, 2020 [1]. The rates of incidence and mortality for most common cancers are shown in Figure 3.1.

Lung cancer is a type of cancer that starts in lungs and is defined as a tumor that is formed by uncontrolled growth of abnormal cells. It can be caused by either environmental factors or aging or family history of the patient, and some other factors include smoking, second-hand smoke, and exposure to radiation therapy. Lung tumor may be either benign or malignant. A lung nodule is a lesion with diameter range from 3 to 30 mm, generally having a regular shape and considered benign, but sometimes it may be malignant also, whereas a tumor with diameter greater than 30 mm is considered malignant tumor and is named as mass, having uneven shape. If a tumor is not able to metastasize or cannot invade the nearby tissues and other parts of the body, then it is known as benign tumor. But a malignant tumor is a type that can reach to other body parts as well, forming new tumors. The two main types of lung cancer are small-cell lung cancer (SCLC) and non-small-cell lung cancer (NSCLC), out of which NSCLC is the most prevailing type because it accounts for around 80%–85% of total lung cancer cases [2]. Both these types are differentiated on the basis of microscopical view, as SCLC is composed of much smaller cells compared to NSCLC. But the treatment to both types is almost similar, only drug may differ in some of the cases, or treatment may differ on the basis of stage of lung cancer. The treatment of lung cancer is decided by the doctors, considering various factors like the type of tumor (whether it is SCLC or NSCLC) and the stage of the tumor.

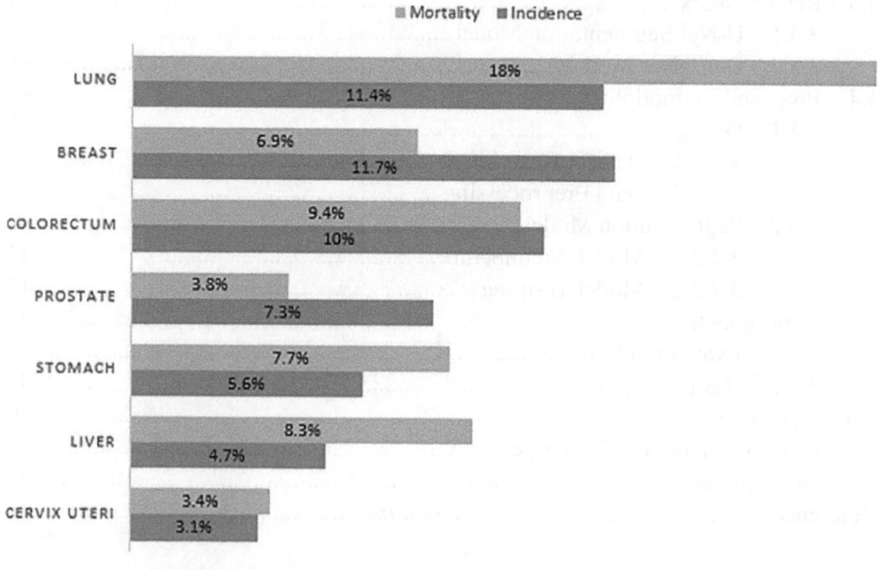

FIGURE 3.1 Incidence and mortality rates of eight top-most common cancers.

The patients' response to a specific treatment is also an important factor. The final decision is taken after consideration of all these aspects.

For diagnosis of a disease or an injury, there are several medical imaging techniques like X-rays, magnetic resonance imaging (MRI), computed tomography (CT) imaging, positron emission tomography (PET) imaging, sonography, mammography and so on. For patients having lung cancer, the diagnosis is usually done by using three medical imaging tests, X-ray images, CT scans and PET scans, out of which CT scan is preferred because this is more reliable as compared to the chest X-ray images in predicting the nature of the tumor, and it can provide more information about certain lung tumor features, including its size, shape and internal density. The CT scan is examined by the radiologists to detect the tumor region in the lungs. But this process is very time-consuming because one CT scan consists of hundreds of slices, and the number of lung cancer patients is also very high, due to which there is a huge burden on radiologists, especially in undeveloped or underdeveloped countries where there are not enough medical experts to examine the cancer imaging tests. To reduce this burden and to provide a second opinion to the doctors, several researchers have proposed different image processing and deep learning techniques for lung cancer detection and analysis. First step in automatic lung cancer detection is tumor segmentation, and if the tumor is segmented correctly then only it can be analyzed in a much better sense.

Computer-aided diagnosis (CAD) plays a vital role to analyze the CT images for lung cancer detection. Initially CAD systems were designed with the help of simple image processing techniques that required hand-crafted features, such as shape, volume, sphericity, eccentricity, solidity, and mean intensity for different tasks of diagnosis like tumor detection, segmentation and classification. But these techniques are not appropriate for working with very large datasets. The main issues with these techniques are the hand-crafted features and vulnerability of these methods to the discrepancy among different CT scans and screening parameters. Deep learning deals with this drawback and can handle large datasets with better performance, and it is proved to be very significant in providing state-of-the-art performance in various computer vision applications, including the field of medical image processing.

In this chapter, the authors have implemented a deep learning architecture based on densely connected convolutional neural network (CNN) for the segmentation of lung tumor using lung CT images. The proposed CNN is based on two networks, U-Net [3] and DenseNet [4]. The network layout is similar to the U-Net architecture consisting of encoder-decoder structure with dense connections between convolutional layers. In the proposed network, the authors have replaced the 2D convolutional and pooling layers with their 3D counterparts and also utilized dense connections between convolutional layers as inspired by DenseNet. The skip connections are provided between the encoder and decoder part to regenerate those features that were lost in encoder operations. The final segmentation masks are generated by the decoder. In data preprocessing, the image is resized from 512×512 to 256×256, and only 16 slices are taken as input at a time to solve the data imbalance problem and also to handle the memory capacity limit of the GPU (graphics processing unit).

The rest of this chapter is organized as follows: in Section 3.2, literature survey is presented. Related work is discussed in Section 3.3; in Section 3.4, the proposed approach is explained in detail. Results and evaluation criteria are provided in Section 3.5; in Section 3.6, discussion is given, and conclusion is made in Section 3.7.

3.2 LITERATURE SURVEY

3.2.1 Traditional vs Deep Learning Approaches

Some of the researchers have developed techniques based on traditional image processing like a technique based on K-Nearest Neighbors (KNN) algorithm lung cancer detection, and Support Vector Machine (SVM) classifier for classification of lung cancer is proposed by Sathishkumar et al. [5]. Singadkar et al. have presented an approach based on fuzzy clustering and non-negative matrix factorization for lung segmentation [6]. Other similar techniques using traditional image processing methods are proposed in [7–10]. Although the simple image processing techniques are capable of providing good results, they require lot of manual processing such as hand-crafted features, and they can process only a small amount of data at a time. Deep learning-based techniques deal with these limitations, as they can extract features automatically and can process a large amount of data. Researchers have designed different deep neural networks for computer vision applications [11–16]. These networks can be used for different purposes including medical image processing. A special architecture called U-Net was proposed by Ronneberger [3], specifically for the segmentation of the biomedical images. It was initially designed as a 2D network, for microscopic cell images which were 2D images, but since then it has been applied for many other applications as well, in various improved forms. Based on encoder-decoder structure of U-Net, numerous deep learning architectures have been designed by the researchers for the processing of medical images including lung CT images, like in [17–22].

3.2.2 Lung Nodule Detection

The lung nodule segmentation is an essential primary step toward early-stage lung cancer detection. For lung nodule detection and to reduce the number of false positives (FP), Setio et al. have developed a deep learning-based approach [23]. For this purpose, they have implemented a multi-view CNN. The proposed network is trained and validated on Lung Image Data Consortium-Image Database Resource Initiative (LIDC-IDRI) dataset [24], and a sensitivity of 90.1% for four FP is achieved. A deep learning architecture inspired from U-Net is proposed by Wenkai Huang et al. [25] for detection of pulmonary nodules using CT scan images. The network is named as noisy U-Net. They have added a noise to the hidden layers of the network during training so as to enhance its sensitivity to small nodules, and the leaky Rectified Linear Unit (ReLU) activation function is used. This network is trained and tested using the LUNA16 dataset [26] and Tianchi Medical AI Competition dataset [27], and it has attained 97.1% sensitivity to small nodules with 3–5 mm diameters. Wangxia Zuo et al. design a multiresolution CNN (MRCNN) for classification of lung nodules using lung CT images [28]. Their proposed MRCNN can capture high-quality features of the lung CT images from different resolutions from various depth layers of the network for lung nodule classification. The LUNA16 dataset was used for the experimentation in this study, and the proposed approach has achieved an accuracy of 0.973, an Area Under the curve (AUC) of 0.995 and a precision of 0.967. A deep deconvolutional residual network-based algorithm is proposed by Ganesh Singadkar et al. [29] for lung nodule segmentation. This network is trained on 2D CT images, and it can segment the lung nodules with different features, from these images. The summation-based long skip connections are implemented between the encoder and decoder part of the network, so as to preserve the spatial information that is destroyed during the pooling process and to capture the high-resolution features. This approach is trained and validated on LIDC-IDRI dataset, and 94.97% average dice score and 88.68% Jaccard index are achieved. An end-to-end CAD system for lung cancer diagnosis is designed by Ozdemir et al. [30]. The proposed system is composed of CADe and CADx, CADe for lung nodule detection and CADx for predicting the nodule malignancy. For nodule detection, they use fully CNN consisting of encoder-decoder structure with instance normalization, and for classification, classification network is used. They have evaluated their approach on LUNA16 and achieved a Competition Performance Metrics (CPM) of 0.921. Masood et al. [31] propose a technique based on multidimensional region-based fully convolutional network for lung nodule detection and classification. They have achieved a sensitivity of 98.1% and classification accuracy of 97.91% on LIDC-IDRI dataset. In [32], the authors have implemented a 2D-3D cascaded network for lung nodule detection, segmentation and classification and achieved 80% dice score. The techniques proposed for the segmentation of lung nodules may not work with significant performance for segmenting large and irregular size tumor. Therefore, for the task of lung tumor segmentation, it is required to design more robust architecture that can capture and segment the large-sized tumors.

3.2.3 Lung Tumor Detection

Various image processing and deep learning techniques have been proposed for lung segmentation, lung nodule detection, segmentation and classification. But for gross tumor segmentation, there is scarcity of literature. An image processing-based technique is used by Uzelaltinbulat et al. for lung tumor segmentation [33]. For image smoothing, they use median filtering, and for the segmentation task, Otsu thresholding is used. They have achieved 97.14% accuracy. Tripathi et al. also utilized simple morphological operations for tumor segmentation [34]. But by using simple image processing techniques, only a small amount of data can be processed. So, to perform segmentation of 3D medical images, now deep learning models are utilized, which once trained using a good amount of data can give a good accuracy on an unseen data. A CNN-based approach with dilated convolutions is utilized by M. Anthimopoulos et al. for the segmentation of lung tumor [35]. This proposed network is named as 2D LungNet. The dilated convolution is a special type of convolution in which kernel is widened by inserting holes in between the kernel elements. It is helpful in capturing more information with same computational cost, which results in high-resolution precision. Another approach by using dilated convolutions, but with a 3D CNN, is used by Hossain et al. They have implemented a 3D CNN architecture for tumor segmentation [36], using lung CT scan images. This architecture is known as 3D LungNet. The performance of 3D LungNet is evaluated on the NSCLC-Radiomics dataset [37], and it has achieved an average dice coefficient of 65.7% and a median dice coefficient of 70.39%. Some of the proposed networks are either computationally more complex or not good enough to segment large irregular shape tumors. Also, class imbalance issue needs to be addressed. To tackle all these issues, the authors have proposed a 3D CNN with dense connections to learn more high-level features of the tumor region, and the data is preprocessed before applying to the neural network by converting it into patches of 16 slices each, and also augmentation techniques are applied to the training data before passing through the deep learning model. This is done to solve the class imbalance problem.

The main objectives of the work proposed in this chapter are:

- To segment lung tumor region, which is an initial task in lung tumor analysis.
- To preprocess the imbalanced lung CT data in an efficient manner.
- To present a deep learning model to perform tumor segmentation.
- To achieve better accuracy for irregular-shaped tumor region.

Lung tumor segmentation is a subtask in lung cancer diagnosis process, and if done manually, it becomes very tedious job. With the help of automatic CAD systems, the burden on radiologists can be reduced. The authors are contributing toward this effort by proposing a deep learning-based approach for lung gross tumor segmentation, which can be used in CAD systems.

3.3 RELATED WORK

3.3.1 U-Net Segmentation Model

U-Net is a popular segmentation network specially designed for biomedical image segmentation, proposed by [3]. It is composed of two parts: down-sampling part and up-sampling part. The down-sampling part is also called encoder and up-sampling as decoder. The encoder is used to extract the features of the input image by using convolutional and pooling layers, while the decoder performs the operation of deconvolution by using up-sampling layers, so as to recover the original image size segmentation mask from the feature maps. And feature maps from encoder are merged with those of decoder to preserve the lost features of the image. Originally U-Net was designed as a 2D network that consists of 2D convolutional and pooling layers and can be used for segmenting 2D images.

Most of the medical images like MRI, CT and PET scans are 3D images. For the segmentation of these images, if 2D network is used, then the individual 2D slices are used as input to the network, which requires a lot of preprocessing, and the information in the adjacent slices may be lost. To deal with this issue, a similar architecture called 3D U-Net has been proposed by [17], in which 3D convolutional and 3D pooling layers are used in encoder and decoder, for volumetric segmentation. It can capture features from consecutive slices to get the volumetric segmentation mask. And similar merge connections are provided between encoder and decoder to access those features that may be lost during down-sampling operation.

3.3.2 DenseNet Model

DenseNet is a densely connected CNN designed by [4]. It was proposed based on the observation that deep CNNs are more efficient if short connections are provided between convolutional layers. This network consists of such connections that each layer is connected to subsequent layers in a feed forward manner. The input feature maps of each convolutional layer are taken from all its preceding layers, and its output feature maps are taken as input to all successive layers. Due to these connections, there are fewer chances that some of the feature will be lost during the convolutional and pooling operations. It has many advantages like, it solves the problem of vanishing gradient, reusing features enhances the parameter efficiency, which leads to better performance.

In this chapter, the authors have utilized both these networks to achieve high performance for lung tumor segmentation task from CT scans.

3.4 PROPOSED METHODOLOGY

The proposed approach consists of two steps: the first is data preprocessing in which data is processed so as to make it cleaner before applying to the CNN, and the second is tumor segmentation using 3D densely connected convolutional network, and the block diagram of this approach is shown in Figure 3.2.

3.4.1 Dataset

3.4.1.1 Dataset Description

The lung tumor dataset used in this study is NSCLC-Radiomics dataset, which is an NSCLC dataset and is taken from the cancer imaging archive website, and this dataset consists of 422 CT scans of patients having NSCLC type tumors, with total 52,073 number of slices [37]. The data is in the form of DICOM and RTSTRUCT files. The segmentation masks of gross tumor and some other anatomical structures (lung, heart, esophagus) are provided by the radiologists. In this study, the authors have utilized the segmentation masks of gross tumor for the task of tumor segmentation. The image size in each CT scan is 512 × 512, and the number of slices differs for each scan. There is a class imbalance issue in this dataset because a large portion of the CT slices does not contain any tumor region. And if such data is directly passed through the CNN, it will affect the performance of the network, because the network cannot learn well from an imbalanced data. So, it is required to perform some data augmentation techniques before applying the data to segmentation model. The data augmentation is done during the training procedure. For augmentation, authors have used different techniques, which are random shifting/rotation and horizontal flip. The description of the dataset is given in Table 3.1.

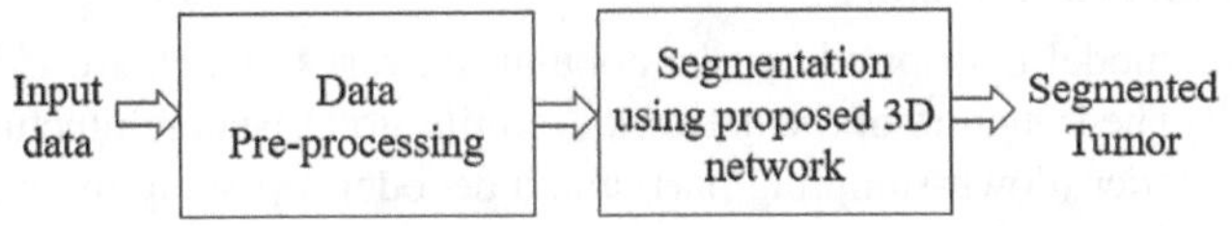

FIGURE 3.2 Block diagram of proposed approach for lung tumor segmentation.

TABLE 3.1
Data Description

Dataset	Number of CT Scans
Training	330
Validation	42
Testing	50
Total	422

3.4.1.2 Data Preprocessing

Out of total 422 CT scans, 330 scans are used for training, 42 scans for validation and 50 scans for testing purpose. Cross-validation by providing the validation data during training helps to control the overfitting problem. Overfitting is a situation where the training accuracy is much larger than the validation accuracy. This is the case when the network learns the training data very well but fails to generalize on validation data. Therefore, rather than testing the overfitted trained model each time, it is better to provide validation data during training, so that if validation accuracy is not improving, the training can be stopped and the model can be improved accordingly. It will save time because training the model each time from scratch till the end is very time-consuming process, so if it is known that the validation accuracy is not increasing, then training can be stopped and then model can be modified by using hyperparameter tuning or in some other way to remove the overfitting or underfitting problem.

First the authors have extracted the data from DICOM and RTSTRUCT files and converted it into NIFTI format using dcmrtrtruct2nii library [38]. This is due to the reason that the NIFTI images are easy to process rather than processing RTSTRUCT file. These NIFTI images are then resized to 256×256 size due to memory limitation of GPU. Then the 3D patches are extracted from all images by selecting only 16 slices in each patch, considering the fact that at least one slice in each patch contains tumor region, which results in a patch with shape $256 \times 256 \times 16$. This will also help to cope up with the computational memory of GPU. The intensity values in the images are clipped into desired Hounsfield range which is −1,000 HU to +400 HU for tumor segmentation task and then scaled between 0 and 1, which is required input range for CNN. The tumor segmentation masks are also resized to the same shape that is $256 \times 256 \times 16$ and are scaled between 0 and 1. Slices of one patch after preprocessing and with their corresponding masks are shown in Figure 3.3.

3.4.2 Segmentation Model

3.4.2.1 Model Architecture

The proposed model is inspired by two popular networks, which are U-Net [3] and DenseNet [4]. The structure of the network is similar to U-Net architecture that consists of an encoder (down-sampling part) and a decoder (up-sampling part). Instead of using dense layers, dense connections are provided between convolutional layers, so as to preserve the high-level features of the image.

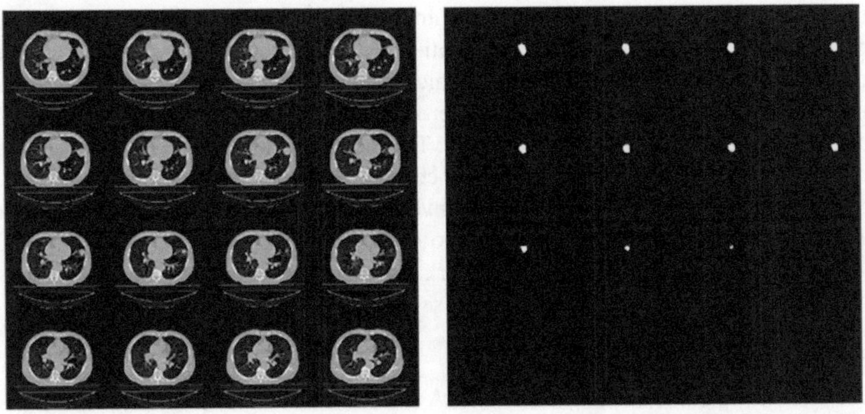

FIGURE 3.3 Visualization of the slices after preprocessing (a) with their corresponding masks (b).

The 3D convolutional layers with kernel size $3 \times 3 \times 3$ are used in both down-sampling and up-sampling part. The activation function ReLU is used in each convolutional layer except the last layer. ReLU is a rectified linear activation function that provides the output if input is positive, otherwise provides zero, and it helps in tackling vanishing gradient problem. Also, batch normalization is used after each convolutional layer to overcome the effect of internal covariate shift, which further helps in reducing the overfitting of the network. A dropout of 0.1 is used in the encoder, which also helps to reduce overfitting. The 3D max-pooling layers with pool size $2 \times 2 \times 2$ are used in the encoder part to down-sample the feature map. And in decoder part, same pool size 3D up-sample layers are used for up-sampling the features. The architecture of proposed network is shown in Figure 3.4. An input image of size $256 \times 256 \times 16$ is given to the network. The high-level features are generated by the encoder block, which are up-sampled by the decoder block to generate the

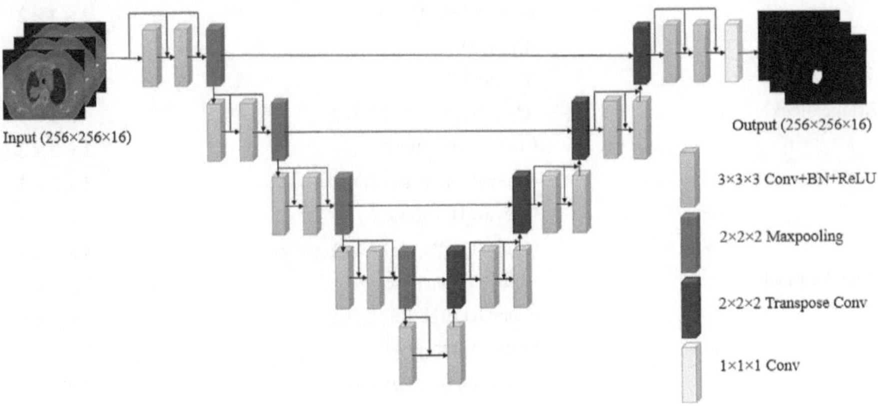

FIGURE 3.4 Architecture of proposed 3D dense convolutional neural network.

final tumor masks. The skip connections are provided between encoder and decoder parts to preserve the high-level information of the tumor. Dense connections are provided between convolutional layer, which means that feature maps from all prior layers will be passed as input to each layer, and each layer's output feature maps will be passed as input to all successive layers. This helps the network to learn the feature maps more accurately, and the performance of the network will be high.

Finally at the end, a pointwise 3D convolutional layer with kernel size $1 \times 1 \times 1$ and sigmoid activation function is used to get the tumor volume masks of the same size as the input provided to the network. The detailed outline of the network is presented in Table 3.2, in which block-wise layer details are given with kernel size.

3.4.2.2 Model Training

The network is trained on 16 GB NVIDIA P100 GPU using cuDNN 7.4 and CUDA 10.0 with 128 GB RAM. For training, 330 CT scans are used with 42 CT scans for validation purpose. As mentioned earlier, the data is having a class imbalance problem, that is, tumor-containing slices are very less as compared to non-tumor slices. Due to this issue, if the data is used without any augmentation, then the network will not be able to learn the tumor region correctly. Therefore, the authors have implemented some augmentation techniques that are random shifting, random rotation and horizontal flip. Input size of $256 \times 256 \times 16$ is set for the network as the data is processed to that size. The training is started with an initial learning rate of 10^{-3}, and Adam

TABLE 3.2
Details of Different Layers and Their Parameters in the Proposed Network

Block	Layers	Kernel Size
	(Conv3D, BN, ReLU)×2	$3 \times 3 \times 3$
	Maxpool3D, Dropout	$2 \times 2 \times 2$
	(Conv3D, BN, ReLU)×2	$3 \times 3 \times 3$
Encoder block	Maxpool3D, Dropout	$2 \times 2 \times 2$
	(Conv3D, BN, ReLU)×2	$3 \times 3 \times 3$
	Maxpool3D, Dropout	$2 \times 2 \times 2$
	(Conv3D, BN, ReLU)×2	$3 \times 3 \times 3$
	Maxpool3D, Dropout	$2 \times 2 \times 2$
Bridge block	(Conv3D, BN, ReLU)×2	$3 \times 3 \times 3$
	Conv3DTranspose	$2 \times 2 \times 2$
	(Conv3D, BN, ReLU)×2	$3 \times 3 \times 3$
	Conv3DTranspose	$2 \times 2 \times 2$
	(Conv3D, BN, ReLU)×2	$3 \times 3 \times 3$
Decoder block	Conv3DTranspose	$2 \times 2 \times 2$
	(Conv3D, BN, ReLU)×2	$3 \times 3 \times 3$
	Conv3DTranspose	$2 \times 2 \times 2$
	(Conv3D, BN, ReLU)×2	$3 \times 3 \times 3$
	Conv3D, Sigmoid	$1 \times 1 \times 1$

optimizer [39] with decay rates β_1 and β_2 with values 0.9 and 0.99, respectively, is used for optimizing the process. The batch size is set to 2, because it can be processed without any memory error by GPU. The dice coefficient is used as accuracy metrics and binary cross-entropy is used for loss calculation. The network is trained for 100 epochs, as after that there is no improvement in validation dice coefficient.

3.5 EXPERIMENTAL RESULTS

3.5.1 Evaluation Criteria

The accuracy is a measure of true positives (TP) as well as true negatives. As the dataset used in this study has more true negatives, accuracy is not a good measure in this case. Therefore, to evaluate the proposed approach, the dice coefficient or dice similarity coefficient (DSC) metrics is used. DSC measures the relative overlap between two samples and is used to compare the ground truth and the predicted mask, and it is considered a good performance measure for image segmentation task. DSC can be defined in two ways as shown in Eqs. (3.1) and (3.2).

$$\text{DSC} = \frac{2\left(|A| \cap |B|\right)}{|A| + |B|} \tag{3.1}$$

$$\text{DSC} = \frac{2\text{TP}}{2\text{TP} + \text{FP} + \text{FN}} \tag{3.2}$$

In Eq. (3.1), A is the predicted mask and B is the ground truth mask. And in Eq. (3.2), DSC is defined in terms of TP, FP and false negatives. DSC ranges between 0 and 1, where 0 means no overlap and 1 means perfect overlap. And dice loss is calculated as shown in Eq. (3.3):

$$\text{Dice loss} = 1 - \text{Dice coefficient} \tag{3.3}$$

Dice loss also ranges between 0 and 1.

3.5.2 Results

The proposed network is trained using a 16 GB NVIDIA P100 GPU with cuDNN v7.4 and CUDA 10.0 with 128 GB RAM. For training, 330 CT scans are used with a cross-validation data of 40 CT scans, and for testing, 50 CT scans are used, and the network was able to obtain a dice coefficient of 0.6734, which is better than the previous-mentioned methods, as given in Table 3.3.

The results of segmentation for three different cases are given in Figure 3.5, which shows that the predicted masks obtained using the proposed approach are much closer to the ground truth masks.

TABLE 3.3
Dice Coefficients of Different Architectures Along with Our Proposed Method

Model	Dice Coefficient (%)
3D U-Net [17]	58.48
2D LungNet [32]	62.67
3D LungNet [33]	65.77
Proposed method	67.34

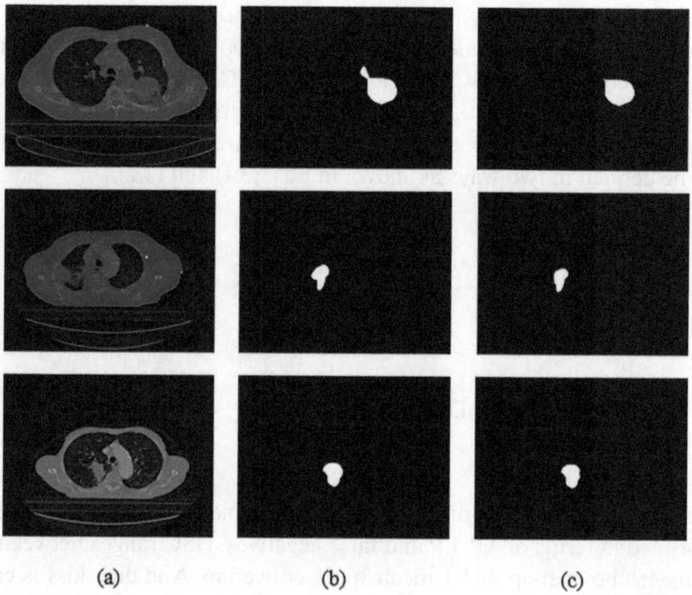

FIGURE 3.5 Segmentation results for three cases: (a) CT images, (b) ground truth masks and (c) predicted masks.

3.6 DISCUSSION

Various approaches have been presented for lung nodule detection, segmentation and classification. To analyze the gross tumor, which is already in its late stages, only a few literatures are available, which are mentioned in Section 3.2. The approach proposed in this chapter is for large-sized tumor segmentation in lung CT images. In later stages of lung cancer, the tumor size becomes large, irregular and it is mostly adjacent to lung walls, which is very difficult to segment. For tumor growth analysis, it is required to segment these large-sized tumor regions. The data used in this research work is highly imbalanced, which is not good for training. To resolve the issue of data imbalance, the authors have implemented a preprocessing technique in which 16 slices of a CT scan image are passed at a time to the network, so that each input contains some suspicious region in lung. The segmentation model is designed using a densely connected CNN, which performs well because high-level features are preserved due to dense connections. But there are some limitations, such as by using preprocessing, the data imbalance problem was resolved up to some extent but not completely. Therefore, some generative adversarial techniques can be utilized for creating false samples, just to balance the positive and negative classes like done in [40], and the segmentation model can be modified accordingly.

3.7 CONCLUSION AND FUTURE SCOPE

Lung cancer diagnosis using the medical images is a very challenging task, especially for large-sized tumor because they are spread over the large portion of the lungs and sometimes adjacent to lung walls. The authors have made an effort to segment the lung tumor using a 3D densely connected CNN consisting of dense connections between convolutional layers. The proposed network is evaluated on NSCLC-Radiomics dataset, which is publicly available, and a dice score of 67.34% is achieved. Many augmentation techniques are utilized and also batch normalization is used to overcome the overfitting problem. The data imbalance problem is resolved by using the proposed preprocessing technique. The proposed approach can be utilized in CAD systems for clinical lung cancer diagnosis, which will help the medical experts in better disease analysis.

Although the proposed approach is able to perform better than some of the previous approaches, still there are many possibilities on the basis of which the approach can further be modified. One such example is tackling the class imbalance problem by using some powerful augmentation techniques like using generative adversarial networks [41] for creating positive samples to balance both tumor and non-tumor classes. The 3D CNN has an advantage that it will provide better volumetric tumor segmentation by considering the consecutive slices, but it comes with large number of training parameters, making it computationally less efficient. So, in future, the authors want to explore generative adversarial networks and also some 2D approaches to increase the performance.

ACKNOWLEDGMENT

The authors are grateful to the Center of Excellence in Signal and Image Processing research lab in Shri Guru Gobind Singhji Institute of Engineering and Technology, Nanded, for providing the support and resources to carry out this research.

REFERENCES

1. *Global Cancer Observatory: Cancer Today.* Lyon: International Agency for Research on Cancer. Retrieved from the Global Cancer Observatory website: https://gco.iarc.fr/today.
2. American Cancer Society. (2021) About lung cancer. Retrieved from the American Cancer Society website: https://www.cancer.org/cancer/lung-cancer/about/what-is.html.
3. Ronneberger, O., Fischer, P. & Brox, T. (2015) U-net: Convolutional networks for biomedical image segmentation. In *Proceedings of International Conference on Medical Image Computing and Computer-Assisted Intervention,* Springer, Munich, Germany, pp. 234–241.

4. Huang, G., Liu, Z., Pleiss, G., Van Der Maaten L. & Weinberger, K. Q. (2019) Convolutional networks with dense connectivity, *IEEE Transactions on Pattern Analysis and Machine Intelligence*, Doi: 10.1109/TPAMI.2019.2918284.
5. Sathishkumar, R., Kalaiarasan, K., Prabhakaran, A. & Aravind, M. (2019) Detection of lung cancer using SVM classifier and KNN algorithm. In *IEEE International Conference on System, Computation, Automation and Networking (ICSCAN)*, Pondicherry, India, pp. 1–7, Doi: 10.1109/ICSCAN.2019.8878774.
6. Singadkar, G., Talbar, S., Sanghavi, P., Jankharia, B. & Talbar, S. (2018) Automatic lung field segmentation based on non-negative matrix factorization and fuzzy clustering. In *Lecture Notes in Networks and Systems,* vol 18. Smart Trends in Systems, Security and Sustainability, Springer, Doi: 10.1007/978-981-10-6916-1_6.
7. Chaudhary, A. & Singh, S. S. (2012, September). Lung cancer detection on CT images by using image processing. In *2012 International Conference on Computing Sciences,* Seattle, Washington, USA, pp. 142–146, IEEE.
8. Makaju, S., Prasad, P. W. C., Alsadoon, A., Singh, A. K. & Elchouemi, A. (2018). Lung cancer detection using CT scan images. *Procedia Computer Science, 125*, 107–114.
9. Nadkarni, N. S. & Borkar, S. (2019, April). Detection of lung cancer in CT Images using image processing. In *2019 3rd International Conference on Trends in Electronics and Informatics (ICOEI)*, Honolulu, Hi, USA, pp. 863–866, IEEE.
10. Singadkar, G., Mahajan, A., Thakur, M. & Talbar, S. (2021). Automatic lung segmentation for the inclusion of juxtapleural nodules and pulmonary vessels using curvature based border correction. *Journal of King Saud University-Computer and Information Sciences, 33*(8), 975–987.
11. Krizhevsky, A., Sutskever, I. & Hinton, G. (2012). ImageNet classification with deep convolutional neural networks. *Neural Information Processing Systems*, 25, Doi: 10.1145/3065386.
12. Liu, S. & Deng, W. (2015) Very deep convolutional neural network-based image classification using small training sample size. In *Proceedings of 3rd IAPR Asian Conference on Pattern Recognition*, Kuala Lumpur, pp. 730–734, Doi: 10.1109/ACPR.2015.7486599.
13. Szegedy, C., Liu, W., Jia, Y., Sermanat, P., Reed, S., Anguelov, D., Erhan, D., Vanhoucke, V. & Rabinovich, A. (2015) Going deeper with convolutions. In *IEEE Conference on Computer Vision and Pattern Recognition (CVPR)*, pp. 1–9, Doi: 10.1109/CVPR.2015.7298594.
14. He, K., Zhang, X., Ren, S. & Sun, J. (2016) Deep residual learning for image recognition. In *2016 IEEE Conference on Computer Vision and Pattern Recognition*, Las Vegas, NV, pp. 770–778, Doi: 10.1109/CVPR.2016.90.
15. Chollet, F. (2017). Xception: Deep learning with depthwise separable convolutions. In *IEEE Conference on Computer Vision and Pattern Recognition (CVPR),* 1800–1807, Doi: 10.1109/CVPR.2017.195.
16. Baheti, B., Innani, S., Gajre, S. & Talbar, S. (2020), Eff-UNet: A novel architecture for semantic segmentation in unstructured environment. In *Proceedings of the IEEE/CVF Conference on Computer Vision and Pattern Recognition (CVPR) Workshops*, Lake Tahoe, Nevada, USA.
17. Cicek, O., Abdulkadir, A., Lienkamp, S. S., Brox, T. & Ronneberger, O. (2018) 3d u-net: learning dense volumetric segmentation from sparse annotation. *IEEE Transactions on Medical Imaging, 37* (12), 2663–2674.
18. Milletari, F., Navab, N. & Ahmadi, S. A. (2016) V-net: Fully convolutional neural networks for volumetric medical image segmentation. In *Proceedings of Fourth IEEE International Conference on 3D Vision (3DV)*, Kuala Lumpur, Malaysia, pp. 565–571.
19. Li, X., Chen, H., Qi, X., Dou, Q., Fu, C. W. & Heng, P. A. (2017) H-denseunet: Hybrid densely connected unet for liver and liver tumor segmentation from CT volumes. arXiv preprint arXiv:1709.07330.

20. Tong, G., Li, Y. Chen, H., Zhang, Q. & Jiang, H. (2018) Improved U-NET network for pulmonary nodules segmentation, *174*, 460–469, Doi: 10.1016/j.ijleo.2018.08.086.
21. Zhao, C., Han, J., Jia Y. & Gou F. (2018). Lung nodule detection via 3D U-net and contextual convolutional neural network. *Proceedings of International Conference on Networking and Network Applications,* 356–361, Doi: 10.1109/NANA.2018.8648753.
22. Xiao, Z., Liu, B., Geng, L., Zhang, F. & Liu, Y. (2020) Segmentation of lung nodules using improved 3D-UNet neural network. *Symmetry, 12*(11), 1787, Doi: 10.3390/sym12111787.
23. Setio, A. A. A., Ciompi, F., Litjens, G., Gerke, P., Jacobs, C., Van Riel, S. J., Wille, M. M. W., Naqibullah, M., Sánchez, C. I. & van Ginneken, B. (2016). Pulmonary nodule detection in CT images: False positive reduction using multi-view convolutional networks. *IEEE Transactions on Medical Imaging, 35,* 1160–1169.
24. Armato III, S. G., McLennan, G., Bidaut, L., McNitt-Gray, M. F., Meyer, C. R., Reeves, A. P., … Clarke, L. P. (2011). The lung image database consortium (LIDC) and image database resource initiative (IDRI): A completed reference database of lung nodules on CT scans. *Medical physics, 38*(2), 915–931.
25. Huang, W. & Hu, L. (2019). Using a noisy U-net for detecting lung nodule candidates, *IEEE Access, 7,* 67905–67915, Doi: 10.1109/ACCESS.2019.2918224.
26. LUNA16, LUNA16 challenge dataset available (online), https://luna16.grand-challenge.org/download.
27. TIANCHI AI, Tianchi medical AI competition dataset, Available (online), https://tianchi.aliyun.com/competition/entrance/231601/information.
28. Zuo, W., Zhou, F. & Wang, L. (2019). Multi-resolution CNN and knowledge transfer for candidate classification in lung nodule detection. *IEEE Access*, Doi: 10.1109/ACCESS.2019.2903587.
29. Singadkar, G., Mahajan, A., Thakur, M. & Talbar, S. N. (2020). Deep deconvolutional residual network based automatic lung nodule segmentation. *Journal of Digital Imaging, 33,* 678–684, Doi: 10.1007/s10278-019-00301-4.
30. Ozdemir, O., Russell, R. L. & Berlin, A. A. (2019). A 3D probabilistic deep learning system for detection and diagnosis of lung cancer using low-dose CT scans. *IEEE Transactions on Medical Imaging, 39*(5), 1419–1429.
31. Masood, A., Sheng, B., Yang, P., Li, P., Li, H., Kim, J. & Feng, D. D. (2020). Automated decision support system for lung cancer detection and classification via enhanced RFCN with multilayer fusion RPN. *IEEE Transactions on Industrial Informatics, 16*(12), 7791–7801.
32. Dutande, P., Baid, U. & Talbar, S. (2021). LNCDS: A 2D-3D cascaded CNN approach for lung nodule classification, detection and segmentation. *Biomedical Signal Processing and Control, 67,* 102527.
33. Uzelaltinbulat, S. & Ugur, B. (2017) Lung tumor segmentation algorithm. In *Proceedings of 9th International Conference on Theory and Application of Soft Computing, Computing with Words and Perception (ICSCCW),* pp. 140–147.
34. Tripathi, P., Tyagi, S. & Nath, M. (2019). A comparative analysis of segmentation techniques for lung cancer detection. *Pattern Recognition and Image Analysis, 29,* 167–173, Doi: 10.1134/S105466181901019X.
35. Anthimopoulos, M., Christodoulidis, S., Ebner, L., Geiser, T., Christe, A. & Mougiakakou, S. (2018) Semantic segmentation of pathological lung tissue with dilated fully convolutional networks. *IEEE Journal of Biomedical and Health Informatics 23*(2), 714–722.
36. Hossain, S., Najeeb, S., Shahriyar, A., Abdullah, Z. R. & Haque, M. A. (2019) A pipeline for lung tumor detection and segmentation from CT scans using dilated convolutional neural networks. In *Proceedings of the IEEE International Conference on Acoustics, Speech and Signal Processing (ICASSP),* Toronto, Canada, pp. 1348–1352.

37. Aerts, H. J. W. L., Wee, L., Rios Velazquez, E., Leijenaar, R. T. H., Parmar, C., Grossmann, P., … Lambin, P. (2019). Data From NSCLC-Radiomics [Dataset] *The Cancer Imaging Archive*. Retrieved from The Cancer Imaging Archive website: Doi: 10.7937/K9/TCIA.2015.PF0M9REI.
38. Phil, T. (2020). Sikerdebaard/dcmrtstruct2nii: v1.0.19 (v1.0.19) [Computer software]. Zenodo, Doi: 10.5281/ZENODO.4037865.
39. Kingma, D. P. & Ba, J. (2014) Adam: A method for stochastic optimization. arXiv preprint arXiv:1412.6980.
40. Gao, C., Clark, S., Furst, J. & Raicu, D. (2019) Augmenting LIDC dataset using 3D generative adversarial networks to improve lung nodule detection. In *Proceedings SPIE 10950, Medical Imaging 2019: Computer-Aided Diagnosis, 109501K,* Doi: 10.1117/12.2513011.
41. Goodfellow, I. J., Pouget-Abadie, J., Mirza, M., Xu, B., Warde-Farley, D., Ozair, S., Courville, A. & Bengio, Y. (2014). Generative adversarial nets. In *Proceedings of the 27th International Conference on Neural Information Processing Systems* – Vol. 2. MIT Press, Cambridge, MA, pp. 2672–2680.

4 Day-Ahead Solar Power Forecasting Using Artificial Neural Network with Outlier Detection

D. Janith Kavindu Dassanayake
and M.H.M.R.S. Dilhani
University of Ruhuna

Konara Mudiyanselage Sandun Y. Konara
University of Ruhuna
University of Agder

Mohan Lal Kolhe
University of Agder

CONTENTS

DOI: 10.1201/9781003220176-4

4.1 INTRODUCTION

In Sri Lanka, the Ministry of Power and Renewable Energy (MPRE) is working towards achieving the status of carbon neutrality by 2050. Hence, there is a substantial increase in the penetration of renewable energy resources such as solar and wind in recent years. Due to the intermittent and uncontrollable nature of renewable energy resources, grid integration of the renewables is challenging. Integration of photovoltaic (PV) systems to the utility grid introduces significant volatility to the grid, which results in system instability [1], electrical power imbalances [2], variation in frequency response [3], etc.

In the modern electric grid, customers are allowed to use electricity in arbitrary quantities at their convenience. However, when aggregating over all the buildings and households, the demand variation is highly predictable. This demand variation is constantly monitored and the generators are dispatched according to the requirement to satisfy the demand. Hence, accurate PV power forecasting can help to plan where the PV systems can be utilized efficiently to satisfy the demand curve. Furthermore, in modern energy management systems, precise PV power forecasting has become a pivotal element to guarantee stable and economic integration of PV systems into the grid [4].

4.2 LITERATURE REVIEW

Many researches have been done on forecasting of PV power generation [4–11]. The two main strategies of forecasting PV power output are direct forecasting and indirect forecasting. The direct approaches predict the PV power output based on the past PV power output data, while the indirect approaches predict solar irradiance based on historical solar irradiance and weather data and thereby convert it to the power output from the PV [4]. The techniques which are commonly used in these strategies are the fuzzy logic method, wavelet analysis, support vector regression and ANN models. The paper [5], consists of a comprehensive study of theoretical forecasting strategies for both solar resource and PV power under four sections; statistical models, AI-based models, physical models and hybrid models. A comparison of the machine learning techniques, linear least squares and support vector machines with multiple kernel functions for generating the prediction model which is capable of analysing day-ahead solar radiation data can be found in [6]. Predicting solar power generation from weather forecasts using machine learning techniques is discussed in [7]. An extensive amount of historical data as well as weather forecast is analysed to correlate the weather parameters present in data with solar intensity as the solar intensity is proportional to solar harvesting. Prediction models are formulated based on linear least squares regression and support vector machines with different kernel functions, and a comparison is made to conclude which model performs better [7]. Day-ahead forecasting of solar power plant output based on meteorological data has been done in literature [9]. A mathematical formulation of the regression model to analyse the statistical significance of the meteorological parameters and improve the prediction accuracy with an empirical clustering approach is discussed in [9]. The algorithm presented in [10] is used to forecast solar irradiance level 2 hours ahead with an artificial neural network (ANN) and present the genetic algorithm to determine the optimum array size and positioning of solar monitoring stations to increase the accuracy of the forecast from the ANN. PV power output forecasting based on day-ahead insolation forecasting using three different neural network approaches has been discussed in [11]. The paper [11] presents a performance comparison of feed-forward neural network, radial basis function neural network and recurrent neural network on forecasting the power generation. Furthermore, solar power forecasting depends primarily on past data. Therefore, it is necessary to pre-process the data by data cleaning, data filling and data transforming techniques if any data inconsistencies and missing data can be found on the dataset. The papers [12–14] discuss the imputation on missing data related to solar researches by various methods and provide a detailed comparison of the imputation methods.

This chapter presents an approach utilizing ANN to forecast day-ahead PV power output of solar arrays installed in the Faculty of Engineering, University of Ruhuna considering the effect of solar irradiance and cell temperature of the solar panels as variables. Interpolation and exponential smoothing are used for the data pre-processing to fill in the missing values of the dataset caused by the faults in the system.

4.3 ELECTRICAL CHARACTERISTICS OF A PV MODULE

4.3.1 Correlation of Temperature and Irradiance to the Output Power of a PV Module

In order to investigate the operational data of the solar panel, the relationship between the PV output power, current, voltage and external environmental variables must be well understood. The short circuit current (I_{SC_m}) and open-circuit voltage (V_{OC_m}) relationships of a PV module at the maximum power point can be found using Eqs. (4.1) and (4.2) [15].

$$I_{SC\_m} = I_{m\_STC}\left[1+\alpha(T_c - T_{STC})\right].\frac{G}{G_{STC}} \tag{4.1}$$

$$V_{OC\_m} = V_{m\_STC}\left[1+\beta(G_{STC})(T_c - T_{STC})\right]\left[1+\delta T_c \ln\left(\frac{G}{G_{STC}}\right)\right] \tag{4.2}$$

V_{m_STC} and I_{m_STC} are voltage and current at the maximum operating points of the PV module under standard test conditions (STC). G_{STC} and T_{STC} are the solar irradiance and cell temperature under STC (which have the values of 1,000 W/m^2 and 25°C, respectively), and G and T_c are irradiance and cell temperature at operating conditions. α, β and δ are solar absorptance of PV layer, temperature coefficient and solar radiation coefficient (compensation coefficients), respectively [15].

The theoretical output power P can be calculated from Eq. (4.5) [15,16]:

$$P = G\tau_{pv}\eta_T A \tag{4.3}$$

$$\eta_T = \eta_{Tref}\left[1-\beta_{ref}\left(T_c-25\right)\right] \tag{4.4}$$

$$P = G\tau_{pv}A\eta_{Tref}\left[1-\beta_{ref}\left(T_c-25\right)\right] \tag{4.5}$$

where G is the irradiance at operating condition, η_T is the conversion efficiency of the solar cell, τ_{pv} is the solar cell transmittance of the outer layer, A is the surface area of the PV module, η_{Tref} is the conversion efficiency of the solar cell at reference conditions, β_{ref} is the value of thermal coefficient of max power for crystalline silicon at reference conditions and has a value of $\beta_{ref} = 0.0045°C^{-1}$and T_c is the operating temperature of the solar cell.

Practically, the cell operating temperatures are not readily available; therefore, it is replaced by T_{NOCT}, i.e., the nominal operating cell temperature. Therefore, cell conversion efficiency is calculated with the use of T_{NOCT}, which can be seen from Eq. (4.6) [15,16]:

$$\eta_T = \eta_{Tref}\left\{1-\beta_{ref}\left[T_a - T_{ref} + \left(T_{NOCT}-T_a\right)\frac{G}{G_{NOCT}}\right]\right\} \tag{4.6}$$

G_{NOCT} is the irradiance at nominal operating conditions and T_a is the ambient temperature. Therefore, the output power of the module can be written as in Eq. (4.7) [16]:

$$P = G\tau_{pv}A\eta_{Tref}\left\{1-\beta_{ref}\left[T_a - T_{ref} + \left(T_{NOCT}-T_a\right)\frac{G}{G_{NOCT}}\right]\right\} \tag{4.7}$$

4.3.2 Variation of Current and Voltage with Irradiance and Temperature

The main external environmental conditions that we are discussing in this chapter are solar irradiance and cell temperature. The solar cell absorbs solar energy and the PV current is generated from the release of photons. This operation continues until the current reaches a maximum point and starts to decrease. The maximum power can be extracted at the saturation point of the output current. The variation of output current with output voltage is shown in Figure 4.1 for different irradiance levels at a constant temperature.

The PV cell's temperature affects the output voltage inversely. The variation of output current with output voltage for different cell temperatures at constant irradiance level is shown in Figure 4.2.

4.3.3 Studied PV System and Data

The PV system selected is the system installed in the Faculty of Engineering, University of Ruhuna, Hapugala, Galle, Sri Lanka, which has the geographical coordinates of 6.0766°N, 80.1959°E. It has an installed peak capacity of 35 kWp and has a total area of 229 m^2. The system was commissioned on 26/07/2017. The PV system is capable of capturing real-time data with an interval of 5 minutes. For training and testing purposes, data within the period of 01/01/2020 to 31/03/2020 was taken.

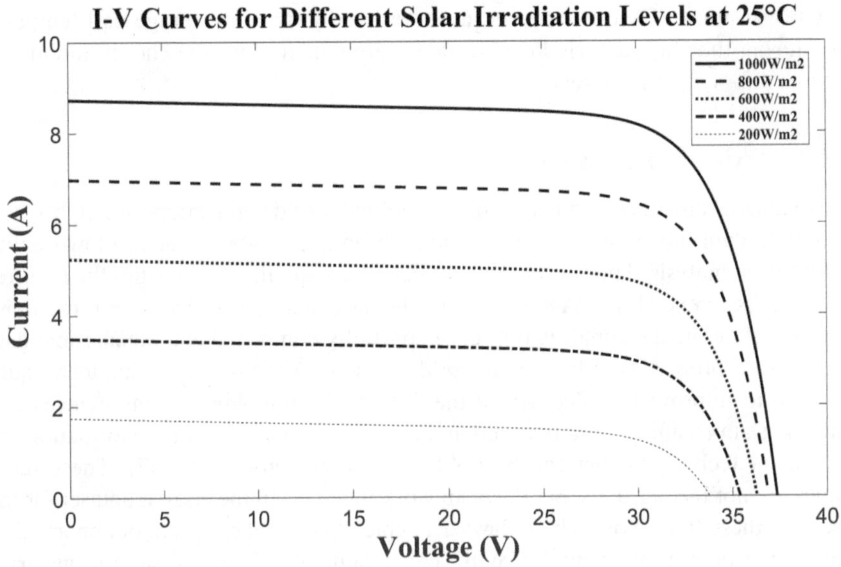

FIGURE 4.1 Variation of output current with output voltage for different cell temperatures at a constant irradiance level.

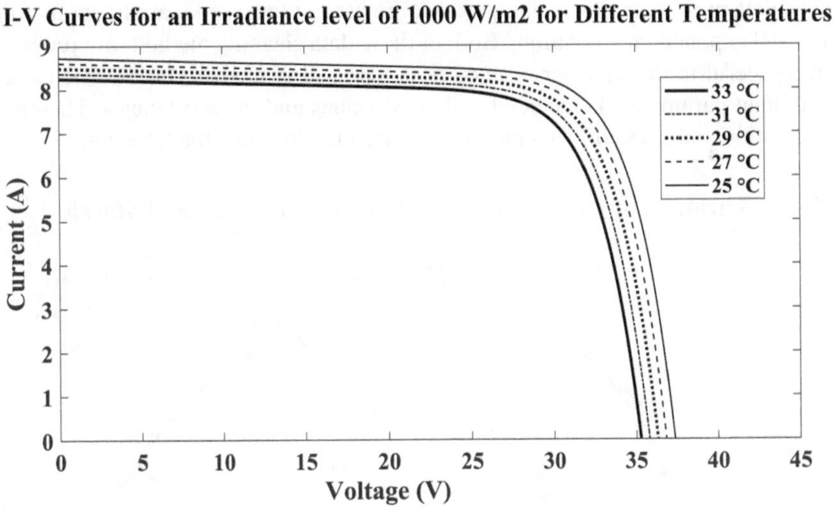

FIGURE 4.2 Variation of output current with output voltage for different cell temperatures at constant irradiance level.

The input parameters that are chosen to train the ANN are output power, irradiance and temperature. The data was taken from 6.35 a.m. to 4.00 p.m. with an interval of 5 minutes. The data records for the months of January and February were taken to train the model, and data records for March were used to check the accuracy and the validity of the model.

The variation of average power, average irradiance and average cell temperature during the chosen period over a month are depicted for the chosen months in Figures 4.3–4.5, respectively.

4.3.4 Data Pre-Processing

The databases created by gathering data from real-world applications are highly vulnerable to data inconsistencies, noises and absent data, which leads to low-quality results after analysis. The chosen data is said to have quality if it satisfies the requirements of its intended use. Accuracy, completeness and consistency are three main factors that define data quality. Imprecise, partially completed and inconsistent data are ordinary properties of large real-world datasets. Data pre-processing techniques are used to improve the accuracy of the datasets by removing inconsistencies and outliers. Data cleaning, data integration, data reduction and data transformation are four major techniques that can be used for data pre-processing [17]. These techniques are not incongruous, but they can work together. In the chosen dataset for the research, these flaws exist. These flaws are typically caused by a fault or abnormality that cannot be recovered immediately. These faults caused the system to generate missing values or outliers. The main reasons for the outliers in the dataset can be the failure of the inverter or the PV array, communication equipment or sensor failure and shutdown of the power unit.

The data pre-processing techniques that are used in the chapter are linear interpolation and exponential smoothing. Both of these data cleaning methods are proven to be resourceful in the short-term time series forecasting models [14,15]. The dataset of the input parameters is included with zero values and inconsistencies. Therefore, interpolation and smoothing techniques are used to eliminate the fallacies.

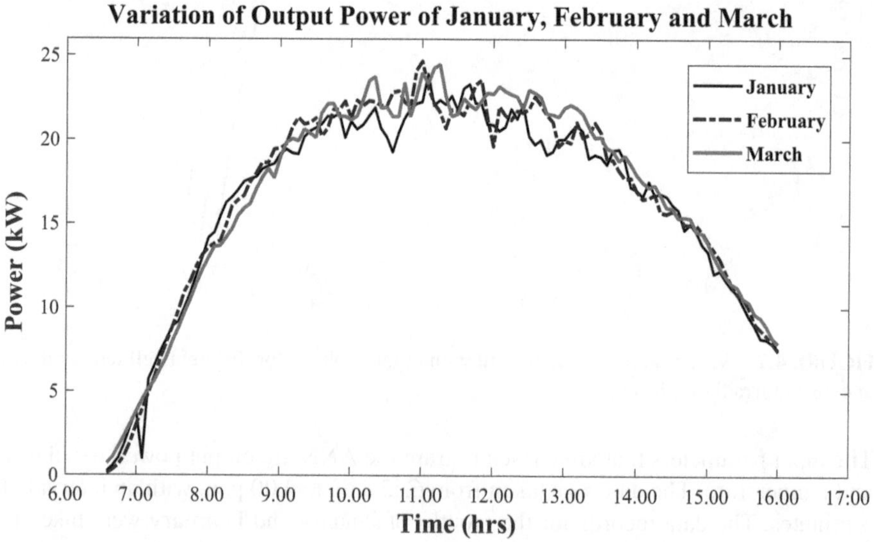

FIGURE 4.3 Variation of average power over the months of January, February and March.

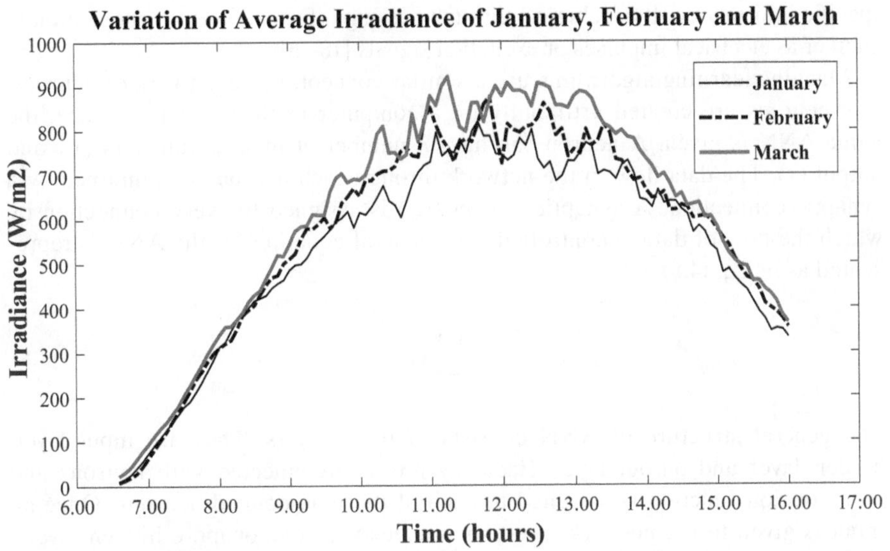

FIGURE 4.4 Variation of average irradiance over the months of January, February and March.

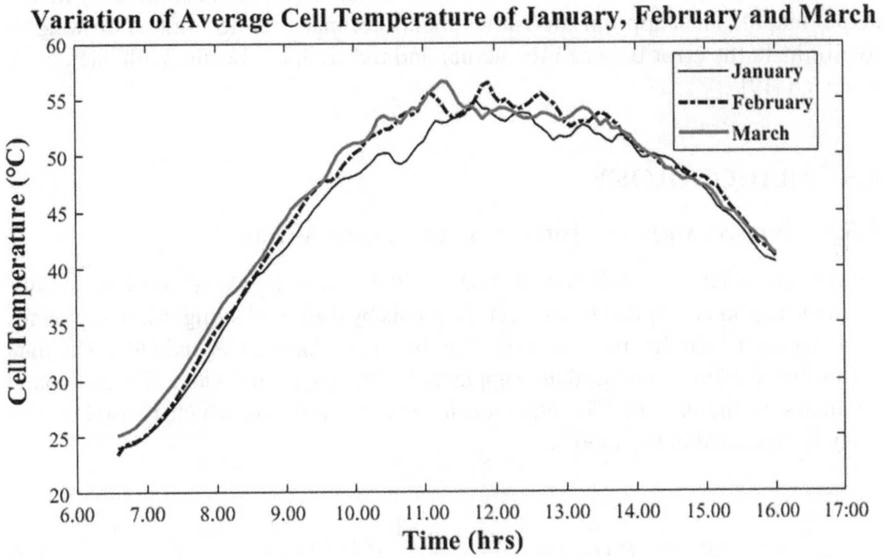

FIGURE 4.5 Variation of average temperature over the months of January, February and March.

4.4 OVERVIEW TO ANN

The human brain is the key to learning. The neurons or nerve cells in the brain are responsible for transmitting and processing information received from our senses. These neurons organize together in our brain to form a nerve network to exchange specific information. It is achieved through passing information from one neuron to another as electrical impulses or excitation signals [18–20].

Machine learning algorithms use a similar concept, i.e., a network of neurons. The neurons are created artificially on a computer in this scenario; hence, the name ANN is given. A neuron can have n number of input parameters (x_n) and output (y). The data flow in the network through each neuron is maintained by a synaptic connection. A synaptic weight (w_n) is assigned to every connection by which the flow of data is controlled. The general equation for the ANN is represented as in Eq. (4.8).

$$y = \sum x_n w_n \tag{4.8}$$

The general structure of ANN consists of three layers. They are input layer, hidden layer and output layer. Each layer is interconnected with neurons and an activation function is assigned to each of these associated neurons. Once an input is given to the network, it communicates with one or more hidden layers. The error between the input and output can be reduced by adjusting the weights of the neurons. This concept is known as the learning rule [20]. Initially, weights are random numbers. After each iteration, the computed output is compared with the actual output. Then appropriated adjustments are made to the connection weights to eliminate the error between the actual and the computed value with each cycle or epoch [19].

4.5 METHODOLOGY

4.5.1 Interpolation for Imputation of Missing Values

Linear interpolation is utilized to discover the values between two data points. A simple method is applied to connect the points by the use of straight-line segments. Each segment bounded by two points can be interpolated independently. The data points that need to be replaced are supplanted with new values with a function found by analysing the dataset. The equation for the interpolation which is used in this study is expressed in Eq. (4.9):

$$P_d(t) = P_d(t-1) + \left[\frac{(m_2 - m_1)}{(m_3 - m_1)}\right]\left(P_d(t+1) - P_d(t-1)\right) \tag{4.9}$$

where

$P_d(t)$ is the power data at the time stamp t of the selected day d, $P_d(t-1)$ is the power data at the time stamp $t-1$ of the selected day d, $P_d(t+1)$ is the power data at the time stamp $t+1$ of the selected day d, m is the time axis positions of power values, where m_1, m_2 and m_3 are the positions of $P_d(t-1)$, $P_d(t)$ and $P_d(t+1)$, respectively.

4.5.2 Exponential Smoothing for Imputation of Missing Values

For the estimation of the regular demand of the identified outliers of a dataset, exponential smoothing can be applied. It is a simple time series forecasting method. Simply the missing values in the selected period are identified and replaced by the exponential smoothing method as given in Eq. (4.10):

$$F_d(t+1) = \alpha A_d(t) + (1-\alpha) F_d(t) \tag{4.10}$$

where $F_d(t+1)$ is the forecasted power demand of day d at time instance $(t+1)$. The actual power demand $A_d(t)$ is selected as they have the same pattern throughout the day. $F_d(t)$ is the forecasted power demand of day d at time period t, and α is an adjustable scalar value. In this research, the optimal value of α is 0.4.

4.5.3 Design of ANN Structure

The designed structure of the ANN is depicted in Figure 4.6.

As illustrated, the input parameters of the ANN are power, irradiance and temperature. The model consists of three hidden layers, and each hidden layer has 36 neurons. The proposed model can be described with Eq. (4.11) to explain the relationship between input and outputs for the ANN.

$$P_t(d) = a_1 P_t(d-1) + a_2 I_t(d-1) + a_3 T_t(d-1) \tag{4.11}$$

The power demand of the forecasting day $P_t(d)$ at time t is calculated using the power demand of the previous day $P_t(d-1)$, irradiance on the previous day $I_t(d-1)$ and cell temperature of the previous day $T_t(d-1)$ at the same period.

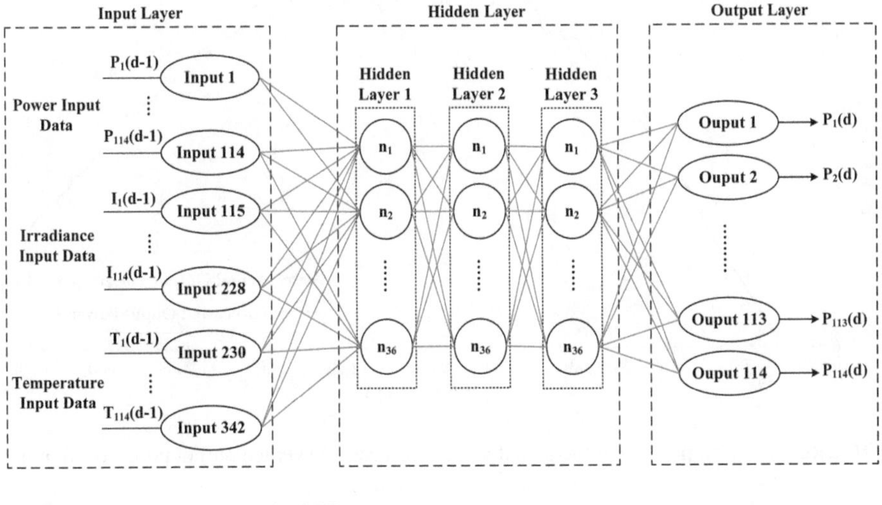

FIGURE 4.6 The proposed ANN structure.

For a chosen day, from 6.35 a.m. to 4.00 p.m., 114 timestamps are available. Therefore, for a given day, a total of 344 data records are used as input to the model as shown in Figure 4.6. $P_1(d) - P_{114}(d)$ represent the 114 power data of the forecasting day *d*.

4.5.4 Evaluation of the Forecasting Model

To verify the performance of the proposed model, Mean Absolute Percentage Error (MAPE) is calculated for each time stamp *t* for the month of March with Eq. (4.12):

$$\text{MAPE} = \frac{P_{d,A}(t) - P_{d,F}(t)}{P_{d,A}(t)} \times 100 \tag{4.12}$$

where $P_{d,A}(t)$ and $P_{t,F}(d)$ are the actual and the forecasted output power at time *t* on the day *d*.

Furthermore, the average value of the MAPE in each timestamp for the whole month is calculated with Eq. (4.13):

$$\text{Average MAPE} = \frac{1}{114}\sum_{t=1}^{114} \frac{P_{d,A}(t) - P_{d,F}(t)}{P_{d,A}(t)} \times 100 \tag{4.13}$$

4.6 Results and Discussion

Figure 4.7 presents the comparison of the actual and the forecasted average power output for the testing month March. It can be observed that the forecasted line follows the actual line closely, but it becomes less accurate at the sudden changes of the output power.

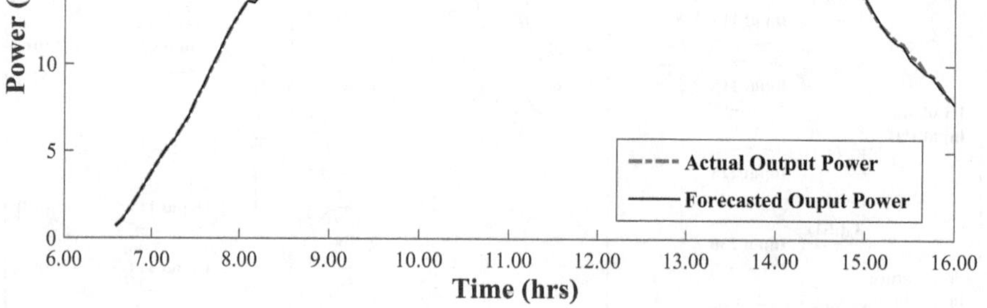

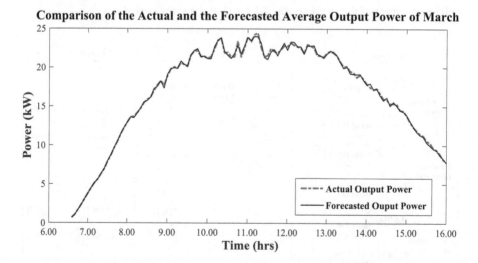

FIGURE 4.7 Comparison of the actual and the forecasted average output power of March.

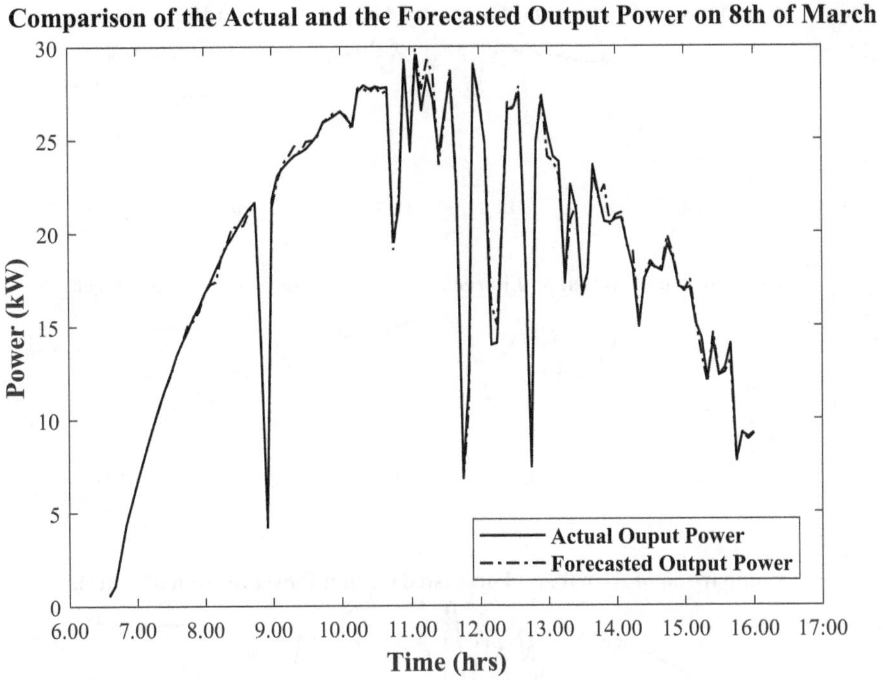

FIGURE 4.8 Comparison of the actual and the forecasted output power on the 8th of March.

In order to verify whether the use of interpolation and exponential smoothing techniques was successful, the best is to consider the 8th of March. A total of five data stamps are missing and it are imputed by the data pre-processing techniques. The average MAPE for that day is 3.71%. The actual and the forecasted values closely follow each other as illustrated in Figure 4.8.

The 10th and 20th of March are the days with minimum daily average MAPE values – 0.06% and 0.01%. The comparison of the power output on the above dates is shown in Figure 4.9a.

When scanning the MAPE values throughout the forecasted month, it is found that maximum MAPE occurred on the days that have sudden variations in the output power. The 16th and 18th of March are two examples of dates where the maximum MAPE values are recorded which are 39.33% and 37.18%, respectively. If we compare the actual and forecasted output power on that day, it can be seen that it is due to rapid fluctuations of the power output that have occurred throughout that day. The comparison of the actual and the forecasted power output for 16th and 18th of March is depicted in Figure 4.9b.

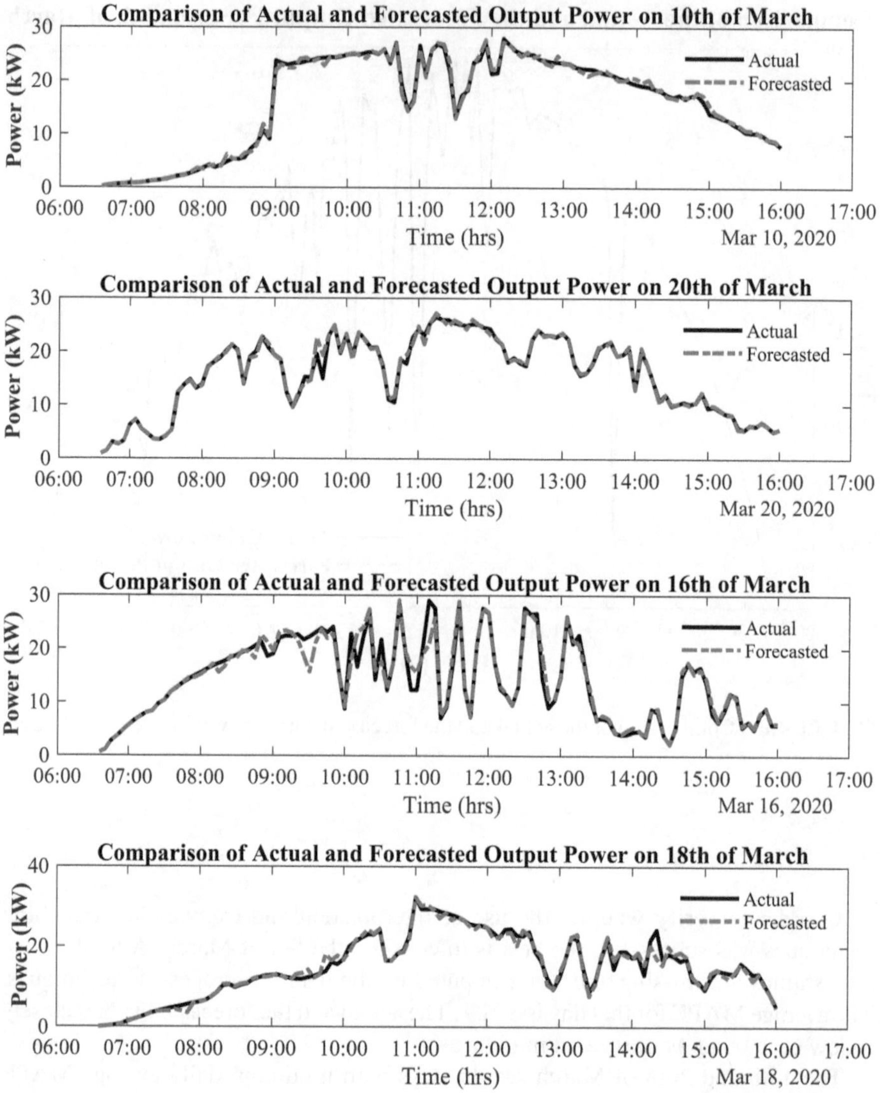

FIGURE 4.9 Comparison of actual and the forecasted output power on four different days.

4.7 CONCLUSION

This chapter presents a day-ahead power forecasting system using an ANN. To train and test the model data for the first 3 months of the year 2020, data records with 5 minutes time intervals are taken. The model is verified with the data for the month of March. The ANN uses three input parameters which are the previous day's output power, irradiance and temperature. All the data are pre-processed with interpolation and exponential smoothing for the purpose of imputation of data. The equation for the ANN model is developed by building a relationship between the previous day's output power and correlating input parameters of the previous day. The results show that the monthly average MAPE of output power is 2.069%. However, during the unexpected rapid variations of the temperature and irradiance, the forecasting accuracy of the model is low.

For further improvement of the model, the research shall use a wider range of datasets and try to improve the accuracy of the model during the sudden fluctuations. In addition, the impact of some additional factors such as humidity, partial shading and weather conditions will be considered for the study.

ACKNOWLEDGEMENT

The authors acknowledge the data and support provided by the Faculty of Engineering, University of Ruhuna, Sri Lanka for this research.

REFERENCES

1. Shah, R., N. Mithulananthan, R.c. Bansal, and V.k. Ramachandaramurthy. "A review of key power system stability challenges for large-scale PV integration." *Renewable and Sustainable Energy Reviews* 41 (2015): 1423–36. Doi: 10.1016/j.rser.2014.09.027.
2. Ueckerdt, F., R. Brecha, and G. Luderer. "Analyzing major challenges of wind and solar variability in power systems." *Renewable Energy* 81 (2015): 1–10. Doi: 10.1016/j.renene.2015.03.002.
3. Nanou, S.I., A.G. Papakonstantinou, and S.A. Papathanassiou. "A generic model of two-stage grid-connected PV systems with primary frequency response and inertia emulation." *Electric Power Systems Research* 127 (2015): 186–96. Doi: 10.1016/j.epsr.2015.06.011.
4. Yang, H.-T., C.-M. Huang, Y.-C. Huang, and Y.-S. Pai. "A weather based hybrid method for 1-day ahead hourly forecasting of PV power output." *IEEE Transactions on Sustainable Energy* 5, no. 3 (2014): 917–26. Doi: 10.1109/tste.2014.2313600.
5. Sharma, N., P. Sharma, D. Irwin, and P. Shenoy. "Predicting solar generation from weather forecasts using machine learning." *2011 IEEE International Conference on Smart Grid Communications (SmartGridComm),* (2011). Doi: 10.1109/smartgridcomm.2011.6102379.
6. Wan, C., J. Zhao, Y. Song, Z. Xu, J. Lin, and Z. Hu. "Photovoltaic and solar power forecasting for smart grid energy management." *CSEE Journal of Power and Energy Systems* 1, no. 4 (2015): 38–46. Doi: 10.17775/cseejpes.2015.00046.
7. Hassan, Md. Z., Md. E.K. Ali, A.B.M. Shawkat Ali, and J. Kumar. "Forecasting day-ahead solar radiation using machine learning approach." *2017 4th Asia-Pacific World Congress on Computer Science and Engineering (APWC on CSE),* (2017). Doi: 10.1109/apwconcse.2017.00050.
8. Serttas, F., F.O. Hocaoglu, and E. Akarslan. "Short term solar power generation forecasting: A novel approach." *2018 International Conference on Photovoltaic Science and Technologies (PVCon),* (2018). Doi: 10.1109/pvcon.2018.8523919.
9. Snegirev, D.A., S.A. Eroshenko, A.I. Khalyasmaa, V.V. Dubailova, and A.I. Stepanova. "Day-ahead solar power plant forecasting accuracy improvement on the hourly basis." *2019 IEEE Conference of Russian Young Researchers in Electrical and Electronic Engineering (EIConRus),* (2019). Doi: 10.1109/eiconrus.2019.8657024.

10. Vanderstar, G., P. Musilek, and A. Nassif. "Solar forecasting using remote solar monitoring stations and artificial neural networks." *2018 IEEE Canadian Conference on Electrical & Computer Engineering (CCECE)*, (2018). Doi: 10.1109/ccece.2018.8447636.
11. Yona, A., T. Senjyu, A.Y. Saber, T. Funabashi, H. Sekine, and C.-H. Kim. "Application of neural network to 24-hour-ahead generating power forecasting for PV system." *2008 IEEE Power and Energy Society General Meeting - Conversion and Delivery of Electrical Energy in the 21st Century*, (2008). Doi: 10.1109/pes.2008.4596295.
12. Layanun, V., S. Suksamosorn, and J. Songsiri. "Missing-data imputation for solar irradiance forecasting in Thailand." *2017 56th Annual Conference of the Society of Instrument and Control Engineers of Japan (SICE)*, (2017). Doi: 10.23919/sice.2017.8105472.
13. Pereira, G.M., R.L. Stonoga, D.H. Detzel, K.K. Küster, R.A. Neto, and L.A.C. Paschoalotto. "Analysis and evaluation of gap filling procedures for solar radiation data." *2018 IEEE 9th Power, Instrumentation and Measurement Meeting (EPIM)*, (2018). Doi: 10.1109/epim.2018.8756358.
14. Demirhan, H., and Z. Renwick. "Missing value imputation for short to mid-term horizontal solar irradiance data." *Applied Energy* 225 (2018): 998–1012. Doi: 10.1016/j.apenergy.2018.05.054.
15. Skoplaki, E., and J.a. Palyvos. "On the temperature dependence of photovoltaic module electrical performance: A review of efficiency/power correlations." *Solar Energy* 83, no. 5 (2009): 614–24. Doi: 10.1016/j.solener.2008.10.008.
16. Hu, A., Q. Sun, H. Liu, N. Zhou, Z.A. Tan, and H. Zhu. "A novel photovoltaic array outlier cleaning algorithm based on sliding standard deviation mutation." *Energies* 12, no. 22 (2019): 4316. Doi: 10.3390/en12224316.
17. Güting, R., S. Chakrabarti, E. Cox, E. Frank, J. Han, X. Jiang, and M. Kamber et al. 2008. *Data Mining*. Burlington: Elsevier.
18. Kim, P. MATLAB Deep Learning, (2017) APress publisher, ISBN: 9781484228449 pages:151.
19. M.H.M.R. Shyamali Dilhani, and C. Jeenanunta. "Daily electric load forecasting: Case of Thailand." *2016 7th International Conference of Information and Communication Technology for Embedded Systems (IC-ICTES)*, (2016). Doi: 10.1109/ictemsys.2016.7467116.
20. Dilhani, M.H.M.R. Shyamali, and C. Jeenanunta. "Effect of neural network structure for daily electricity load forecasting." *2017 Moratuwa Engineering Research Conference (MERCon)*, (2017). Doi: 10.1109/mercon.2017.7980521.

5 Fuzzy-Inspired Three-Dimensional DWT and GLCM Framework for Pixel Characterization of Hyperspectral Images

K. Kavitha
Vellammal College of Engineering and Technology

D. Sharmila Banu
Ultra College of Engineering and Technology

CONTENTS

DOI: 10.1201/9781003220176-5

5.1 INTRODUCTION

Remote sensing is the technology that is capable of collecting information about the area under investigation, without establishing direct contact. To construct the image of the landscape, the sensors need to acquire the reflectance information from the Earth's surface. In active remote sensing system, the sensor antenna itself emits energy toward the Earth's surface and measures the energy that is being reflected back to it. The energy measured by the sensor is collected in several spectral bands.

The spectral resolution is the spectral range of a single band. The area covered by the sensor defines geometrical resolution. The advances in imaging sensor technologies and satellite technologies resulted in the evolution of very high-resolution imaging systems. The most prominent advances are multispectral systems, which can acquire multispectral images that exhibit good geometrical resolution and hyperspectral systems, which can acquire hyperspectral images that exhibit good spectral resolution. As hyperspectral sensors produce rich spectral information for the fine identification of the land cover classes, they occupy a significant place in remote sensing research like urban planning, vegetation identification, and crop classification [1].

To simultaneously exploit the spectral and spatial features from the hyperspectral data, it is necessary to decompose the data using three-dimensional wavelets. Qian et al. (2012) decomposed the hyperspectral cube using 3D DWT with different scales and frequencies, thereby capturing the joint spatial and spectral characteristics of the data [2]. Structured sparse logistic regression–based classification of pixels and extraction of features and 3D discrete wavelet transform (3D DWT) was reported in this work.

Ye et al. (2014) utilized the 3D DWT algorithm for extracting the spatial-spectral features and used the extracted coefficient for feature grouping and reported an algorithm for hyperspectral image analysis. The hyperspectral data spectral-spatial features are extracted and selected using the wavelet-coefficient correlation matrix (WCM) [3]. The adjacent wavelet-coefficient subspaces were grouped and hyperspectral image analysis was done.

Guo et al. (2014) proposed the classification of urban multi/hyperspectral imagery by texture feature extraction by using 3D wavelet transform and proved the ability of the method [4]. Three types of approaches, namely, pixelwise, nonoverlapping, and overlapping cube for 3D DWT texture extraction, were proposed and reported. Cao et al. (2020) proposed an enhanced 3D DWT approach to extract the features along with CNN model for classification [5].

For single-band images, 2D texture analysis is adequate. The downsides of 2D texture analysis cannot treat the stacked data, such as multispectral and hyperspectral data. Tsai et al. (2007) extended the GLCM (gray level cooccurrence matrix) computation using 3D form in which the cubes of hyperspectral image are considered volumetric data and use the same for the extraction of discriminant texture features for the classification of hyperspectral image [6]. Tsai and Lai (2013) extracted the three-dimensional cooccurrence features from the hyperspectral cube. They formulated the three-dimensional textural features, which are the extension of two-dimensional cooccurrence features [7]. Su et al. (2014) reported an approach that uses spectral features that are reduced and volumetric texture features for the classification of hyperspectral images. By using this method, the volumetric GLCMs are used for the extraction of volumetric texture features. The reported Volumetric Gray Level Co-occurrence Matrix (VGLCM) method outperforms the GLCMs method and yielded better performance in the classification of hyperspectral images [8].

In the above-mentioned works, classification is enabled based on the per-pixel classification approach, in which a pixel is purely assigned to the class. But most of the hyperspectral images contain pixels that are mixed, i.e., the pixels that contain cover classes that are more than one. So, it is necessary to decompose the mixed pixels and to identify their class belonging by any of the unmixing techniques. For the identification of classes in which the pixels are mixed, a membership function is required. So, fuzzy-based approaches are useful in this regard.

Moraes et al. (2002) proposed an architecture that is expert in fuzzy systems for the classification of images, and its implementation of rules is done through some mathematical invariant operators of morphology [9]. The proposed architecture use is demonstrated by a system that is expert and used in the hyperspectral image area classification. Zhang et al. (2012) proposed a hyperspectral image classification algorithm that integrates an unsupervised neural network and a fuzzy system for the classification of hyperspectral images [10]. Mylonas et al. (2014) developed a spectral-spatial feature and fuzzy membership–based method and performed the classification of a land cover of a hyperspectral image that is of high resolution [11].

5.2 EXPERIMENTATION

5.2.1 3D DWT and 3D GLCM-Based Approach for Hyperspectral Image Classification

In this methodology, hyperspectral cube is decomposed by using 3D DWT, and 3D GLCMs is formed. From the GLCM, the statistical and cooccurrence features like entropy, energy, homogeneity, contrast, maximum probability, mean, standard deviation are extracted. On concatenating the extracted features, classification is done and classification maps are generated. Parameters such as class-wise accuracies, overall accuracies, and kappa values are calculated and some other DWT-based methods are compared with the proposed method performance and cooccurrence-based methods. The proposed 3D DWT and 3D gray level cooccurrence matrix (3D GLCM) combination extracts the joint spatial-spectral features and yields better result. The workflow is shown in Figure 5.1.

5.2.1.1 3D DWT Decomposition

3D DWT decomposition is the extended computation of 1D DWT proposed in [12]. In 1D DWT, by using low-pass (L) and high-pass (H) filter banks, DWT retrieves

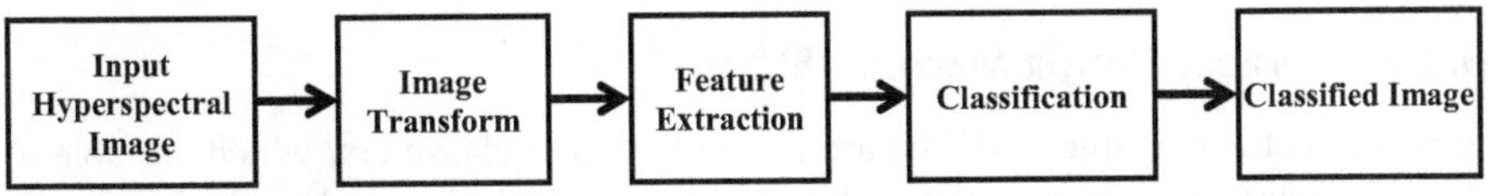

FIGURE 5.1 General classification methodology.

the approximation and details of the given information, respectively. In 1D, low-pass sequences, i.e., $h(n)$, and high-pass sequence i.e., $g(n)$, can be obtained by converging the impulse responses and by using scaling and translation parameters.

$$\varphi(t) = \sum_{n=0}^{N} l(n)\varphi(2t - n). \tag{5.1}$$

where $\varphi(t)$. denotes the scaling function and t and n represents the scaling parameter and translation parameter, respectively as given in Eq. (5.1).

The wavelet function is given in Eq. (5.2).

$$\Psi(t) = \sum_{n=0}^{N} h(n)\Psi(2t - n) \tag{5.2}$$

Conventional 1D DWT modeling is carried out along the three-dimensional hyperspectral axes (x, y, z). In practice, the spatial coordinates of an image are denoted by x and y directions, and the spectral axis is denoted by z axis. The tensor product is used for the construction of 3D DWT as given in Eq. (5.3).

$$\begin{aligned} I^{(x,y,z)} &= \left(L^x \oplus H^x\right) \otimes \left(L^y \oplus H^y\right) \otimes \left(L^z \oplus H^z\right) \\ &= L^x L^y L^z \oplus L^x L^y H^z \oplus L^x H^y L^z \oplus L^x H^y H^z \oplus \\ &\quad H^x L^y L^z \oplus H^x L^y H^z \oplus H^x H^y \oplus L^z H^x H^y \end{aligned} \tag{5.3}$$

where $\oplus$ denotes space direct sum and $\otimes$ denotes tensor product, while L represents the low-pass filters and H high-pass filters along the x, y and z axes, respectively. After first-level processing, the decomposition of volume data into eight subbands, i.e., *LLL*, *LLH*, *LHL*, *LHH*, *HLL*, *HLH*, *HHL*, and *HHH* will take place.

5.2.1.2 3D GLCM Feature Extraction

Two-dimensional GLCM calculation is extended to 3D GLCM for calculating the textural properties of any three-dimensional data like hyperspectral data (Tsai et al. 2007). 3D GLCM is also called voxel cooccurrence matrix that collects the neighborhood relation along three axes of the three-dimensional data, and it can be used to interpret and visualize the three-dimensional data. Figures 5.2 and 5.3 depict the decomposition method and decomposed subbands of a hyperspectral image.

5.2.2 Support Vector Machine (SVM)

Support vector machines (SVMs) are the widely used classifiers, which are able to classify linear and nonlinear data. The nonlinear separability can be achieved by the kernel trick [13].

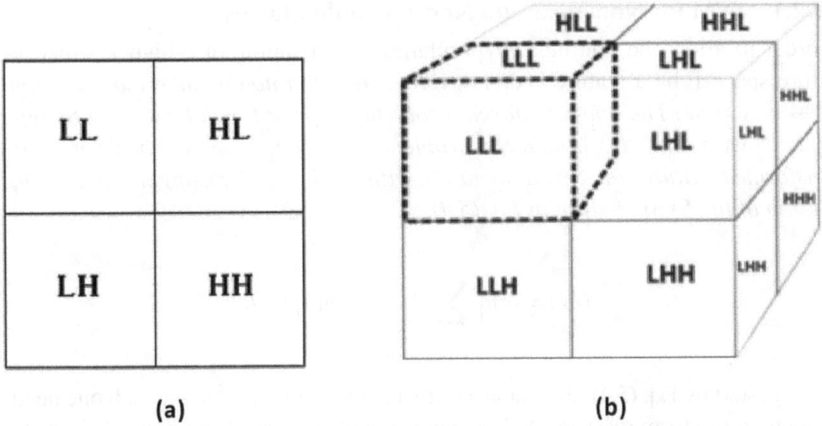

FIGURE 5.2 Decomposition of data (a) decomposed subbands of 2D image and (b) decomposed subbands of the hyperspectral cube.

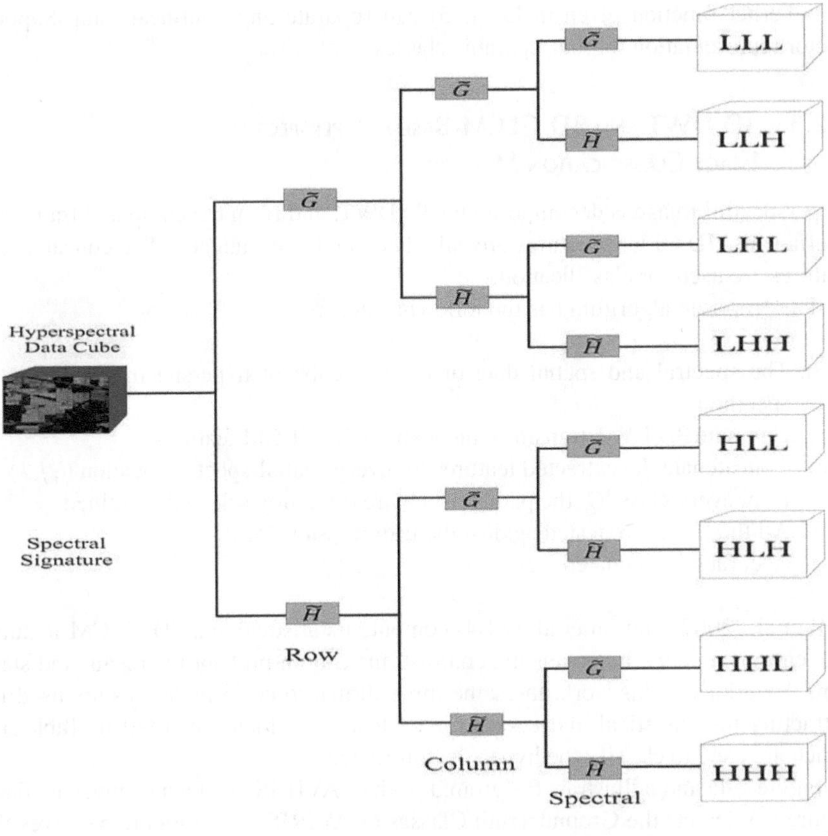

FIGURE 5.3 One-level 3D DWT decomposition of hyperspectral data.

5.2.2.1 SVM for Nonlinear and Nonseparable Classes

In order to allow the nonlinear hyperplanes, the mapping of a higher dimensional feature space F by a feature vector, $\boldsymbol{x} \varepsilon \boldsymbol{R}^N$ *through a nonlinear vector function* Φ: $R^N \rightarrow F$ *is done. The optimal margin problem in space F can be written by replacing* $\boldsymbol{x}_i$. $\boldsymbol{x}_j$ *with* $\Phi(x_i)\cdot\Phi(x_j)$, *then the problem of optimization is solved for* λ_i *in the transformed feature space by association with the* $\lambda_i > 0$. *By using this mapping, the solution of the SVM is given in Eq. (5.4).*

$$f(x) = \text{sgn}\left[\sum_i \lambda_i \varphi(x) \cdot \varphi(x_i) + b\right] \tag{5.4}$$

As suggested by Eq. (5.4), the scalar products are the quantities in which one needs to compute in the form $\Phi(x)$. $\Phi(y)$. So the concept of the kernel function k is introduced for the convenience in such a way that

$$k(x_i, x_j) = \varphi(x_i).\varphi(x_j) \tag{5.5}$$

This kernel function given in Eq. (5.5) can separate any nonlinear data. Support vector representation for nonseparable classes.

5.2.3 3D DWT and 3D GLCM-Based Hyperspectral Image Classification Method

Hyperspectral image is decomposed by 3D DWT, and from the computed transform coefficients, 3D GLCM features are calculated and concatenated. The concatenated features are used for classification.

The proposed algorithm has the following steps:

i. The spectral and spatial data of the test cube of hyperspectral image is specified.
ii. Compute 3D DWT transform and extract 3D GLCM features.
iii. Concatenate the extracted features for every spatial-spectral location (*i*, *j*, *k*) from every class '*C*' the pixels which are randomly selected is trained.
iv. All the pixels are tested against the training sample.
v. Accuracy is evaluated.

Tsai et al. (2007) and Su et al. (2014) computed statistical and 3D GLCM features like entropy, energy, homogeneity, contrast, maximum probability, mean, and standard deviation. In this work, the same three-dimensional formulations are used for extracting the statistical and cooccurrence features which are listed in Table 5.1, which are used to classify the hyperspectral image.

Figure 5.4a and b illustrate the ground truth of AVIRIS and Pavia University data. Figure 5.5 depicts the Ground Truth Classes of AVIRIS, while Figure 5.6 gives the same for Pavia University.

TABLE 5.1
Extracted 3D Wavelet Statistical and Cooccurrence Features

Feature	Formula
Mean	$\left(\frac{1}{3}\right)\sum_i\sum_j\sum_k ip(i, j, k)+\sum_i\sum_j\sum_k jp(i, j, k)+\sum_i\sum_j\sum_k kp(i, j, k)$
Standard deviation	$\sigma_i=\sqrt{\sigma_i^2},\quad \sigma_j=\sqrt{\sigma_j^2},\quad \sigma_k=\sqrt{\sigma_k^2}$
Energy	$\sum_i\sum_j\sum_k p^{2(i,j,k)}$
Entropy	$-\sum_i\sum_j p(i, j)\log_2 p(i, j)+\sum_j\sum_k p(j, k)\log_2 p(j, k)+\sum_i\sum_k p(i, k)\log_2 p(i, k)$
Contrast	$\sum_i\sum_j (i-j)^2\ p(i,j)+\sum_i\sum_j (j-k)^2\ p(j,k)+\sum_i\sum_j (i-k)^2\ p(i, k)$
Homogeneity	$\sum_i\sum_j (p(i-j)/(1+\|i-j\|)+\sum_j\sum_k (p(j-k))/(1+\|j-k\|)+\sum_i\sum_k (p(i-k))/(1+\|i-k\|)$
Maximum probability	$\max\{p(i,j,k)\}$

$p(i, j, k)$ is the value of the element of 3D GLCM at the coordinates (i, j, k).

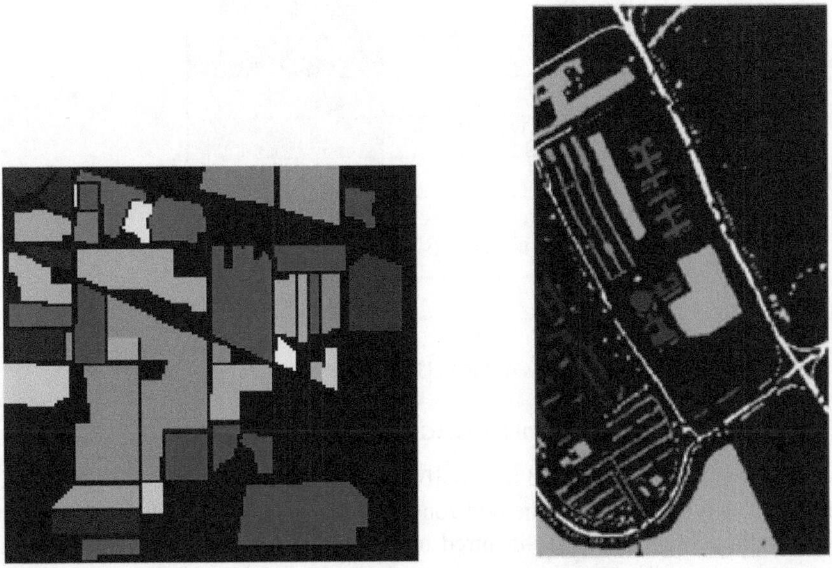

FIGURE 5.4 Groundtruth map of AVIRIS and groundtruth map of Pavia University.

CLASS NAME	THEME
ALFALFA	
CORN NOTILL	
CORN MINTILL	
CORN	
GRASS PASTURE	
GRASS TREES	
GRASS PASTURE MOWED	
HAY WINDROWED	
OATS	
SOYBEAN NOTILL	
SOYBEAN MINTILL	
SOYBEAN CLEAN	
WHEAT	
WOODS	
BUILDING GRASS TREES DRIVES	
STONE STEEL TOWERS	

FIGURE 5.5 Groundtruth classes of AVIRIS.

CLASS NAME	THEME
ASPHALT	
MEADOWS	
GRAVEL	
TREES	
PAINTED METAL SHEET	
BARE SOIL	
BITUMEN	
SELF BLOCKING BRICKS	
SHADOWS	

FIGURE 5.6 Groundtruth classes of Pavia University.

5.2.4 Proposed Fuzzy-Inspired Image Classification Method

Like other works, SVM is used to classify the pixels. Fuzzification and reclassification modules of the proposed method convert the normal multiresolution transform-based methodology into fuzzy-inspired methodology. The edges of the classes are identified by using Canny edge detector, and the pixels at the edges are assumed as mixed pixels. By applying fuzzy-inspired approach, overall, average, and class-wise

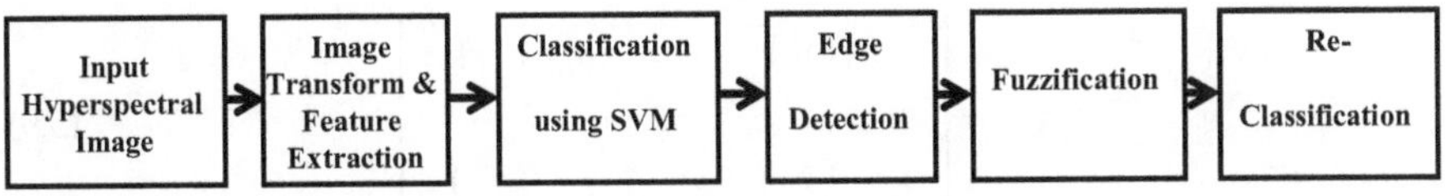

FIGURE 5.7 Block diagram of the proposed fuzzy-inspired classification method.

accuracies are observed and kappa statistics are computed. Results before and after applying fuzzy framework are obtained and compared. Figure 5.7 shows the steps involved in the proposed fuzzy-based methods.

5.2.4.1 Mixed Pixel Identification

Mixed pixels are the representatives of more than one class due to the overlapping reflectance of the neighboring pixels. It is implied that mixed pixels are available at the edges where overlapping of more than one classes is possible. Hence the pixels at the edge are identified by using the conventional edge detection algorithms. As reported by previous researchers, Canny edge detector is used [14].

5.2.4.2 Fuzzification

The fuzzy sets are the representation of the hyperspectral image bands (i.e., total number of bands 'b' which is available in the hyperspectral cube). The fuzzy subsets are the consideration of land cover classes of hyperspectral image (i.e., number of classes available in the hyperspectral cube) [15]. The membership function f_b, $c(x_b)$ is used for the definition of fuzzy sets and subsets in which x_b is the representation of the gray level of X in b and b. In B-dimensional space, the pixel vector X is given in Eq. (5.6):

$$X = \left[x_1, x_2, \ldots x_b \ldots x_B\right]^T \tag{5.6}$$

The fuzzy requirement is given in Eq. (5.7).

$$\forall x_b \in [0,255], \quad \sum_{i=1}^{N} f_{b,i}(x_b) = 1 \tag{5.7}$$

N indicates the number of fuzzy subsets.

5.2.4.3 Membership Function

The class 'c' in and 'b' membership function is given in Eq. (5.8).

$$f_{b,c}(x_b) = \exp\left(-\left(x_b - \mu_{b,c}\right)^2 \middle| 2\sigma^2_{b,c}\right) \tag{5.8}$$

where $\mu_{b,c}$ is the representation of mean of class c in band b. The process of fuzzification is the computation of the $\mu_{b,c}$ membership degrees for every class c and b from the membership functions for a given pixel. $f_{i,p}$ is the matrix of fuzzy inputs of order $B \times N$ which is derived, where N denotes the number of classes and B denotes the number of bands which is given in Eq. (5.9).

$$f_{i,p} = \begin{bmatrix} f_{1,1}(x_1) & \cdots & f_{1,N}(x_1) \\ \vdots & \ddots & \vdots \\ f_{B,1}(x_B) & \cdots & f_{B,N}(x_B) \end{bmatrix} \tag{5.9}$$

For each pixel, the class extent is given by this matrix. From the matrix, the membership value of classes for any individual pixels can be inferred.

5.2.4.4 Reclassification

With the reference to the membership value matrix, the class which is having the high membership value will be assigned to the pixels. Thus pixels that are mixed at the edges are converted into the unique class pixel. The pixels have 'one pixel-one class' property is ensured after applying the membership function and the class memberships are properly identified. When the class belonging to the mixed pixel is determined, at a spatial location (*i, j*), the average gray level is replaced for the gray level at that particular spatial location for the determined class in that band. After this replacement, feature extraction and reclassification are carried out and accuracies are calculated.

5.2.4.5 Fuzzy-Inspired Process

The overall algorithm for the fuzzy-inspired process suggested in this paper is as follows:

i. The edges are detected and the pixels which are mixed are identified.
ii. The fuzzy set $\{b_1, b_2, \ldots, b_N\}$ is defined where N represents the number of bands in the hyperspectral image.
iii. The fuzzy subset $[c_1, c_2, \ldots, c_n\}$ is defined where n is the number of land cover classes in the hyperspectral image.
iv. Observe 'x_b' the gray level or reflectance value of mixed pixel 'x' in band b.
v. The class 'c' mean gray level value in band 'b' is calculated and denoted as $\mu_{b,c}$.
vi. The class 'c' standard deviation $\sigma_{b,c}$ in band 'b' is calculated.
vii. The membership function $f_{b,c}(x_b)$ for pixel 'x' is calculated in band 'b' by using Eq. (2.8).
viii. The class which has a high membership value is identified.
ix. For all mixed pixels the steps from iv to ix will be repeated.
x. Extract the feature for the pixels with received values.
xi. Reclassify the pixels.

5.3 RESULTS AND DISCUSSION

5.3.1 Results Obtained for Simple 3D DWT and GLCM Method

Two types of dataset are used for the evaluation of the proposed method. Indian Pines data which is taken over Northwestern Indiana captured by AVIRIS (Airborne Visible Infrared Imaging spectrometer) and University of Pavia dataset acquired by ROSIS (Reflective Optics System Imaging Spectrometer) is used. The Indian Pines dataset has

220 bands with 145×145 pixels each and 20 m of spatial resolution of 20 m. Among 220 bands, 180 bands are used for testing, and the remaining bands are not used since they are affected by noise and absorption of water. University of Pavia data contains nine land cover classes captured with a geometrical resolution of 1.3 m. The data has 103 bands with the size of 610×340 pixels each. All the bands are used to test the algorithm.

One level of 3D DWT yields eight subbands. As recommended by previous researchers, the conventional Db2 wavelet is selected for the experiment as Db2 wavelet provides good energy compaction. Other Daubechies wavelets are avoided to minimize the computational time, as they have more number of vanishing movements. Using the wavelet coefficients, three-dimensional cooccurrence features are extracted which are listed in Table 5.1 from the eight decomposed subbands and concatenated to classify the hyperspectral data. To extract these joint spatial-spectral cooccurrence features, two different overlapping windows of size $3 \times 3 \times 3$ are used for the Indian Pines experiment, as it contains small classes like oats, grass pasture mowed, and alfalfa. As the University of Pavia data contains large classes, medium size overlapping sliding window of $7 \times 7 \times 7$ is used. As the contextual information can be better represented by an overlapping window than a nonoverlapping window, the overlapping sliding window is preferred.

Unlike other three-dimensional data like other solid data and MRI, the spectrum is the third axis (z) of hyperspectral image which is used as an alternative to a geometric axis. Therefore, the variance of the same pixel is calculated by using the direction of 0° at two wavelengths without any other spatial direction that is suggested by Tsai et al. (2007). Hence, the direction of 0° and distance of 1 (adjacent neighborhood) is used for extracting the features. Concatenating all the extracted features, the algorithm is executed and classification results are observed. While training the samples, 5% of the randomly selected labeled samples are used and for testing all samples were used. MATLAB package and OSU SVM are used for conducting the experiment. In SVM, kernel type of radial basis function (RBF) is used. For classification, the strategy of one against one is used.

In order to evaluate the performance of different classification methods, the following quantitative measures are used:

1. Class-wise accuracy (CA): This metric shows the value of the ratio when the correctly classified pixel number is divided by the total number count of pixels in a class.

$$\text{CA} = \frac{\text{Number of correctly classified pixels in the class}}{\text{Total number of pixels in the class}}$$

2. Overall accuracy (OA): This value is defined as the ratio of correctly classified pixels number to the total number of pixels in an image.

$$\text{OA} = \frac{\text{Number of correctly classified pixels}}{\text{Total number of pixels in the image}}$$

3. Kappa coefficient (k): This value provides information about the agreement corrected by the level of agreement that could be expected due to chance alone. The kappa statistic is a measure of closeness of the instances that is classified by the machine learning classifier matched with the ground truth data. It is defined in Eq. (5.10).

$$K = \frac{N\sum_{i=1}^{n} m_{i,i} - \sum_{i=1}^{n} (G_i G_i)}{N^2 - \sum_{i=1}^{n} (G_i C_i)} \tag{5.10}$$

where the class number is represented by i

The total count of the pixels classified when compared with the ground truth is denoted by N

The number of pixels that belongs to the ground truth class i is represented by $m_{i,i}$, which is also been classified with a class i

The total count of the pixels classified which belongs to class i will be denoted by C_i

The total count of pixels of ground truth image which belongs to class i will be denoted by G_i

From the obtained results of 3D DWT and 3D GLCM method, it is observed that Db2 wavelet and the sliding window of size $3 \times 3 \times 3$ produce better classification accuracies for Indian Pines data. For University of Pavia data, fuzzy framework is implemented on the features of Db2 wavelet and sliding window of size $7 \times 7 \times 7$.

5.3.2 Results Obtained for Fuzzy-Inspired 3D DWT and 3D GLCM Method

The class-wise accuracies and overall accuracies for the Indian Pines dataset and the University of Pavia dataset for fuzzy-inspired 3D DWT and 3D GCLM are shown in Tables 5.2 and 5.3, respectively. Even though the overall accuracies of the fuzzy-based method are slightly increased, the class-wise accuracies are reduced for a few classes. On observing the obtained results for fuzzy-inspired and nonfuzzy approaches, it is noted that the fuzzy-inspired approach increases the OA, AA, and kappa statistics for both the data. For smaller classes like oats and grass pasture mowed, the fuzzy-inspired method performs better. Few classes like corn notill, corn mintill, soy clean, and woods yield lesser class-wise accuracies in Indian Pines data. In University of Pavia data, Gravel, Self-Blocking Bricks and Shadows show accuracy increment. For other classes, marginal class-wise accuracy difference is observed. However, the class-wise accuracies of some of the classes like Asphalt, Meadows, Painted Metal Sheets, and Bare Soil exhibit slightly lesser than the fuzzy-inspired process. From both the datasets, it is observed that the classes that are less prone to mixed pixel problems exhibit a slight reduction in accuracies after applying fuzzy-inspired model. In order to overcome this limitation, it is decided to select the

TABLE 5.2
Overall Average Accuracies and Kappa (in %) of 3D DWT and 3D GLCM Features Before and After Applying the Proposed Fuzzy-Based Algorithm for Indian Pines Data

		Class-Wise Accuracies (in %)		
Sr. No.	**Class**	**Before Applying Fuzzy**	**After Applying Fuzzy**	**Adaptive Fuzzy**
1	Alfalfa	83.3	87.04	98.15
2	Corn Notill	94.2	93.65	97.49
3	Corn Mintill	97.7	96.88	97.72
4	Corn	88.4	94.44	97.86
5	Grass pasture	88.9	96.58	98.99
6	Grass trees	96.3	95.05	99.33
7	Grass pasture mowed	100	100	100
8	Hay windrowed	98.5	99.59	100
9	Oats	35	55.00	100
10	Soybean Notill	94.5	96.59	98.97
11	Soybean Mintill	94.1	96.96	98.54
12	Soybean clean	95.9	89.58	95.93
13	Wheat	99.5	96.23	99.53
14	Woods	99.0	98.76	99.61
15	BGTD	83.4	96.05	95.26
16	Stone-steel-towers	84.2	83.16	100
Overall accuracy		94.62	95.86	**98.39**
Average accuracy		89.56	92.22	**98.53**
Kappa		92.18	93.24	**95.94**

nonfuzzy and fuzzy approaches in a class-specific manner, and this method is called the adaptive fuzzy method.

In the adaptive fuzzy method, nonfuzzy approach is chosen for the classes that exhibit better results than fuzzy-inspired methods. For the rest of the classes, the fuzzy-inspired method is chosen and experimentation is repeated, and hence it is called adaptive fuzzy approach. Results are obtained and parametric analysis is done. While applying adaptive fuzzy along with 3D DWT and 3D GLCM, overall and average accuracies and kappa statistics are increased. Improvement in class-wise accuracies is observed for all the classes except Building Grass Tree Drives, in which a slight accuracy reduction is observed for Indian Pines data. The same is observed for the class Shadows in University of Pavia data. The observations imply that the adaptive fuzzy method finds difficulty in classifying such classes. For the above-mentioned classes, fuzzy-inspired approach performs better. Classification maps of the proposed methods for before and after applying fuzzy-inspired algorithm for Indian Pines and University of Pavia data are shown in Figures 5.8 and 5.9, respectively. The accuracies are listed in Table 5.4.

TABLE 5.3
Overall Average Accuracies and Kappa (in %) of 3D DWT and 3D GLCM Features Before and After Applying the Proposed Fuzzy-Based Algorithm for University of Pavia Data

		Class-Wise Accuracies		
Sr. No.	Class	Before Applying Fuzzy	After Applying Fuzzy	Adaptive Fuzzy
1	Asphalt	98.79	98.3	99.28
2	Meadows	99.40	99.2	99.95
3	Gravel	96.86	99.1	99.62
4	Trees	93.31	93.9	97.62
5	Painted metal sheets	98.51	97.9	99.85
6	Bare Soil	97.67	96.7	99.22
7	Bitumen	99.10	99.3	99.85
8	Self-blocking bricks	97.18	99.2	99.38
9	Shadows	97.25	99.8	98.52
Overall accuracy		98.27	98.36	**99.49**
Average accuracy		97.56	98.38	**99.25**
Kappa		96.48	96.92	**97.63**

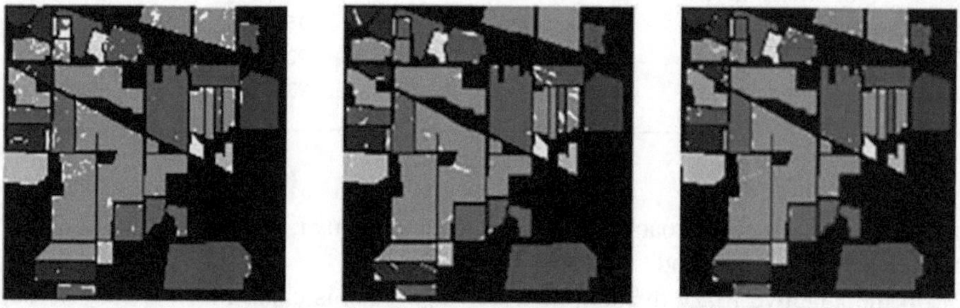

(a) (b) (c)

FIGURE 5.8 Classification maps of fuzzy-based 3D DWT and 3D GLCM method for Indian Pines (a) Before applying fuzzy, (b) After applying fuzzy, and (c) Adaptive fuzzy.

From the comparison in Table 5.5, it is evident that the proposed method yields better results than the conventional spectral feature based classification methods other often addressed joint spectral-spatial features and competitive to the convolutional neural network-based classification methods.

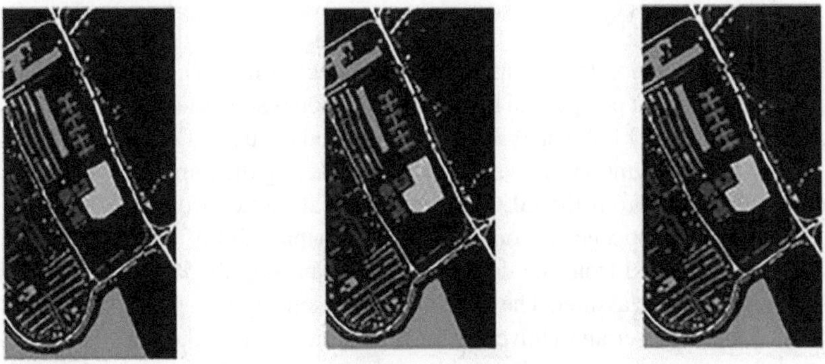

FIGURE 5.9 Classification maps of fuzzy-based 3D DWT and 3D GLCM method for university.

TABLE 5.4
Performance Comparison of Proposed Method with Remaining 3D DWT Methods for Indian Pines Data

	Class-Wise Accuracies (in %) Using 5% Training Samples Using Features				
	Existing Method				
Parameter/Methods and Authors	**LDA Features**	**Spectral Features**	**Attribute Profile**	**CNN Active Learning**	**Proposed Fuzzy Method**
Overall accuracy	65.22	80.56	91.13	94.28	98.39
Average accuracy	73.67	77.81	95.07	89.79	98.53
Kappa	60.61	85.86	89.87	NA	95.94

TABLE 5.5
Performance Comparison of Proposed Method with Remaining 3D DWT Methods for University of Pavia Data

	Class-Wise Accuracies				
	Existing Method				
Parameter/ Methods and Authors	**LDA Features (Li et al. 2012)**	**Spectral Features (Li et al. 2012)**	**Attribute Profile (Ghamisi et al. 2014)**	**CNN Active Learning (Cao et al. 2020)[a]**	**Proposed Fuzzy Method**
Overall accuracy	75.59	80.99	94.77	99.15	99.49
Average accuracy	75.88	88.28	95.41	97.45	99.25
Kappa	68.16	76.16	93.17	NA	97.63

[a] Used only 1% of training samples.

5.4 CONCLUSION

Different multiresolution transform-based hyperspectral image classification methods and a fuzzy-inspired hyperspectral image classification is suggested. In the first method, the combination of 3D DWT and 3D GLCM methods is used. Different wavelets and various overlapping window sizes are used for extracting the joint spatial-spectral and contextual features. From the GLCM, the statistical and cooccurrence features like entropy, energy, homogeneity, contrast, maximum probability, mean, standard and deviation are extracted from the decomposed subbands. Only 5% of the samples are used for training the classifier. The classification is done by using the SVM.

Indian Pines dataset and University of Pavia dataset are taken for the evaluation of the performance of presented algorithms. From the obtained results, it is inferred that the joint spatial-spectral features are extracted and are able to classify the pixels in the hyperspectral images better. The results obtained from this method are superior to the conventional spectral feature–based methods and other two-dimensional DWT-based methods. The same is true while comparing the results with other two-dimensional cooccurrence-based feature extraction methods too.

In order to improve the accuracies further, mixed pixel problem is considered. By fuzzy-based framework, the mixed pixels class membership is evaluated and they are assigned with the labels of the class with the highest class membership. Fuzzy-based method is combined with the multiresolution transform-based methods. Experiments are conducted before and after applying the fuzzy frameworks. In all three experiments, the overall accuracies of the fuzzy-based method are better than the non-fuzzy-based method. But the class-wise accuracy of a few classes is less. Hence an adaptive technique is suggested in which one can choose the fuzzy and nonfuzzy methods in a class-specific manner.

5.5 SCOPE

A few possible future works are proposed in this section. Three different wavelet transforms are used in this work. Some other multiresolution transforms can be used for decomposing the hyperspectral data and classification can be done. Fuzzy-based method is proposed to handle the mixed pixel issue. Some other artificial intelligence method can be used in the future for resolving the same.

REFERENCES

1. D. Landgrebe, G.E. Cohn, J.C. Owicki. (1999). Some fundamentals and methods for hyperspectral image data analysis. *SPIE Photon.* 1(1), 1–15.
2. Y. Qian, M. Ye, J. Zhou. (2012). Hyperspectral image classification based on structured sparse logistic regression and three-dimensional wavelet texture features. *IEEE Trans. Geosci. Remote Sens.* 51(4), 2276–2291.
3. Y. Ye, Z.S. Prasad, W. Li, J.E. Fowler, M. He. (2014). Classification based on 3D DWT and decision fusion for hyperspectral image analysis. *IEEE Geosci. Remote Sens. Lett.* 11(1), 173–177.
4. H. Guo, X. Huang, L. Zhang. (2014). Three-dimensional wavelet texture feature extraction and classification for multi/hyperspectral imagery. *IEEE Geosci. Remote Sens. Lett.* 11(12), 2183–2187.

5. X. Cao, J. Yao, Z. Xu, D. Meng (2020). Hyperspectral image classification with convolutional neural network and active learning. *IEEE Trans. Geo. Remote Sens.* 58(7), 4604–4616.
6. F. Tsai, C.K. Chang, J.Y. Rau, T.H. Lin, G.R. Liu. (2007). 3D computation of gray level co-occurrence in hyperspectral image cubes. *Spring. Lect. Notes Comp. Sci.* 4679, 429–440.
7. F. Tsai, J.S. Lai. (2013). Feature extraction of hyperspectral image cubes using three-dimensional gray level co-occurrence. *IEEE Trans. Geosci. Remote Sens.* 51(6), 3504–3513.
8. H. Su, Y. Sheng, P. Du, C. Chen, K. Liu. (2014). Hyperspectral image classification based on volumetric texture and dimensionality reduction. *Front. Earth Sci.* 9, 225–236.
9. R.M. Moraes, G.J. Banon, S. Sandri. (2002). Fuzzy expert systems architecture for image classification using mathematical morphology operators. *Info. Sci.* 142(1), 7–21.
10. L. Zhang, D. Tao, X. Huang. (2012). On combining multiple features for hyperspectral remote sensing image classification. *IEEE Trans. Geosci. Remote Sens.* 50(3), 879–893.
11. S.K. Mylonas, D.G. Stavrakoudis, J.B. Theocharis, P. Mastorocostas. (2014). Spectral-spatial classification of remote sensing images using a region-based genesis segmentation algorithm. *Proc. IEEE Int. Conf. Fuzzy Syst.* 1976–1984.
12. S.G. Mallat. (1989). A theory for multi resolution signal decomposition: the wavelet representation, *IEEE Trans. Pat. Ana. Machine Intell.* 2(7), 674–693.
13. V.N. Vapnik. (1999). An overview of statistical learning theory. *IEEE Trans. Neural Net.* 10(5), 988–999.
14. J. Canny. (1986). A computational approach to edge detection. *IEEE Trans. Pat. Ana. Machine Intell.* 8(6), 679–698.
15. F. Melgani, A.R. Bakir, A. Hasheny, S.M.R. Taha. (2000). An explicit fuzzy supervised classification method for multispectral remote sensing images. *IEEE Trans. Geosci. Remote Sens.* 38(1), 287–302.

6 Painless Machine Learning Approach to Estimate Blood Glucose Level with Non-Invasive Devices

Altaf O. Mulani
SKN Sinhgad College of Engineering

Makarand M. Jadhav
NBN Sinhgad School of Engineering

Mahesh Seth
TECH-CITY Research and Consulting (OPC) Pvt. Ltd.

CONTENTS

6.1 INTRODUCTION

Presently, there is no permanent medicine to cure diabetes. Patients with diabetic symptoms are kept within the range by controlling their blood glucose levels. Various techniques are available to measure and monitor the regulation of glucose. However, available glucose meters offered to a customer has their own disadvantages. These meter types need a blood sample extracted by pecking a needle into the

DOI: 10.1201/9781003220176-6

patient's fingers, which is a painful process. This results in the formation of copious calluses on the fingertips and causes more pain to lure blood again and again for repetitive measurements. The second type, which is a continuous glucose meter, provides incessant monitoring of glucose levels based on measuring interstitial fluid. This meter requires added calibration to blood samples and is not cost-effective. The third type of glucose meter is comparatively cheap and uses disposal electrode strips. However, strips are expensive for patients who require to monitor and measure their glucose level several times a day.

In general, it is observed that factors such as quality of strips and inadequate knowledge of clinicians are affecting the measurement of glucose level. Further, clinicians, as well as patients, must be aware of test strip physical and chemical performance under different environmental conditions. It is observed that 91%–97% of overall inexactness in blood glucose errors are operator-dependent. The maximum discrepancies are due to unknowingly applied mechanical stress on strips, use of non-calibrated as well as faulty meters and adequate blood quantity applied on the strip for glucose measurement. It is concluded that clinicians need to instruct patients to read operating manuals of instruments. For example, glucose oxidase meters misjudge glucose levels at low temperatures, whereas glucose-dehydrogenase meters give unpredictable measurements in increased humidity conditions.

Figure 6.1 demonstrates the generation of glucose in the human body. Diabetes mellitus in India is known as type 1 and type 2 diabetes. A glucose concentration in the range of 72–134 represents non-diabetic patients, whereas the range of 111–330 mg/dL represents diabetic patients. This diabetes damages the eyes, kidneys and nerves. Further, prediabetes hypertension due to kidney failure. In type-1 diabetes, insulin deficiency in patients destroys pancreatic beta-cell. This, in turn, doesn't uphold glucose homoeostasis in patients. Hence, 50%–60% of patients under 18 years of age are infected by type-1 diabetes [1]. Day-to-day life cycle against the biological clock is one significant factor in gaining weight. This has resulted in attaining type 2 diabetes. In connection with this, the patient is suggested to perform daily physical

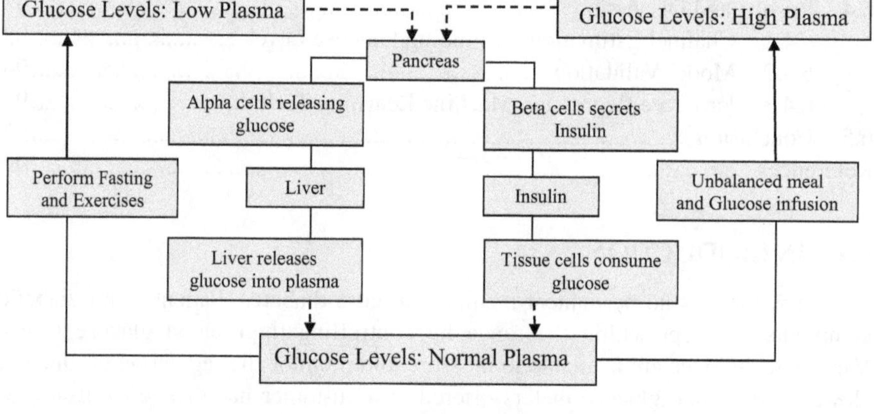

FIGURE 6.1 Typical blood glucose generation system [22].

exercise and adopt a good diet plan. This helps to control chronic diseases like vascular as well as neurological illnesses [2]. In 2003, the cost of treating diabetes was estimated to be $132 billion.

The survey has shown that the estimated count of diabetes patients detected will be 37.3 million by 2021 costing an estimated $192 billion [3,4]. Red blood cells in human body have a life of 90–120 days and produce haemoglobin. The process of glucose getting mixed in cells generates glycated haemoglobin called HBA1C. This is an adopted method in clinic/home to monitor and control glucose level as well as insulin quantity of the patient. Based on this, diet plans and yoga exercises are suggested to patients.

6.2 TYPES OF GLUCOSE MONITORING TECHNIQUES

Presently, there are two types of glucometers available in the market. One is based on an invasive method and the other on non-invasive method to measure the glucose level in the blood cells.

6.2.1 Invasive Method for Glucose Measurement

The first glucose monitoring meter was designed in the late 1960s. The process to measure glucose level in blood remains the same in the current available meters. Factors such as device miniaturization, daily easiness to measure as well as monitor, and mechanism to acquire record data are changes found in present meters. Presently, blood meters use glucose oxidase and platinum-coated strips for glucose measurement. Here, initially, blood is pecked from the patient's finger and placed on enzyme strips. Further, amperometry analysis is carried out on the yield of hydrogen peroxide engendered from the chemical reactions of glucose and oxygen [5,6].

6.2.2 Non-Invasive Method for Glucose Measurement

Non-invasive method for measuring glucose is classified into ten categories. This method is intensely investigated over the past several decades. The working principle of each method is elaborated in this section.

Interstitial Fluid Chemical Method: This is the first traditional monitoring system designed and approved by FDA [9,10] in 2001. This meter was able to measure blood glucose levels accurately in several trials. It was made available in the form of a watch called gluco-watch [8]. It works on the principle of enzymatic reaction. Here, interstitial fluid is expatriated by pressing the disposal pad hard-pressed on the skin. The cost of pads has shut down the manufacturing and use of this meter [11].

Breath Chemical Method: This method was designed to measure glucose in the blood with the help of acetone level in an exhaled breath. It has been shown that the level of acetone abruptly rises in diabetic patients with increase in glucose [12].

Infrared Spectroscopy Method: This method is classified into two groups – near-infrared and mid-infrared. It works on the principle of light characteristics to illustrate the content of glucose in blood [15]. Here, relatively thin tissue of body skin is exposed for glucose measurement. These tissues are available on the earlobe, lips and fingers of the human body [14]. It is observed that such measurements are sensitive to skin structure and are not a reliable method.

Optical Coherence Tomography Method: In this method, the principle of light scattering mechanism based on refractive index values is used to measure glucose in the blood. Here the light is incident on the human arm, and the phase component of reflected light is used to measure refractive index using Snell's law. It is observed that an increase in glucose level increases incident angle and in turn changes the refractive index [16].

Temperature Modulated Localized Reflectance Method: This is another method based on the light scattering mechanism. It is observed that soft skin tissue of humans is sensitive to environmental conditions as well as to concentrations of glucose in the blood. Thus, modifying skin temperature in the range of 22°–38°C is resulting in change of the reflected angle of light. The refractive index is measured thereafter to determine glucose levels in the blood [17].

Raman Spectroscopy Method: In this method, a laser device is used to stimulate glucose molecules in the ocular fluid. Thus, an oscillation in a fluid causes changes in light properties [18]. The amount of light scattered is used as a metric to measure the glucose level in blood.

Polarization Change Method: This method works on the principle of light polarization. It is observed that polarization angle of light source is a function of glucose content in the blood [19]. However, pH and interfering compounds have prevented its feasibility for glucose measurement.

Ultrasound Method: This method works on the principle of ultrasound to monitor as well as measure glucose levels in the blood. Here the frequency of laser is varied to determine glucose contents in the blood. A short laser ray is used to heat a small tissue area of human body. This causes a pulse of ultrasonic to propagate into this tissue [20].

Fluorescence Method: In this method, soft tissues of the skin are excited with ultraviolet laser operated at 380 nm wavelength. This in turn generates fluorescence that is used as a performance metric to measure glucose level in the blood. It is concluded that intensity of fluorescence depends on levels of glucose in the blood and is affected by skin thickness as well as pigments [13].

Thermal Spectroscopy Method: In this method, infrared radiation emission from the human body is used for glucose measurement. Here, the concentration of glucose shows an absorptive effect on the volume of infrared emission [7,8].

Ocular Spectroscopy Method: In this method, human tears are chemically evaluated with the help of spectrometer to divulge glucose levels in the blood. A contact laser lens is designed [21] that retorts with the glucose contents extracted from tears. It is observed that reflected light from the lens is able to change the wavelength based on the glucose contents in the blood.

Impedance Spectroscopy Method: In this method, radio-frequency radiation is used to measure impedance of skin tissue to character glucose levels in the blood. It is observed that tissue impedance is a function of signal wavelength as well as of radiation-matter interaction properties.

6.3 PAINLESS NON-INVASIVE GLUCOMETER USING MACHINE LEARNING APPROACH

Imagine diabetics who no longer have to suffer the frequent pain of drawing blood to measure their blood sugar levels. Frequent monitoring results in better disease management, less suffering, fewer medical complications and lower healthcare costs.

In day-to-day life, an embarrassing approach of pricking blood from the human body is used to measure and record glucose levels of diabetic patients. This process is carried out either in the clinic or at home. However, obtained glucose levels in the blood are at that specific time only. Hence, there is a need to monitor and measure unobserved glycaemic changes in blood several times a day to avoid complications. Therefore, designing painless and low-cost non-invasive meters for continuous glucose measurement has always been an area of interest. Such personal device is found better for glucose control and taking necessary preventive actions at the right time to save lives.

Figure 6.2 represents the block diagram of proposed non-invasive glucometer. The channel estimation signal response transmitted from Wi-Fi transmitter to

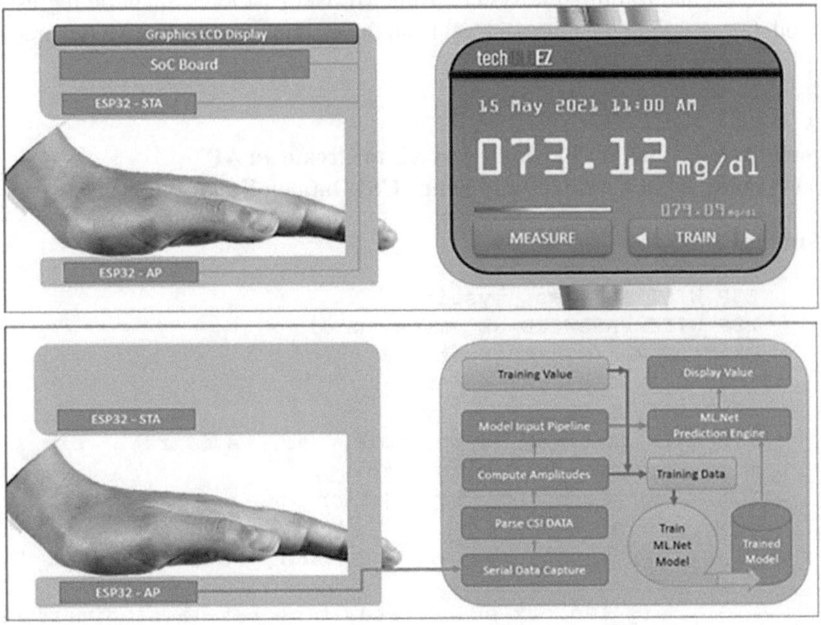

FIGURE 6.2 Block diagram of a proposed non-invasive glucometer.

receiver while transferring a data packet is analysed for the dielectric properties of the sample (hand) kept between transmitter and receiver. A dielectric characterization is used to estimate the dielectric property of the sample with low-cost commodity Wi-Fi devices. The channel state information (CSI) is captured and the amplitude values for respective subcarriers are computed. To reduce the dimension of the dataset, some key frequency data is taken. The model is trained using ML.Net Fast Forest Regression. Using the reduced dataset, the model is trained for different glucose concentrations. Using water-soluble glucose solution, the accuracy obtained is around 90%. Further, real patient data is to be collected and the machine-learning approach is used to train the system based on the data values collected/experimented. The description of non-invasive glucometer to measure glucose level in human blood using Wi-Fi CSI data and machine learning is described in Algorithms 1–5.

Practically, signal propagation between source and destination is affected due to physical properties of the wireless medium. Further, resource allocation at the physical layer is done based on received signal strength indicator (RSSI) as well as CSI values obtained from Wi-Fi devices. RSSI provides information on wireless channel properties and packet delivery status. On the other hand, CSI gets amplitude and phase information of subcarriers used for communication. All this information is used to estimate the channel properties of the communication link.

We have used the CSI data to train and measure the blood glucose levels keeping a human hand between STA (Transmitting Station) and AP (Receiving Access Point). The variations in dielectric properties of blood glucose affect the amplitudes for the subcarriers. The system is trained with multiple individuals by taking actual blood glucose levels and training the system using ML.Net Fast forest quantile regression method. For checking the glucose level from the system, the ML.Net Model predicts the glucose level in Algorithm 1.

Algorithm 1: ESP32 Wi-Fi Module as AP to Create an AP (Access Point) That Captures and Sends CSI Data on Serial Port

Definitions

```
ESP_WIFI_SSID as 'myssid'
ESP_WIFI_PASSWORD as 'mypassword'
EXAMPLE_MAX_STA_CONN as '16'
SHOULD_COLLECT_CSI as 'y'
SEND_CSI_TO_SERIAL as 'y'
CSI_DATA as array [type, role, mac, rssi, rate, sig_
mode, mcs, bandwidth, smoothing,
      not_sounding, aggregation, stbc, fec_coding, sgi,
      noise_floor,
      ampdu_cnt, channel, secondary_channel, local_
      timestamp, ant,
      sig_len, rx_state, real_time_set, real_timestamp,
      len, Non-
      HT_128byte_CSI_DATA ]
```

Register Events

```
( s_wifi_event_group )
SYSTEM_EVENT_AP_STACONNECTED : Print To Serial 'STA_MAC
Connected'
SYSTEM_EVENT_AP_STADISCONNECTED : Print To Serial 'STA_
MAC Disconnected'
```

Program

```
nvs_init : ESP32 Library Function to initialise Non
Volatile Storage softap_init:
      initialise the s_wifi_event_group
      initialise the tcpip_adapter
      initialise the wifi_config.ap with parameters
            ssid=ESP_WIFI_SSID;password=ESP_WIFI_PASS-
            WORD;
            max_connection = EXAMPLE_MAX_STA_CONN;
            authmode = WIFI_AUTH_WPA_WPA2_PSK; channel
            = 8;
      if ESP_WIFI_PASSWORD is empty
            set wifi_config.ap.authmode as
            WIFI_AUTH_OPEN
      esp_wifi_set_mode to WIFI_MODE_AP : Print Error
      to Serial
      esp_wifi_set_config to wifi_config if mode is
      'AP' : Print Error to Serial
      esp_wifi_start : Print Error to Serial
      esp_wifi_set_ps as WIFI_PS_NONE to disable Power
      Saving Mode
      Print to Serial : 'softap_init finished'
csi_init : Print To Serial CSI DATA on socket request
event
```

The ESP32 is configured and initialised as an access point. Packets will move from transmitter to receiver. On receipt of an incoming request from STA, the CSI values are filled in CSI_DATA array and printed on serial port. The frequency responses of subcarriers are measured as CSI values. It is stored in registers as two bytes of signed characters. One-byte stores imaginary value, whereas the second byte stores the real value of CSI. At the end, the bequest long training field (BLTF) of 128 fields from the CSI information received.

Algorithm 2: ESP32 Wi-Fi Module as STA to Create a STA (Station) Which Sends 100 Requests/Sec to AP to Generate CSI_DATA

Definitions

```
ESP_WIFI_SSID as 'myssid'
ESP_WIFI_PASSWORD as 'mypassword'
```

```
PACKET_RATE as '100'
SHOULD_COLLECT_CSI as 'y'
SEND_CSI_TO_SERIAL as 'y'
CSI_DATA as array   [ type, role, mac, rssi, rate,
                    sig_mode, mcs, bandwidth,
                    smoothing,
                    not_sounding, aggregation, stbc,
                    fec_coding, sgi, noise_floor,
                    ampdu_cnt, channel, secondary_
                    channel, local_timestamp, ant,
                    sig_len, rx_state, real_time_set,
                    real_timestamp, len, Non-
                    HT_128byte_CSI_DATA]
```

Register Events

```
( s_wifi_event_group )
SYSTEM_EVENT_STA_START          : esp_wifi_connect( )
SYSTEM_EVENT_STA_GOT_IP         : Print To Serial 'STA_IP'
SYSTEM_EVENT_STA_DISCONNECTED : Print To Serial 'STA
                                Disconnect',
                                retry esp_wifi_connect( )
```

Program

```
nvs_init   : ESP32 Library Function to initialise Non
Volatile Storage
station_init  :
      initialise the s_wifi_event_group
      initialise the tcpip_adapter
      initialise the wifi_config.ap with parameters
            ssid = ESP_WIFI_SSID; password = ESP_WIFI_
            PASSWORD;
            channel = 0;
      esp_wifi_set_mode to WIFI_MODE_STA        : Print
      Error to Serial
      esp_wifi_set_config to wifi_config if mode is
      'STA'  : Print Error to Serial
      esp_wifi_start                            : Print
      Error to Serial
      esp_wifi_set_ps as WIFI_PS_NONE to disable Power
      Saving Mode
      Print to Serial : 'softsta_init finished'
task_socket_transmitter_sta_loop :
      while(1) :
            close sockets
            if wifi_not_connected delay for 1000ms
            open socket to IP 192.168.4.1     : Print
            Error to Serial
                  while(1) :
```

```
                    if wifi_not_connected break  : Print
                    Error to Serial
                    send data '1\n' over socket
                    delay 1ms
                delay 2ms
```

The ESP32 is configured and initialised as a station. The access point SSID/Password is configured & Wi-Fi connection is established. At each second, an http request is sent to the AP, which enables the AP to generate the CSI data.

Algorithm 3: Capture CSI_DATA over Serial Port

Definitions

```
sp as Serial Port, 115200 baud
```

Register Events

```
spi_DataRecieved  :
            Check if Data is from AP & verify
            AP_MAC_Address
            Split & Transfer the captured CSI_DATA
            line to GUI Text Box
```

Program

```
open_serial   : opens the Serial Port
                Set GUI button text as 'STOP'
close_serial  : close the Serial Port
                Set GUI button text as 'START'
```

The serial port is initialised at 115,200 baud. A serial data receive event and its delegate function are defined. On serial data receive event the function verifies if the data is from the AP & confirms by checking its MAC address. On verification, the required data is parsed & forwarded to the GUI layer of the application.

Algorithm 4: Parse CSI_DATA and Plot Graph

Definitions

```
CSIDATASET 2D array of [64][1024]
list_im, list_real, list_amp
list_amp_ch8, list_amp_ch16, list_amp_ch24, list_amp_ch32,
list_amp_ch40, list_amp_ch48, list_amp_ch56
chart1
```

Program

```
loop for i=0:32
        clear list_im, list_real, list_amp
        set cd as CSIDATASET [ i ]
        split cd values in list_im, list_real
        loop for j=0:64
              calculate amplitude :
                        list_amp[ j ] = sqRoot( list_im[ j
]^2 + list_real[ j ]^2 )
        hampel filter on list_amp
        plot the list_amp[ i] of 64 amplitudes on chart1
if textBoxTrain value is set
        save the data to trainCSV for 8,16,24,32,40,48,56
        spaced frequencies
else
        generate input_pipeline for 8,16,24,32,40,48,56
        spaced frequencies
        Predict result from the input_pipeline using
        ML.Net
        Display result on GUI
```

Algorithm 5: Train Model with ML.Net

Definitions

```
TRAIN_DATA_FILEPATH
MODEL_FILEPATH
mlContext object of ML.Net
```

Program

```
load training data from TRAIN_DATA_FILEPATH to ML.Net &
generate trainingDataView
build trainingPipeline from mlContext
Train Model mlModel with ML.Net:Regression Training
[ FastForest ]
Evaluate quality of model & Print Metrics for
Regression Model
Save Model to MODEL_FILEPATH
```

A Hampel filter can be applied to our time series to identify outliers and replace them with more representative values. The filter is basically a configurable-width sliding window that we slide across the time series. For each window, the filter calculates the median and estimates the window's standard deviation σ. For any point in the window, if it is more than 3σ out from the window's median, then the Hampel filter identifies the point as an outlier and replaces it with the Windows Median.

6.4 RESULTS AND DISCUSSION

This work used Wi-Fi sensing methodology to develop a painless non-invasive device for measuring glucose level. It provides abundant signals for accurate detection and estimation of signals. Table 6.1 highlights the specifications of proposed non-invasive glucometer device. Here, Wi-ESP is used as a tool to extract CSI information, whereas ESP32 works as an access point that implements various layer protocol to communicate with different routers in client mode. All these accurate 52 subcarriers channel information along with CSI measurements in the frequency domain helps to enhance channel estimation.

6.4.1 Channel Estimation for Finding Glucose Level

CSI is a metric used for recitation amplitude and phase disparities among subcarrier frequencies as shown in Figure 6.3. This is the result of signal fading which is available on the path from transmitter to receiver. In this work, a codebase for ESP-32s microcontrollers in C is used with Express as development framework. It runs two applications. One is for access point to initialize on-board Wi-Fi stack to connect devices for communication. Second is an active station to send requests to the server seriatim on the access point. This application helps to collect CSI data for processing. It is printed repeatedly to a serial port of the device. It can be concluded that the tool gives access to 64 subcarriers. The attained resolution of individual imaginary and real number is 8 bits. It is on-par with any other tools.

The system design and configuration consist of two ESP32 Wi-Fi devices, which are installed on a personal computer. Two devices are utilized to emit and receive signals. An off-loading task is well-thought-out to detect real-time signals between access point and station. The receiver collects raw CSI signals with the help of serial communication from the channel. In this work, additional header information per CSI frame for all subcarriers is transmitted. This information consists of MAC address, RSSI as well as CSI. The device measure CSI data for a set of 32 transmissions for all subcarriers. Further, captured output resulted in formation of frame size

TABLE 6.1
Specifications of Painless Non-Invasive Glucometer Device

Application: To determine blood glucose in whole blood by non-invasive method
Operating conditions:
Temperature: 4°C–45°C
Humidity: <98%
Operating altitude: 0–3,094 m
Display: Large display with 3.5 inch LCD
Measuring range: 40–400 mg/dL
Measuring time: less than 3 seconds
Memory capacity: 2,000 blood glucose test results with time, date and daily as well as weekly average report for random or before or after meal
Connectivity: USB interface

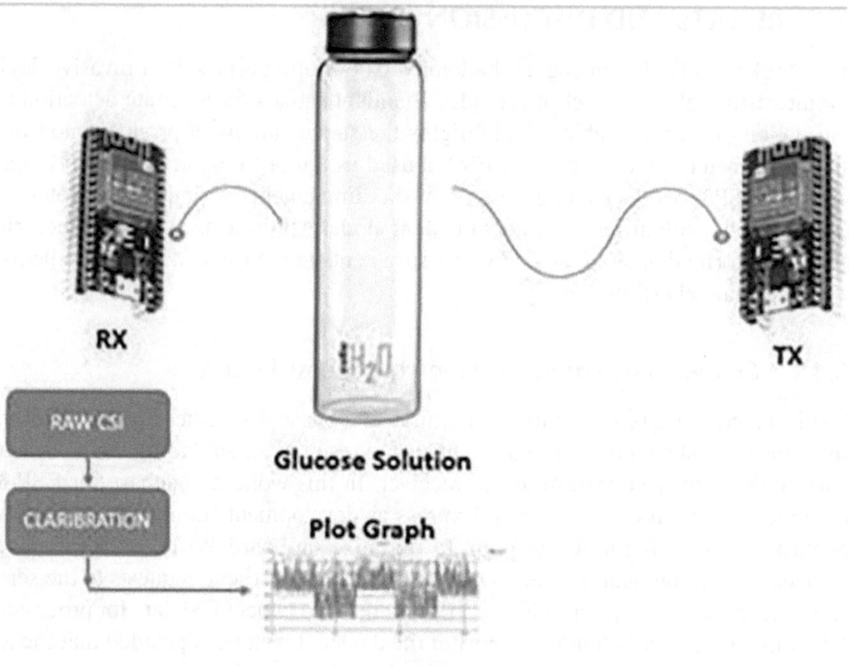

FIGURE 6.3 Extraction of channel state information.

equal to 1 kB. Further, serial communication speed is set to 115,200 bits/s. At this rate, the highest throughput is achieved by retaining serial reliability. The results are validated by placing a dielectric material to obtain a change in CSI for signal processing. A theoretical model in Figure 6.4 quantitatively characterizes the relationship between signal characteristics and material dielectric properties.

CSI is a metric used in orthogonal frequency-division multiplexing (OFDM) for describing amplitude and phase variations across multiple subcarrier frequencies as

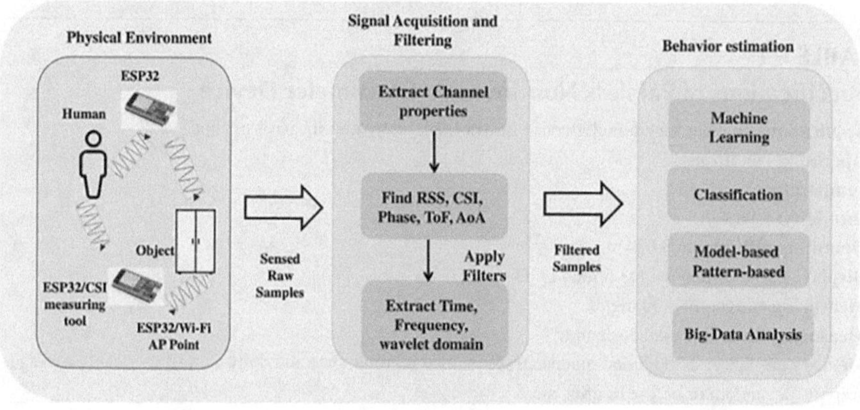

FIGURE 6.4 Conventional block diagram of glucometer.

wireless signals are transmitted between a transmitter and receiver. A pilot symbol helps to detect variations across subcarriers. The estimates vector $\boldsymbol{H}$ is represented as fitness function in terms of CSI with

$$Y = H * X + N \tag{6.1}$$

where Y is the detected receiver vector, X represents the transmitter vector and N is the noise vector. It is observed that the information collected for each subcarrier is a complex value. Thus with the help of real as well as imaginary value, amplitude $A^{(i)}$ and phase $\varphi^{(i)}$ is measured by

$$A^{(i)} = \sqrt{\left(h_{im}^{(i)^2} + h_r^{(i)^2}\right)}$$

$$\begin{aligned} A^{(i)} &= \sqrt{\left(h_{im}^{(i)^2} + h_r^{(i)^2}\right)} \\ \emptyset^{(i)} &= a\tan 2\left(h_{im}^{(i)}, h_r^{(i)}\right) \end{aligned} \tag{6.2}$$

All these measurements are taken by assuming the transmitter section is in line of sight with the receiver section. Thus, locating ESP32 devices becomes an important factor to enhance estimation accuracy. The amplitude depends upon the received signal strength and is captured by the parameter RSSI. The collected CSI frame is calibrated as per the RSSI value to minimize the effect of the automatic gain control of the transmitted signal.

When building models for forecasting time series, we generally want "clean" datasets. Usually, this means we don't want missing data and we don't want outliers and other anomalies. But real-world datasets have missing data and anomalies. We use the Hampel filter to deal with these problems. A Hampel filter is applied to identify outliers and replace them with more representative values as explained in Figure 6.5. The filter is basically a configurable-width sliding window that we slide across the time series. For each window, the filter calculates the median and estimates

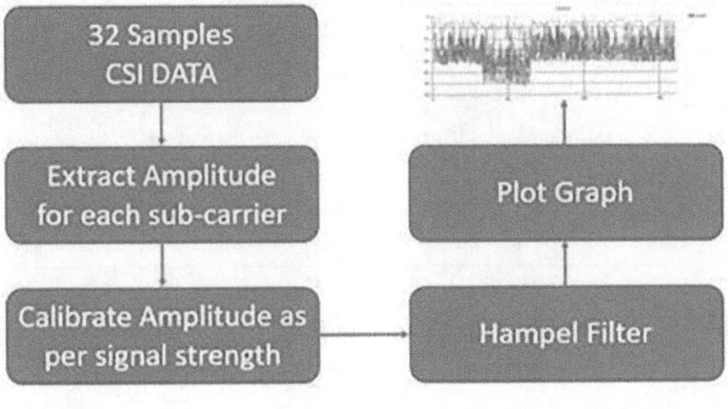

FIGURE 6.5 Operation of Hampel filter.

the window's standard deviation σ. For any point in the window, if it is more than 3σ out from the window's median, then the Hampel filter identifies the point as an outlier and replaces it with the window's median.

6.4.2 Model Validation

This section describes model validation in the presence of a liquid with specific glucose content. In this work, total three experiments are performed by placing three containers comprising air, water and 5% glucose solution between the access point and station. Here antenna is assumed to be in the line of sight to avoid signal loss. It is found that an amplitude attenuation occurs with different concentrations of glucose liquid placed between the access point and station. Figure 6.6 shows the plot of amplitudes for all subcarriers against different glucose levels. When the container is empty, the amplitude received is represented for all the subcarriers. The amplitude significantly attenuates when a 5% glucose solution is placed between access point and station. In practice, high variations in amplitude is observed due to different propagation mechanism.

The second absurdity is the phase shift obtained because of the initial phase of the vacant space, signal replications as well as variation in human movement velocity. For simulation, attenuation is presumed persistent over a period, whereas the distance between access point and station is reduced to minimize the phase shift variation.

In Figures 6.6 and 6.7, an experiment was carried out by placing three containers containing honey, DNS and 5% dextrose solution. The amplitude for 'Honey' is highly attenuated than for 5% dextrose. DNS as well as 5% dextrose show similar attenuation as both solutions contain the same amount of glucose. In Case Study 1, we found out that there is an attenuation in CSI amplitudes if a bottle filled with glucose solution is kept in between access point and station. Thereafter, ML.Net is used to take blood samples of various humanoid to train the device. It is accomplished in three steps: combines data loading, data transformations and creating machine-learning

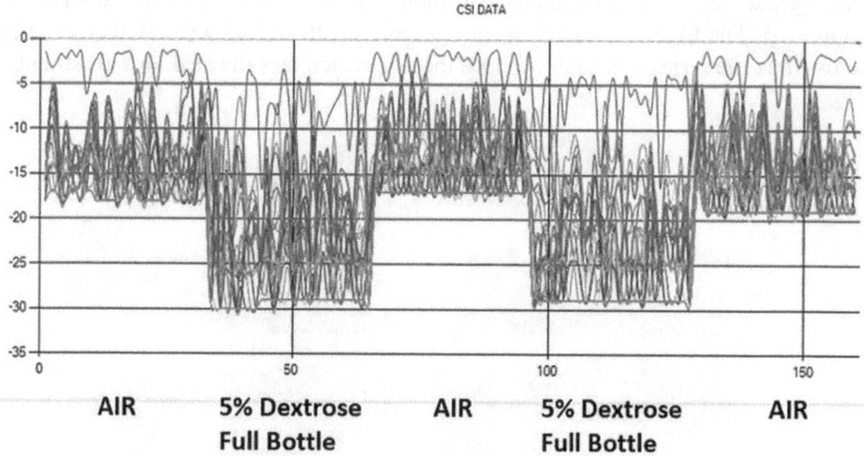

FIGURE 6.6 Experimental results for training the model.

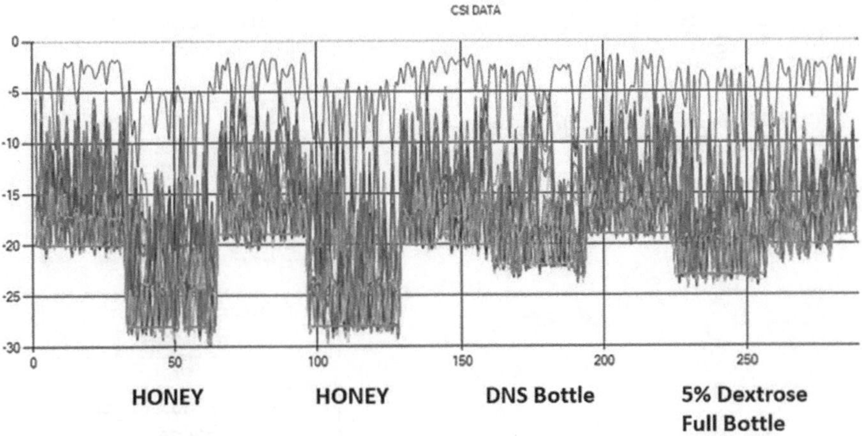

FIGURE 6.7 Experimental results of keeping a human hand to measure glucose level.

models. It helps in selection of superlative algorithm for learning task. It is used to find patterns to make predictions on new and unknown data. In machine learning, feature attribute is to make predictions and the label for actual prediction.

A fast-tree regression trainer is used as an estimator to detect glucose content in the blood. In the end, the trained model is saved as binary file. It is amalgamated into. NET applications to make predictions as shown in Figure 6.8.

6.4.3 Fast-Tree Regression Machine Learning Technique

Fast-Tree is an effective execution of the gradient boosting algorithm to solve regression problems. It is a machine-learning technique. Regression is a binary tree flow chart. Thereafter, predefined loss fitness function is established in a step-wise manner. It measures errors for individual steps and helps to correct them in the succeeding steps. The decision is based on the applied test when the feature value in the input sample is less than or equal to one of the possible values. The output of the algorithm at any instance is the summation of the tree outputs.

6.5 CONCLUSION

A glucose concentration in the range of 111–330 mg/dL represents a diabetic patient. A blood glucose meter is used most at home to measure glucose content in the blood as per the doctor's prescription. Thus, a portable painless and cost-effective non-invasive glucose monitoring device is developed to monitor the glucose level of a person at home as well as at the office using Wi-Fi module.

A realistic reusable low-cost non-invasive glucose monitoring system is developed using 802.11a Wi-Fi module. A variation in amplitudes as well as phases of received packets helps to measure blood glucose levels. A Hampel filter is used to suppress abrupt amplitude variations occurring due to the environmental effects. A total of three containers is placed between transmitter and receiver containing air, water and 5%

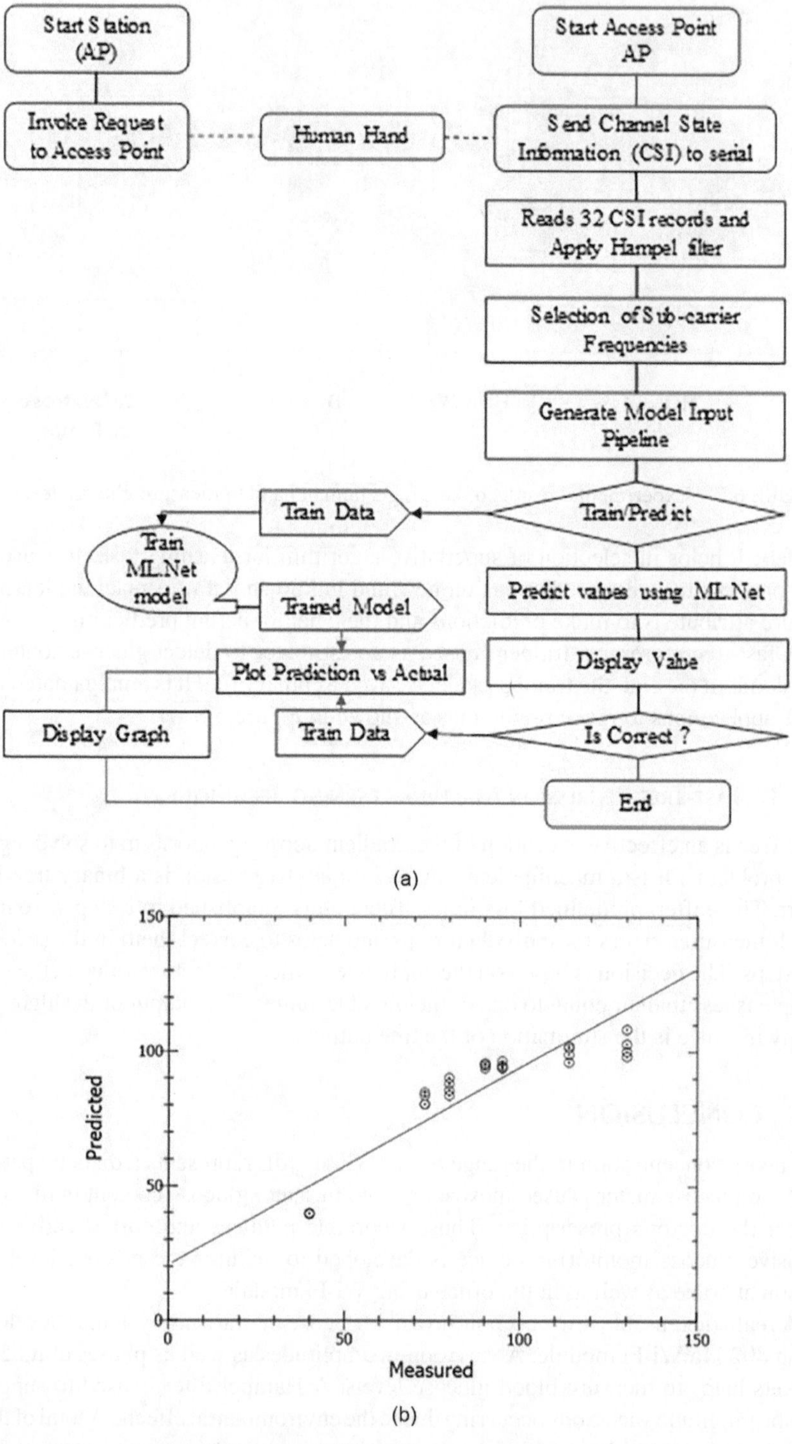

FIGURE 6.8 (a) Flow chart of proposed non-invasive glucometer (NIG) and (b) Graph of predicted vs measure CSI values.

glucose solution to record variations in subcarrier amplitudes. Further, Fast-Tree Regression algorithm is used as a machine-learning technique to train the model for different glucose concentrations for accurate prediction and detection of diabetes. It also helps to reduce dataset dimension for minimizing the time for training the system. The accuracy can be increased further above 95% by training the system with a few more factors such as skin pigment, blood flow, epidermis, and bone density.

REFERENCES

1. Masoud, B., Zahra, A., Mojgan, D., and Light, P. E. (2020). Non-invasive continuous-time glucose monitoring system using a chipless printable sensor based on split ring microwave resonators. *Scientific Reports (Nature Publisher Group), 10*(1), 1–15.
2. Islam, M., Ali, M. S., Shoumy, N. J., Khatun, S., Karim, M. S. A., and Bari, B. S. (2020). Non-invasive blood glucose concentration level estimation accuracy using ultra-wide band and artificial intelligence. *SN Applied Sciences*, *2*(2), 1–9.
3. Hernandez, S. M., and Bulut, E. (2020, August). Lightweight and standalone IoT based WiFi sensing for active repositioning and mobility. In *2020 IEEE 21st International Symposium on" A World of Wireless, Mobile and Multimedia Networks"(WoWMoM)*, Ireland (pp. 277–286). IEEE.
4. Wang, X., Yang, C., and Mao, S. (2020). On CSI-based vital sign monitoring using commodity WiFi. *ACM Transactions on Computing for Healthcare*, *1*(3), 1–27.
5. Tan, S., Zhang, L., and Yang, J. (2018, July). Sensing fruit ripeness using wireless signals. In *2018 27th International Conference on Computer Communication and Networks (ICCCN)*, China (pp. 1–9). IEEE.
6. Song, H., Wei, B., Yu, Q., Xiao, X., and Kikkawa, T. (2020). WiEps: Measurement of dielectric property with commodity Wifi device—an application to ethanol/water mixture. *IEEE Internet of Things Journal*, *7*(12), 11667–11677.
7. Islam, M., Ali, M. S., Shoumy, N. J., Khatun, S., Karim, M. S. A., and Bari, B. S. (2020). Non-invasive blood glucose concentration level estimation accuracy using ultra-wide band and artificial intelligence. *SN Applied Sciences*, *2*(2), 1–9.
8. Baghelani, M., Abbasi, Z., Daneshmand, M., and Light, P. E. (2020). Non-invasive continuous-time glucose monitoring system using a chipless printable sensor based on split ring microwave resonators. *Scientific Reports*, *10*(1), 1–15.
9. Li, F., Valero, M., Shahriar, H., Khan, R. A., and Ahamed, S. I. (2021). Wi-COVID: A COVID-19 symptom detection and patient monitoring framework using WiFi. *Smart Health*, *19*, 100147.
10. Wang, F., Han, J., Zhang, S., He, X., and Huang, D. (2018). CSI-Net: Unified human body characterization and pose recognition. *arXiv preprint arXiv:1810.03064.*
11. Jain, P., Joshi, A. M., Agrawal, N., and Mohanty, S. (2020). iGLU 2.0: A new non-invasive, accurate serum glucometer for smart healthcare. *arXiv preprint arXiv:2001.09182.*
12. Hanna, J., Bteich, M., Tawk, Y., Ramadan, A. H., Dia, B., Asadallah, F. A., and Eid, A. A. (2020). Noninvasive, wearable, and tunable electromagnetic multisensing system for continuous glucose monitoring, mimicking vasculature anatomy. *Science Advances*, *6*(24), eaba5320.
13. Forbes, G., Massie, S., and Craw, S. (2020, November). WiFi-based human activity recognition using raspberry Pi. In *2020 IEEE 32nd International Conference on Tools with Artificial Intelligence (ICTAI)*, Baltimore (pp. 722–730). IEEE.

14. Islam, M., Min, C. L., Shoumy, N. J., Ali, M. S., Khatun, S., Karim, M. S. A., and Bari, B. S. (2020, May). Software module development for non-invasive blood glucose measurement using an ultra-wide band and machine learning. *Journal of Physics: Conference Series, 1529*(5), 052066. IOP Publishing.
15. Hernandez, S. M. and Bulut, E. (2020, March). Performing WiFi sensing with off-the-shelf smartphones. In *2020 IEEE International Conference on Pervasive Computing and Communications Workshops (PerCom Workshops)*, Atlanta, Asia (pp. 1–3). IEEE.
16. Atif, M., Muralidharan, S., Ko, H., and Yoo, B. (2020). Wi-ESP—A tool for CSI-based device-free Wi-Fi sensing (DFWS). *Journal of Computational Design and Engineering*, *7*(5), 644–656.
17. Geng, Z., Tang, F., Ding, Y., Li, S., and Wang, X. (2017). Noninvasive continuous glucose monitoring using a multisensor-based glucometer and time series analysis. *Scientific reports*, *7*(1), 1–10.
18. Zhang, Y., Zhang, Y., Siddiqui, S. A., and Kos, A. (2019). Non-invasive blood-glucose estimation using smartphone PPG signals and subspace kNN classifier. *Elektrotehniski Vestnik*, *86*(1/2), 68–74.
19. Mulani, A. O. and Mane, P. B. (2016, October) Area efficient high speed FPGA based invisible watermarking for image authentication. *Indian Journal of Science and Technology*, *9*(39). Doi: 10.17485/ijst/2016/v9i39/101888.
20. Mulani, A. O. and Mane, P. B. (2017). An efficient implementation of DWT for image compression on reconfigurable platform. *International Journal of Control Theory and Applications*, 10(15), 1006–1011.
21. Jadhav, M. M., Durgude, Y., and Umaje, V. N. (2019). Design and development for generation of real object virtual 3D model using laser scanning technology. *International Journal of Intelligent Machines and Robotics*, *1*(3), 273–291.
22. Jadhav, M. M. Chavan, G. H., and Mulani, A. O. (2021). Machine learning based autonomous fire combat Turret. *Turkish Journal of Computer and Mathematics Education*, *12*(2), 2372–2381.

7 Artificial Intelligence and Machine Learning in Biomedical Applications

Vaibhav V. Dixit
RMD Sinhgad School of Engineering Pune

Mayuresh B. Gulame
G. H. Raisoni College of Engineering and Management

CONTENTS

DOI: 10.1201/9781003220176-7

7.1 INTRODUCTION

Artificial intelligence (AI) is a branch of engineering that has been around since the 1940s and specialises in using data, probabilities, and various types of uncertainties to solve problems that conventional computer scientists find difficult [1]. AI is classified as computer intelligence, which is different from humans or other living class intellect. AI is also known as the analysis of intelligent robots, or any object or device that can recognise and know its environment and take suitable action to improve its chances of succeeding in its goals. In both computer vision and medical imaging, machine learning (ML) is a methodology that fosters many AI applications. However, using this technique blindly, particularly for medical applications, may result in poor results. As a result, potential pitfalls and associated challenges in ML phases such as pre-processing, learning, and evaluation must be understood. There's a lot of hope that applying AI to healthcare and biomedical application will result in significant changes in all fields, from diagnosis to treatment. Biomedical engineering is in high demand and many countries are facing a physician shortage. Healthcare organisations are still struggling to keep up with all of the recent technical advances and the high demands of customers in terms of service levels [2].

7.1.1 Innovations of Technology

There have been many technological advances in the fields of AI and ML over the last decade. The rapid growth of AI tools and technologies, including the field of biomedical and healthcare, has been facilitated by a perfect arrangement of improved computer processing speed, superior data collection, and a strong AI talent pool. This will result in a significant change in the step of AI technology, as well as its acceptance. ML algorithms can identify trends associated with diseases and health conditions. ML techniques have the potential to improve healthcare access in developed countries as well as innovation in cancer detection and treatment [3].

Deep learning (DL) is a comparatively novel and rapidly expanding outlet of ML. It uses multi-layered deep neural networks (DNNs) to model abstraction from significant data, allowing it to make sense of data such as images, sounds, as well as texts. DL enables the discovery of relations that were previously difficult to discover using traditional ML algorithms [4]. In the 1980s, artificial neural networks (ANNs) were used to construct the early architecture for DL, and the true impact of DL was seen in 2006. DL has been used in an extensive range of applications since then. ANN is one of the techniques which is used in biomedical engineering applications. In comparison to previous neural networks, which only had three to five layers of connections, ANNs and DL networks have more than 10 layers of connections [5].

Figure 7.1 shows overall ML techniques used in world of AI with year. As the fame of AI, ML and DL are growing due to their reliability and flexibility. There are several methods and technics in the area of biomedical imaging, which are popular over time. Nowadays, the most popular tool in biomedical application is convolution neural network (CNN). CNN can be used with other combinations of DL architecture like CNN with auto-encoder, support vector machine (SVM) for classification, and K-means algorithm in image segmentation; likewise, there are numerous approaches

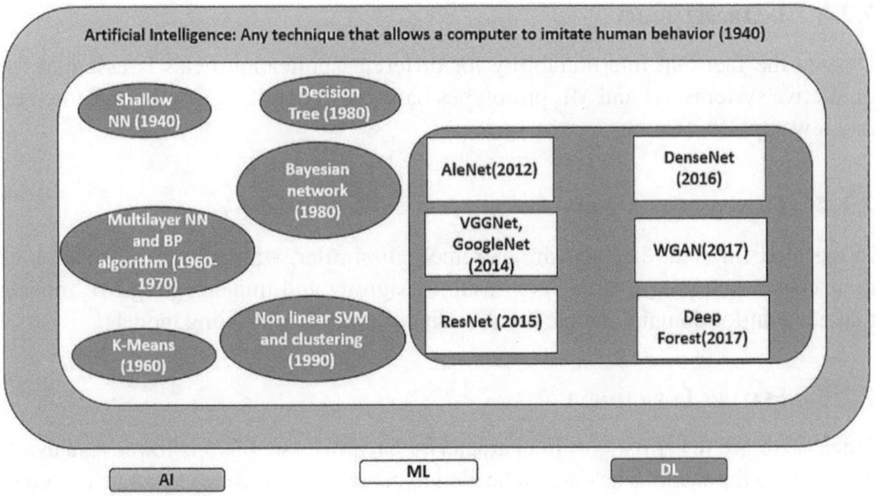

FIGURE 7.1 Overview of techniques used in artificial intelligence [14].

and the architecture available for resolving the real-life and other research problems. There are certain CNN models existing with various layers and structures, like Visual Geometry Group (VGG) [6], AlexNet [7], GoogLeNet [8], Residual Networks (ResNet) [9], Highway nets [10], DenseNet [11], Wasserstein Generative Adversarial Network (WGAN) [12] and many more [13,14].

7.2 CHALLENGES AND ISSUES

There are several problems and challenges that are essential to be addressed in various application domains, especially in biomedical applications, which are enlisted below.

7.2.1 Data Collection

AI and ML are capable to manage a large amount of data because it is highly computational. A minimum of ten sample parameters are needed as a common rule. We can get a huge amount of data for various applications like natural language and computer vision. Since we are all aware that the world's population is growing day by day, the count of disease cases is also growing, making data collection easier.

7.2.2 Poor Quality of Data

Data quality is a very important act in AI and ML because in certain application domains, heterogeneous, raw, noisy, and incomplete data can lead to incorrectly interpreted results. As a result, maintaining data quality for such a large and heterogeneous database while training a DL model has many concerns to consider, such as data shortage, data duplication, and lost values.

7.2.3 Interpretability

Despite the fact that interpretability for different application fields is essential for predictive systems, AI and ML prototypes have only been successfully implemented in a few application areas.

7.2.4 Domain Complexity

In the medical field, data sets are extremely dissimilar, with partial information of their causes and progression. As a result, designing and implementing ML models that account for domain complexity are critical aspects of training models.

7.2.5 Feature Enrichment

Since there are fewer patients to characterise each disease, there is fewer data available around the world. The data set needed to generate the features is not restricted to a single data source, such as social media. Another research issue facing the research community is the convergence of data sources with DL models.

7.2.6 Temporal Modelling

Time is important in the health, in biomedical sector, and in real-life issues where machines such as computer-aided design (CAD) systems, electronic health records (EHRs), and other monitoring devices are involved. To improve the efficiency of any application, DL training should be quicker, more precise, and more efficient.

7.2.7 Balancing Model Accuracy and Interpretability

In ML, model efficiency is important, but model reliability and interpretability are equally important. AI and ML are becoming more common due to their promising results and high efficiency, but making the outcomes more understandable is also a challenge. Investigators must concentrate on model efficiency and algorithms in order to improve the system's ability to predict outcomes.

7.2.8 Legal Issues

One of the most important and newest challenge factors in AI is the legal concern. If the public perceives the data collected as invading their privacy, then you might have some trouble in your research. So while doing data collection, legal factors need to be considered [15,16].

7.3 ARTIFICIAL INTELLIGENCE AND MACHINE LEARNING APPLICATIONS IN BIOMEDICAL

It is usually said that AI and ML technologies can enhance rather than fully replace the effort of physicians and other healthcare specialists. AI is ready to support healthcare employees and radiologists with extensive activities, which contain clerical

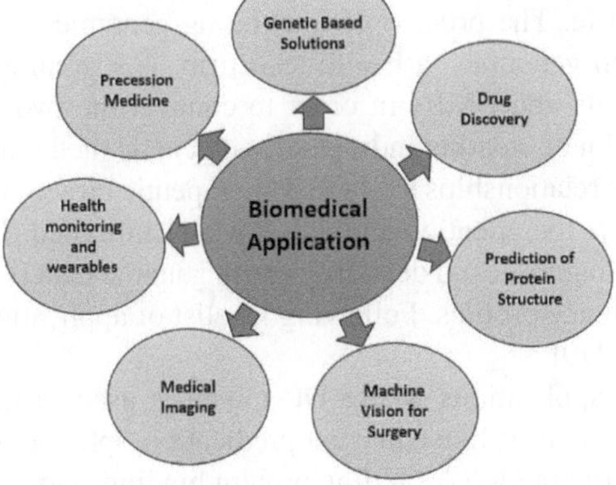

FIGURE 7.2 Different biomedical applications using AI and ML.

responsibilities, medical reporting, and community outreach, moreover technical assistance in regions including image processing, automation of medical system, and patient tracking. Figure 7.2 illustrates different biomedical applications in various areas.

7.3.1 Precision Medicine

AI has been used in the area of biomedicine for years, and its potential to advance medicine and science continues to evolve exponentially each year. Even precision medicine would be impossible to achieve without the use of ML algorithms to help in the practise. Precision medicine is a new approach to clinical science and patient care that focuses on better understanding and managing disease by combining multi-modal or multi-omics data from a person to make patient-specific decisions. Precision medicine permits healthcare treatments to be personalised to specific patients based on their disease history. The personalised treatment option will take into account genomic variations as well as medical treatment contributing aspects like age, gender, topography, ethnicity, history of family, resistant profile, metabolic summary, and microbiome. Precision medicine has a number of benefits, including lower healthcare costs, less adverse drug reactions, and improved drug action efficacy. Precision medicine innovation is expected to benefit patients greatly and transform the health facilities [17]. One popular way through which precision medicine works is omics-based tests. ML algorithms are used to identify associations and predict care responses for individual patients using genetic data from a population pool. Other biomarkers, such as protein expression, sequence of DNA, and metabolic profile, are used with ML to facilitate personalised treatments [18,47].

7.3.2 Genetics-Based Solutions

Genomics is a system of techniques for analysing DNA sequences in order to learn about the structure and function of genomes, gene regulation, and genetic changes linked to a variety of diseases. Within the next decade, it is expected that a significant portion of the world's populace will be given complete genome sequencing, either at birth or later in life. The process of integrating genomic and phenotype data is still underway. Deep genomics, a health-tech firm, is searching for trends in both large genetic data sets and EMRs in order to connect the two in terms of disease markers. With the aim of creating individualised genetic medicines, Deep Genomics Company uses these relationships to classify therapeutic targets, either current therapeutic targets or new therapeutic candidates. Many inherited disorders have signs that are difficult to diagnose, and decoding entire genome data is also difficult due to the numerous hereditary profiles. Following is a list of applications for genetics and genomics using ML [19].

There are many applications where DL model is used in genomic sequencing and gene expression analysis like enhancer predication, splicing pattern prediction in separate tissues, predicting RNA-binding protein binding sites, modelling structural binding preferences, miRNA precursor and miRNA target prediction, etc. [19].

7.3.3 Drug Improvement and Discovery

Pipelines for drug discovery and production are lengthy, complicated, and reliant on a variety of aspects. ML methods offer a range of tools for improving exploration and decision-making for well-defined problems with lots of good data. Any disaster in this phase has a significant economic effect, and most drug aspirants fail at some point throughout production and never reach the marketplace.

Figure 7.3 shows the ML strategies that have been used to address the discovery of drugs. Unsupervised methods are used to build algorithms that permit data clustering, while supervised learning methods, i.e. regression and classifier methods, are used to address the prediction of data categories. The following mentioned ML techniques have been used for drugs development: absorption, distribution, metabolism

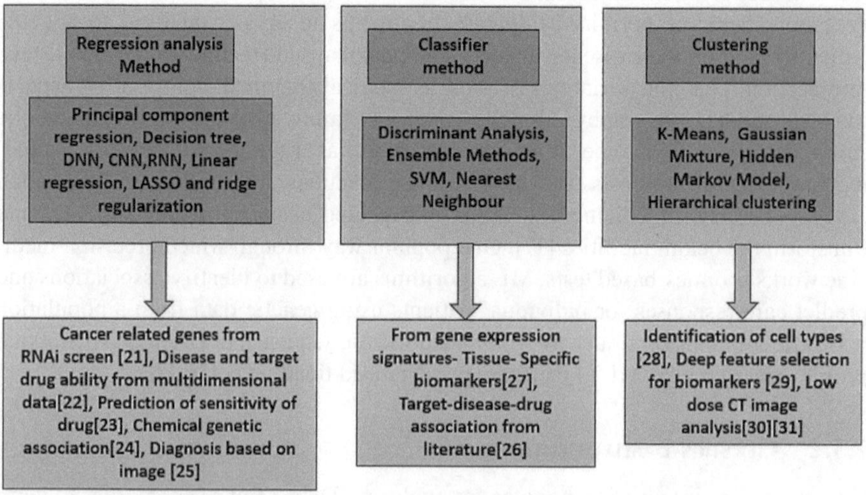

FIGURE 7.3 Illustrations of drug discovery applications using machine learning tools [20].

TABLE 7.1
Applications of Protein Structure Prediction

Sr. No.	DL/ML/ Architecture	Application
1	DBN	Predicting binding sites of RBPs and modelling structural binding preferences [32]
2		Estimation of protein disorder [33]
3		Secondary structures estimation and solvent accessible surface area of proteins [34]
4	CNN	Protein structure properties predication like disorder regions, solvent accessibility, etc. [35]
5		Prediction of protein secondary structures [36]
6		Protein disorder regions prediction [37]
7	SAE	Backbone Cα angles and dihedrals for sequence-based prediction [38]
8	RNN	Estimation of protein secondary structure [39]
9		Protein contact map protection [40]

and excretion (ADME), CNN; deep auto-encoder neural network (DAEN), DNN, generative adversarial network (GAN), recurrent neural network (RNN), SVM, support vector regression (SVR), etc.

7.3.4 Prediction of Protein Structure

The amino acid sequence that makes up a protein determines its 3D structure. The computational prediction of 3D protein structure from 1D sequences is a very tough task. The accurate 3D construction of a protein is critical to its function, and incorrect structures could lead to an extensive variety of diseases. Table 7.1 shows different DL techniques used for a variety of applications.

7.3.5 Medical Image Recognition

DL states the multi-layered nature of ML, and among all DL methods, CNNs have proven to be the most capable in the field of image recognition. A convolutional layer is formed by convolving an image with dissimilar weights and creating a stack of filtered images which is shown in Figure 7.4. Pooling is then applied to all of these filtered images, resulting in a smaller representation of the original stack of images and the removal of all negative values by a rectified linear unit (ReLU). After that, deep stacking is the process of stacking all of these operations on top of one another to construct layers. This practice can be replicated several times, with the image being increasingly filtered and smaller each time. The final layer is known as a fully connected layer, in which each attribute allocated to all layers contributes to the final outcomes.

Humans choose machine because machines are quicker and more reliable than humans. CAD and automated medical image processing are the preferred, if not crucial, options in medical sciences. Another common CAD activity in medical imaging

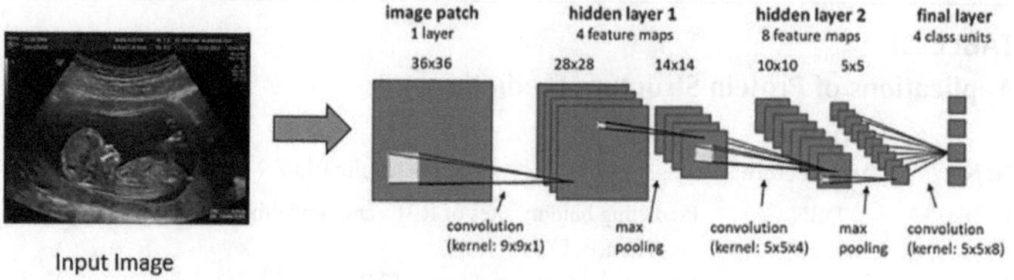

FIGURE 7.4 Overview of convolution neural network [42].

TABLE 7.2
Medical Imaging Application Using AI and ML

Sr. No.	ML/DL Architecture	Applications
1	CNN	Brian Lump segmentation [42]
2		Pancreas in CT segmentation [43]
3		Segmentation of knee cartilage [44]
4		Segmentation of hippocampus [45]
5		Histopathological cancer classification [46]
6		Grading of coronary artery calcium in CT images [47]
7	Stacked auto-encoder (SAE)	Hippocampus segmentation from infant brains [48]
8		From 4D patient data organ detection [49]
9		Scoring of mammographic density [50]
10	Deep Belief Network (DBN)	Heart ventricle segmentation by MR records [51]
11		Categorise retinal-based diseases [52]
12	Deep Neural Network (DNN)	Gland instance segmentation [53]
14	Recurrent Neural Network (RNN)	EEG-based epileptic seizure propagation prediction by time-delayed NN [54]

is malignancy diagnosis as well as computing the strength of lesions. CNNs have become more common in current years as a result of their outstanding success and dependability (Table 7.2). The steps in disease detection process which are normally used by researchers are pre-processing, object and lesion detection, segmentation, classification, and registration.

7.3.6 Health Monitoring and Wearables

Individuals have relied on doctors for centuries to educate them about their own bodies, and this practise continues to some degree today. Wearable health devices (WHDs) are a new tool that allows for continuous monitoring of humans in various situations. Their application versatility is the secret to their early adoption, and users can now monitor their behaviour while exercising, meditating, or even underwater [55]. The aim is to give people control of their own fitness by giving them the ability to evaluate data and monitor their own health. Figure 7.5 shows how WHD creates individual empowerment.

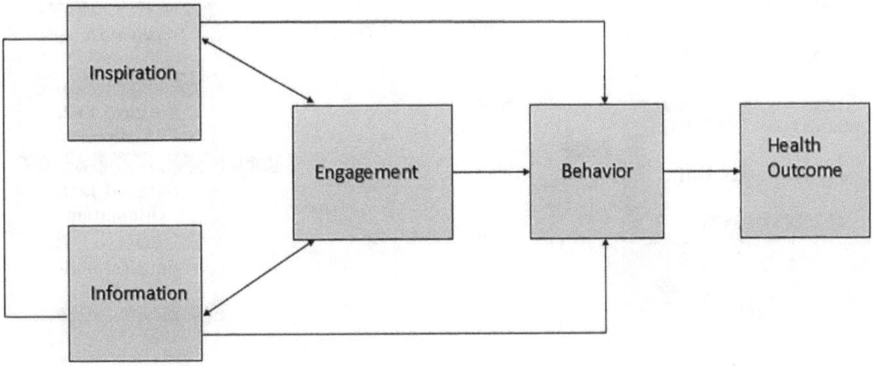

FIGURE 7.5 The flow of wearable health device application [56].

A wearable device may appear to be a simple band or watch at first glance; however, these machines link the gap among the number of technological disciplines, like biomedical engineering, materials science, computer software design, and data science. Wearable system design and implementation using ML are feasible for a variety of applications, including real-time identification of heart attacks, irregular heart sounds, blood pressure tracking, and gait examination for diabetic foot checking. The signal from these prototypes is fed into an ML algorithm, which uses it to identify the wearer's disease. Garmin wearables are an excellent model of this; with an emphasis on staying healthy, they cover a wide range of sports and offer a significant amount of data on their Garmin connect application, which allows operators to analyse and notice their everyday activities. Gamification is constantly being used in conjunction with these initiatives [56].

7.3.7 Minimally Invasive Surgery (MIS)

ML and AI learning has sparked a lot of attention in computer-assisted models for minimally invasive surgery (MIS) in the last 5 years. DNNs are extremely effective for resolving classification difficulties in complex surgical circumstances due to the easy access to images in surgical interventions. While there have been many advancements in the field of surgery as calculated by surgical procedure results, the core practise of surgery is a relatively low-tech process for the most part, relying on hand tools and instruments for cutting and stitching. Figure 7.6 shows tasks to be performed in MIS application using CNN by using surgical and kinematics data sets. Initially, data set is used to feed the DL models. The DL prototypes are then equipped to accomplish one of four tasks: analysis of surgical image, analysis of surgical work, surgical skill assessment, or automation of surgical tasks.

Surgical Image Analysis: Analysing the surgical picture is important to know the surgical situation, for taking proper decision. For the same, depending on the predicted performance of the model, the methods for object recognition in an image can be distributed into categories like classification, detection, and segmentation.

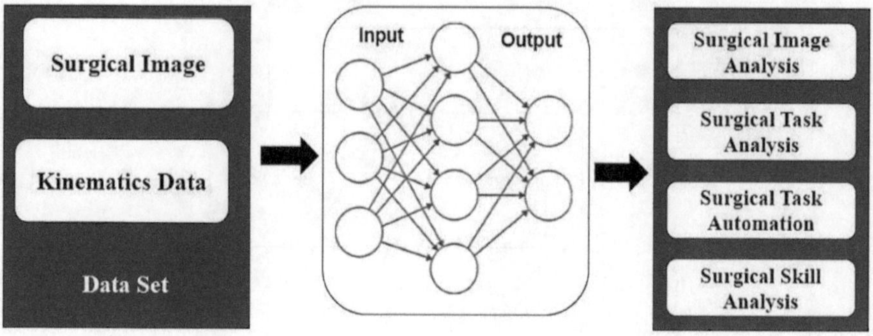

FIGURE 7.6 Different surgical operations performed by using CNN [57].

Surgical Task Analysis: Operating task analysis is a crucial task in the area of MIS because of its numerous applications, which include the creation of context-aware systems, automated grading of medical video records, self-directed robotic applications, and so on.

Assessment of Surgical Skill: The evaluation of clinical skills to rate trainees' concert and track their progress during the training process is a key task in health training.

Automation of Tasks: Over the past few decades, significant progress has been ready in the mechanisation of whole operating tasks or the semi-automation of specific sections, allowing robotic systems to collaborate with surgeons performing complex manoeuvres [57].

7.3.8 Monitoring by Biosensor

The most pressing necessity for AI in biomedicine is disease diagnosis. In this field, there have been a number of notable breakthroughs. In vitro diagnostics by biosensors or biochips is one big class of diagnosis that AI helps health practitioners to provide earlier and more precise diagnoses for a variety of diseases. Biosensors and associated point-of-care testing (POCT) systems can identify cardiovascular disorder in the primary stages using integrated AI. In addition to diagnosis, AI may aid in the prediction of cancer patient survival rates, like colon cancer patients [58]. Figure 7.7 shows how POCT models work.

Additional important type of disease diagnosis is the one that depends on (two-dimensional) medical imaging and (one-dimensional) signal processing. In the diagnosis, treatment, and prognosis of diseases, AI technique is applied to one-dimensional signal processing, extraction of biomedical signal features like Electroencephalography (EEG), electromyography (EMG), and electrocardiography (ECG). Guessing epileptic seizure is an essential application of EEG. It is important to anticipate seizures in order to reduce the severity of their effects on patients. To improve image quality and analysis efficiency, AI has been used in image segmentation, multidimensional

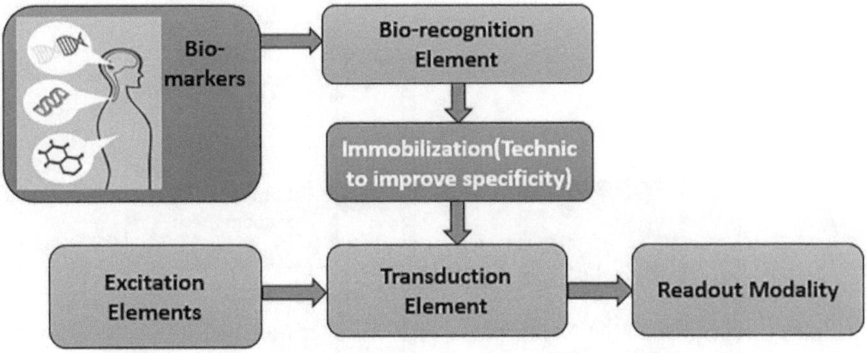

FIGURE 7.7 Schematic of PCOT system [58].

imaging, and thermal imaging. AI may even be embedded in portable ultrasonic systems, allowing untrained individuals in underdeveloped areas to use ultrasound as an effective tool for diagnosing a variety of illnesses.

In addition to the above uses, AI can help Decision Support Systems (DSSs) to increase diagnostic accuracy and ease disease management, reducing staff workload. For example, in thyroid nodules classification, AL and ML play very important roles in identifying whether nodules come under benign or malignant case [59]. These applications show that AI can be useful for diagnosing, managing, and even predicting diseases and patient conditions early and accurately.

7.4 SUCCESS ELEMENTS FOR AI IN BIOMEDICAL ENGINEERING

In biomedical applications, AI can help physicians, patients, and other healthcare staff in the following ways. These ideas act as motivations and will enlarge their involvement toward a fruitful execution of AI in biomedical application (Figure 7.8) [14]:

1. Assessment of disease
2. Management of difficulties
3. Patient-care support throughout a treatment
4. Investigation directed at discovery or management of disorder.

7.4.1 Assessment of Condition

Mood disorders and mental health issues are hot topics in today's world. According to the WHO, one out of every four people in the world suffers from such illnesses, which can hasten their road to ill-health. ML systems have recently been created to identify words and intonations in a person's speech that may signify a mood disorder. An MIT-based lab has shown investigation into the identification of initial indications of depression using speech by using neural networks.

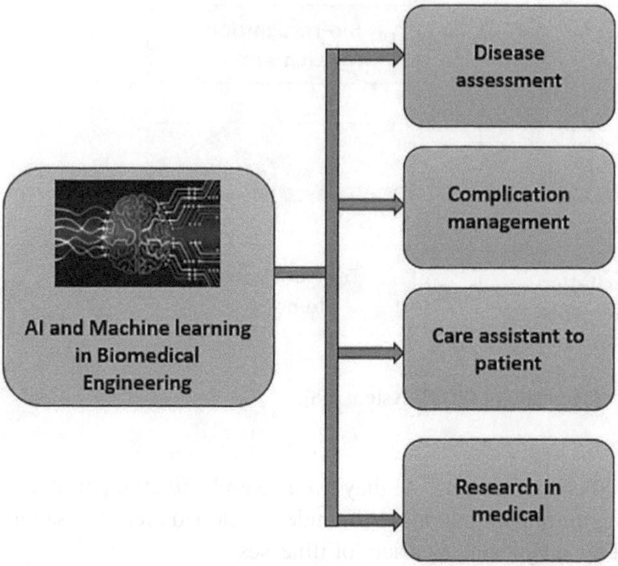

FIGURE 7.8 Illustrations of success factors of AL and ML in biomedical application [14].

7.4.2 Managing Complications

Patients normally handle the general feeling of being unwell as well as many problems that come with mild illnesses. However, in certain cases, it is essential to treat these signs in order to avoid extra development and, eventually, to relieve additional complex symptoms. ML methods can also help predict serious problems like neuropathy, which can develop in people with type 2 diabetes, as well as early cardiovascular abnormalities.

7.4.3 Patient-Care Assistance

Healthcare support systems will help physicians function more efficiently while also promoting patient autonomy and well-being. If every patient is viewed as a separate system, a bespoke solution may be applied depending on the various data available. This is particularly critical for the elderly and disabled in our communities. Digital health assistants, for example, may remind people to take their prescribed drugs at specific times or suggest different workout routines for the best results. Affective computing may make a major contribution in this field. This environment may be used at home or in a clinic to ease job pressure and improve service for healthcare staff.

7.4.4 Medical Research

AI has the potential to speed up diagnosis and medical investigation. Nowadays, a growing number of collaborations amongst biotech, MedTech, and medicine firms have emerged to speed up the development of different drugs. These collaborations are mostly born out of societal need and desire, rather than curiosity-driven science. Another example is recent studies on the COVID-19 pandemic that has affected people all over the world. Predictive oncology, a precision medicine company, has announced the launch of an AI platform that will use more than 12,000 computer simulations per system to speed up the development of new diagnostics and vaccines.

7.5 CONCLUSION

This chapter explains the importance and role of AI and ML in different biomedical applications like precision medicine, MIS system, drug discovery system, healthcare, disease detection system like cancer and PCOT, prediction of protein structure, etc. Moreover, this chapter explains different DL techniques like ANN, CNN, RNN, models which have been used in healthcare and in biomedical application. This chapter also focuses on challenges and issues in biomedical research. Lastly, the chapter demonstrates success factors in biomedical application to guide and motivate the researchers.

REFERENCES

1. Chan, Y.-K., Chen, Y.-F., Pham, T., Chang, W., and Hsieh, M.-Y. "Artificial intelligence in medical applications" *Hindawi J. Healthcare Eng.* Vol. 2018, Article ID 4827875.
2. Rong, G., Mendez, A., Assi, E.B., Zhao d, B., and Sawan, M. *Artificial Intelligence in Healthcare: Review and Prediction Case Studies*, China, Elsevier, 2020, 6(3), 291–301.
3. Zemouri, R., Zerhouni, N., and Racoceanu, D. "Deep learning in the biomedical applications: Recent and future status" *Appl. Sci.* 2019, 9, 1526.
4. Cao, C., Liu, F., Tan, H., Song, D., Shu, W., Li, W., Zhou, Y., Bo, X., and Xie, Z. "Deep learning and its applications in biomedicine" Genomics Proteomics Bioinformatics 2018, 6 (1), 17–32.
5. Nayak, R., Jain, L.C., and Ting, B.K.H. "Artificial neural networks in biomedical engineering: A review" In *Proceedings Asia-Pacific Conference on Advance Computation*, India, 2001.
6. Simonyan, K., and Zisserman, A. "Very deep convolutional networks for large-scale image recognition" arXiv 2014, arXiv:1409.1556.
7. Krizhevsky, A., Sutskever, I., and Hinton, G.E. "ImageNet classification with deep convolutional neural networks" In *International Conference on Neural Information Processing Systems*, Lake Tahoe, NV, USA, 3–6 December 2012; Curran Associates Inc.: Lake Tahoe, NV, 2012, pp. 1097–1105.
8. Szegedy, C., Liu, W., Jia, Y., Sermanet, P., Reed, S.E., Anguelov, D., Erhan, D., Vanhoucke, V., and Rabinovich, A. "Going deeper with convolutions" arXiv 2014, arXiv:1409.4842.
9. He, K., Zhang, X., Ren, S., and Sun, J. Deep residual learning for image recognition. https://doi.org/10.48550/arXiv.1512.03385, 2016.
10. Srivastava, R.K., Greff, K., and Schmidhuber, J. "Training very deep networks" In *Conference on Neural Information Processing Systems*, India, 2015.
11. Huang, G., Liu, Z., and van der Maaten, L. "Densely connected convolutional networks" *IEEE*, 2018 1, 2262–2269.
12. You, C., Yang, Q., Shan, H., Gjesteby, L., Li, G., Ju, S., Zhang, Z., Zhao, Z., Zhang, Y. (Member, IEEE), Cong, W., and Wang, G. "Structurally-sensitive multi-scale deep neural network for low-dose CT denoising 2169–3536", *IEEE* 2018 3.

13. Shadi, N. and Albarqouni, M. *Machine Learning for Biomedical Applications: From Crowdsourcing to Deep Learning*, Chine, 2017.
14. Bohr, A., and Sonohaler, K.M. *Artificial Intelligence in Healthcare* Copenhagen: ChemoMetec, Lillerød 2020.
15. Panayides, A.S., Amini, A., Filipovic, N.D., Sharma, A., Tsaftaris, S.A., Young, A., Foran, D., Do, N., Golemati, S., Kurc, T., and Huang, K. "AI in medical imaging informatics: Current challenges and future directions" *IEEE J. Biomed. Health Info.* 2020 24(7), 1837–1857.
16. Weinrich, M. "Special challenges and opportunities for application of bio-medical sensors" In *Smart Biomedical and Physiological Sensor Technology XIII*, edited by B.M. Cullum, D. Kiehl, E.S. McLamore, Proc. of SPIE, India, Vol. 9863, 986300.
17. Uddin, M., Wang, Y., and Woodbury-Smith, M. "Artificial intelligence for precision medicine in neurodevelopmental disorders" *NPJ Digital Med.* 2019 2, 112. Doi: 10.1038/s41746-019-0191-0.
18. Afzal, M., Islam, S.R., Hussain, M., and Lee, S. "Precision medicine informatics: Principles, prospects, and challenges" *IEEE Access* 2020, 8, 13593–13612. Doi: 10.1109/ACCESS.2020.2965955.
19. Libbrecht, M.W., and Noble, W.S. "Machine learning in genetics and genomics" *Nat Rev Genet.* 2015 June 16(6), 321–332. Doi: 10.1038/nrg3920.
20. Vamathevan, J., Clark, D., Czodrowski, P., Dunham, I., Ferranl, E., Lee, G., Li, B., Madabhushi, A., Shah, P., Spitzer, M., and Zhao, S. "Applications of machine learning in drug discovery and development" *Nat. Rev. Drug Discover.* 2019, 18, 463–477. Doi: 10.1038/s41573-019-0024-5.
21. Godinez, W.J., Hossain, I., Lazic, S.E., Davies, J.W., and Zhang, X. "A multi- scale convolutional neural network for phenotyping high-content cellular images" *Bioinformatics* 2017, 33, 2010–2019.
22. Costa, P.R., Acencio, M.L., and Lemke, N. "A machine learning approach for genome-wide prediction of and druggable human genes based on systems- level data" *BMC Genom.* 2010, 11, S9–S9.
23. Li, B., Shin, H., Gulbekyan, G., Pustovalova, O., Nikolsky, Y., Hope, A., Bessarabova, M., Schu, M., Kolpakova-Hart, E., Merberg, D., and Dorner, A. "Development of a drug- response modeling framework to identify cell line derived translational biomarkers that can predict treatment outcome to erlotinib or sorafenib" *PLOS One* 2015, 10, e0130700.
24. McMillan, E.A., Ryu, M.J., Diep, C.H., Mendiratta, S., Clemenceau, J.R., Vaden, R.M., Kim, J.H., Motoyaji, T., Covington, K.R., Peyton, M., and Huffman, K. "Chemistry-first approach for nomination of personalized treatment in lung cancer" *Cell* 2018, 173, 864–878.
25. Sharma, H., Zerbe, N., Klempert, I., Hellwich, O., and Hufnagl, P. "Deep convolutional neural networks for automatic classification of gastric carcinoma using whole slide images in digital histopathology" *Comput. Med. Imaging Graph.* 2017, 61, 2–13.
26. Mamoshina, P., Volosnikova, M., Ozerov, I.V., Putin, E., Skibina, E., Cortese, F., and Zhavoronkov, A. "Machine learning on human muscle transcriptomic data for biomarker discovery and tissue- specific drug target identification" *Front. Genet.* 2018, 9, 242.
27. Bravo, A., Pinero, J., Queralt- Rosinach, N., Rautschka, M., and Furlong, L.I. "Extraction of relations between genes and diseases from text and large- scale data analysis: Implications for translational research" *BMC Bioinform.* 2015, 16, 55.
28. Kim, J., Kim, J.-j., and Lee, H. "An analysis of disease- gene relationship from Medline abstracts by DigSee" *Sci. Rep.* 2017, 7, 40154.
29. Ding, J., Condon, A., and Shah, S.P. "Interpretable dimensionality reduction of single cell transcriptome data with deep generative models" *Nat. Commun.* 2018, 9, 2002.

30. Wang, B., Zhu, J., Pierson, E., Ramazzotti, D., and Batzoglou, S. "Visualization and analysis of single- cell RNA- seq data by kernel- based similarity learning" *Nat. Methods* 2017, 14, 414.
31. Chen, H., Zhang, Y., Kalra, M.K., Lin, F., Chen, Y., Liao, P., Zhou, J., and Wang, G. "Low- dose CT with a residual encoder decoder convolutional neural network (RED-CNN)" Preprint at arXiv https://arxiv.org/abs/1702.00288 2017.
32. Zhang, S., Zhou, J., Hu, H., Gong, H., Chen, L., and Cheng, C. "A deep learning framework for modeling structural features of RNA-binding protein targets" *Nucleic Acids Res.* 2016, 44, e32.
33. Wang, S., Weng, S., Ma, J., and Tang, Q. "DeepCNF-D: Predicting protein order/disorder regions by weighted deep convolutional neural fields" *Int. J. Mol. Sci.* 2015, 16, 17315–17330.
34. Heffernan, R., Paliwal, K., Lyons, J., Dehzangi, A., Sharma, A., Wang, J., Sattar, A., Yang, Y., and Zhou, Y. "Improving prediction of secondary structure, local backbone angles, and solvent accessible surface area of proteins by iterative deep learning" *Sci. Rep.* 2015, 5, 11476.
35. Wang, S., Li, W., Liu, S., and Xu, J. "RaptorX-property: A web server for protein structure property prediction" *Nucleic Acids Res.* 2016, 44.
36. Troyanskaya, O.G. "Deep supervised and convolutional generative stochastic network for protein secondary structure prediction" *Proc. 31st Int. Conf. Mach. Learn.* 2014, 32, 745–753.
37. Wang, S., Weng, S., Ma, J., and Tang, Q. "DeepCNF-D: Predicting protein order/disorder regions by weighted deep convolutional neural fields" *Int. J. Mol. Sci.* 2015, 16, 17315–17330.
38. Lyons, J., Dehzangi, A., Heffernan, R., Sharma, A., Paliwal, K., Sattar, A., Zhou, Y., and Yang, Y. "Predicting backbone Cα angles and dihedrals from protein sequences by stacked sparse auto-encoder deep neural network" *J. Comput. Chem.* 2014, 35, 2040–2046.
39. Baldi, P., Pollastri, G., Andersen, C.A.F., and Brunak, S. "Matching protein beta-sheet partners by feedforward and recurrent neural networks" *Proc Int. Conf. Intell. Syst. Mol. Biol.* 2000, 25–36.
40. Pollastri, G., and Baldi, P. "Prediction of contact maps by GIOHMMs and recurrent neural networks using lateral propagation from all four cardinal corners" *Bioinformatics* 2018.
41. Hashimoto, D.A., Rosman, G., Rus, D., and Meireles, O.R. "Artificial intelligence in surgery: Promises and perils" *Ann. Surg.* 2018, 268(1), 70.
42. Pereira, S., Pinto, A., Alves, V., and Silva, C.A. "Brain tumor segmentation using convolutional neural networks in MRI images" *IEEE Trans. Med. Imag.* 2016, 35, 1240–1251.
43. Roth, H.R., Farag, A., Lu, L., Turkbey, E.B., and Summers, R.M. "Deep convolutional networks for pancreas segmentation in CT imaging" ArXiv1504.03967.
44. Prasoon, A., Petersen, K., Igel, C., Lauze, F., Dam, E., and Nielsen, M. "Deep feature learning for knee cartilage segmentation using a triplanar convolutional neural network" *Med Image Comput. Assist. Interv.* 2013, 8150, 246–253.
45. Kim M, Wu G, Shen D. Unsupervised deep learning for hippocampus segmentation in 7.0 tesla MR images. In: Wu, G., Zhang, D., Shen, D., Yan, P., Suzuki, K., Wang, F., editors. New York: Springer-Verlag New York Inc., 2013, 1–8.
46. Xu, Y., Zhu, J.Y., Chang, E., and Tu, Z. "Multiple clustered instance learning for histopathology cancer image classification, segmentation and clustering" In *Proceedings of the IEEE Computer Society Conference on Computer Vision and Pattern Recognition*, UK, 2012.

47. Martorell-Marugán, J., Tabik, S., Benhammou, Y., del Val, C., Zwir, I., Herrera, F., and Carmona-Sáez, P. "Deep learning in omics data analysis and precision medicine" 10.15586/computationalbiology, Computational Biology: Codon Publications, Chapter 3, 2019.
48. Guo, Y.R., Wu, G.R., Commander, L.A., Szary, S., Jewells, V., Lin, W.L., and Shen, D. "Segmenting hippocampus from infant brains by sparse patch matching with deep-learned features" *Med. Image Comput. Assist. Interv.* 2014, 8674, 308–315.
49. Shin, H.C., Orton, M.R., Collins, D.J., Doran, S.J., and Leach, M.O. "Stacked autoencoders for unsupervised feature learning and multiple organ detection in a pilot study using 4D patient data" *IEEE Trans. Intell.* 2013, 35, 1930–1943.
50. Kallenberg, M., Petersen, K., Nielsen, M., Ng, A.Y., Diao, P.F., Igel, C., Vachon, C.M., Holland, K., Winkel, R.R., Karssemeijer, N., and Lillholm, M. "Unsupervised deep learning applied to breast density segmentation and mammographic risk scoring" *IEEE Trans. Med. Imaging* 2016, 35, 1322–1331.
51. Ngo, T.A., Lu, Z., and Carneiro, G. "Combining deep learning and level set for the automated segmentation of the left ventricle of the heart from cardiac cine magnetic resonance" *Med. Image Anal.* 2017, 35, 159–171.
52. Arunkumar, R., and Karthigaikumar, P. "Multi-retinal disease classification by reduced deep learning features" *Neural Comput. Appl.* 2015, 1–6.
53. Xu, Y., Li, Y., Liu, M., Wang, Y., Fan, Y., Lai, M., and Chang, E.I. "Gland instance segmentation by deep multichannel neural networks" arXiv160704889.
54. Mirowski, P.W., Madhavan, D., and Lecun, Y. "Time-delay neural networks and independent component analysis for EEG-Based prediction of epileptic seizures propagation" *Proc. Conf. AAAI Artif. Intell.* 2007, 234, 1892–1893.
55. Randhawa, P., Shanthagiri, V., and Kumar, A. "A review on applied machine learning in wearable technology and its applications" *International Conference on Intelligent Sustainable Systems (ICISS)*, Palladam, India, 07-08 December 2017, 10.1109/ISS1.2017.8389428.
56. MacDermott, Á., Lea, S., Iqbal, F., Idowu, I., and Shah, B. "Forensic analysis of wearable devices" Fitbit, Garmin and HETP Watches, 2019 10th IFIP (NTMS), India.
57. Rivas-Blanco, I., Pérez-Del-Pulgar, C.J., García-Morales, I., and Muñoz, V.F. "A review on deep learning in minimally invasive surgery" *IEEE Access* 2016, 124, 6 Doi: 10.1109/ACC ESS.2021.3068852.
58. Vasan, A.S.S., Doraiswami, R., and Mahadeo, D.M. "Point-of-care biosensor systems" *IEEE*, 2017, 4.
59. Gulame, M.B., Dixit, V.V., and Suresh, M. "Thyroid nodules segmentation methods in clinical ultrasound images: A review" Doi: 10.1016/j.matpr.2020.10.2592214-7853/_2020 Elsevier Ltd.

8 The Use of Artificial Intelligence-Based Models for Biomedical Application

Sharad Mulik and Nilesh Dhobale
RMD Sinhgad School of Engineering

Kanchan Pujari
Smt. Kashibai Navale College of Engineering

Kailash Karande
SKN Sinhgad College of Engineering

CONTENTS

DOI: 10.1201/9781003220176-8

8.1 INTRODUCTION

In today's modern world, healthcare industries are growing very fast. The shortage of professionals and spending on this industry is more; hence, this industry's revolution is needed. So, to overcome this problem, the artificial intelligence (AI)-based solution needs to be adopted; thus, it reduces unnecessary costs and provides the best solution for present issues in the healthcare industry.

The healthcare industry before 2010 was emphasizing treatment based on history or evidence-based. After 2010, these healthcare industries emphasized result-based treatment. Since 2020 technology in the medical field is changing very fast based on intelligent solutions. It includes AI, Robotics, and Virtual or Augmented reality [1]. In a recent year's survey, it was found that 69% of businesses are adopting AI in their business, and 22 % of companies are thinking of adopting AI in their business for smooth flow of operation [2]. In the United States of America, up to 2026, the AI-based healthcare industry has the potential to save $150 and because of the use of AI, healthcare industries are increasing day by day [3]. AI has been used in the healthcare industry for health monitoring, surgery, primary diagnosis, nursing assistant, organization workflow, treatment design, and medical consultation.

Recently, AI-based technologies and their application have improved the quality and efficiency of the healthcare industry. Improvement in the healthcare industry increases the life of the human being, which is a great achievement [4,5]. As AI includes machine learning (ML), deep learning (DL), natural language processing (NLP), and smart robots, it is possible to innovate the healthcare industry; hence, it will be better than conventional technologies [6,7]. Many researchers, physicians, technology-based product developers have been attracted to AI and its implementation in the healthcare industry. Additionally, from various studies, it was found that AI in the healthcare industry can diagnose complicated human diseases more efficiently [8,9]. The AI-based technologies in the healthcare industry need a large dataset of patients for diagnostic purposes; hence, these technologies play a crucial role in decision-making for doctors and healthcare professionals [10–12]. In one study for the detection of breast cancer, AI-based technologies were applied, and it was found that the AI-based system is more suitable than a conventional system which is nothing but the diagnosis with healthcare professionals [13]. The diagnosis is based on the AI process the treatment efficiently and more quickly. Due to the large dataset of disease gained hence, it works more quickly and efficiently. Due to a more quick and efficient process, it is possible to give fast treatment [14]. An advanced virtual avatar has been used to diagnose patients with mental diseases, which is the best solution [15].

In England, Wi-Fi-based armbands were developed for monitoring the patient's respiratory rate, body temperature, blood pressure, pulse, oxygen level. With this system, over 50,000 people's treatment was carried out. Additionally, due to the AI-based real-time analysis in the hospital, it is observed that the emergency room visits were reduced significantly. Here 96% improvement on a hospital is observed compared with conventional 50% improvement. In another example of the United States of America, the AI-based tool has reduced $4 billion saving with a 31% reduction rate in readmission [16]. The AI has enormous potential for emergency treatment in the healthcare industry which is more helpful for healthcare professionals [7]. The application of AI in the healthcare industry has a huge scope, but at the same time, it has a lot of challenges cyber security, the privacy of data, ownership of data, sharing of data, medical error responsibility, risk of system failure, issues of medical ethics, etc. [5,7,17]. In healthcare industry, treatment process is changing due to use of AI, but at the same time there is issue of patients safety and privacy [4,5].

The AI-based technologies in the healthcare industry have flexibility in problem-solving as well as human-oriented values. The AI-based technologies in the healthcare industry are more controversial because it is not available for all care providers. Therefore, it is necessary to analyze the AI-based technologies and their application for future patients' diagnoses and also needs to emphasize the quality of service and the operational policies.

8.2 AI METHODS AND APPLICATIONS

In the healthcare industry, different types of AI are used. The use of AI in the healthcare industry has been categorized in Figure 8.1. From the market analysis companies and research articles, data has been collected in this writing [3]. AI can be used in the healthcare industry for primary diagnosis, virtual assistants for nurses and doctors, error reduction in dosage, medical trials, surgery, medical treatment design, etc. The implementation of AI is helpful for decision-making for doctors and nurses [18]. Different methods such as ML, NLP, neural network (NN), DL, machine vision (MV)/computer vision have been used in the healthcare industry, which is briefly explained below. Different AI methods and their uses are shown in Figure 8.2.

	Human interaction	No humans involved
Specific systems	**ASSISTED INTELLIGENCE** • AI systems that assist humans in making clinical decisions actions • Hard-wired systems that do not learn from their interactions	**AUTOMATED INTELLIGENCE** • Automation of manual and cognitive clinical tasks that are routine task • Does not involve new ways of doing things -automates existing clinical tasks
Adaptive systems	**AUGMENTED INTELLIGENCE** • AI systems that augmented human clinical decision making and continuously learn from their interactions with humans and the environment	**AUTONOMOUS INTELLIGENCE** • AI systems that can adapt to different situations and can act autonomously without human assistance

FIGURE 8.1 AI in healthcare industry categorization [19].

8.2.1 Machine Learning (ML)

It is the type of AI that gives an intelligent solution. ML includes supervised learning algorithm, unsupervised learning algorithm, and reinforcement learning algorithm. The supervised learning algorithm generally builds the mathematical model with the help of desired inputs and outputs [20]. Unsupervised learning algorithms typically take a data set that comprises only inputs and finds structure present in the data, like data points grouping. The test dataset, which is not labeled, classified, and categorized unsupervised learning, is learned from it [21]. Reinforcement learning algorithm algorithms do not consider an exact mathematical model of the discrete-time stochastic control process.

8.2.2 Natural Language Processing (NLP)

It is a type of AI that can allow computers to read, understand, and develop meaning from human languages. Because of this, natural interaction between humans and computers takes place. To perform the desired task, computer systems generally understand and manipulate the natural language. With the patient's health history and the patient's own speech, NLP can predict the future disease [22].

8.2.3 Neural Network (NN)

NN is an AI type that includes an input layer, an output layer, and a hidden layer. In the input layer, the input can be image or speech, and the hidden layer process this data, and the output layer gives output. The deep neural network is a subset of the NN that includes convolutional, recurrent, reinforcement, transfer NN, which can be implemented in the healthcare industry skin lesions, retinal, and pathology problems.

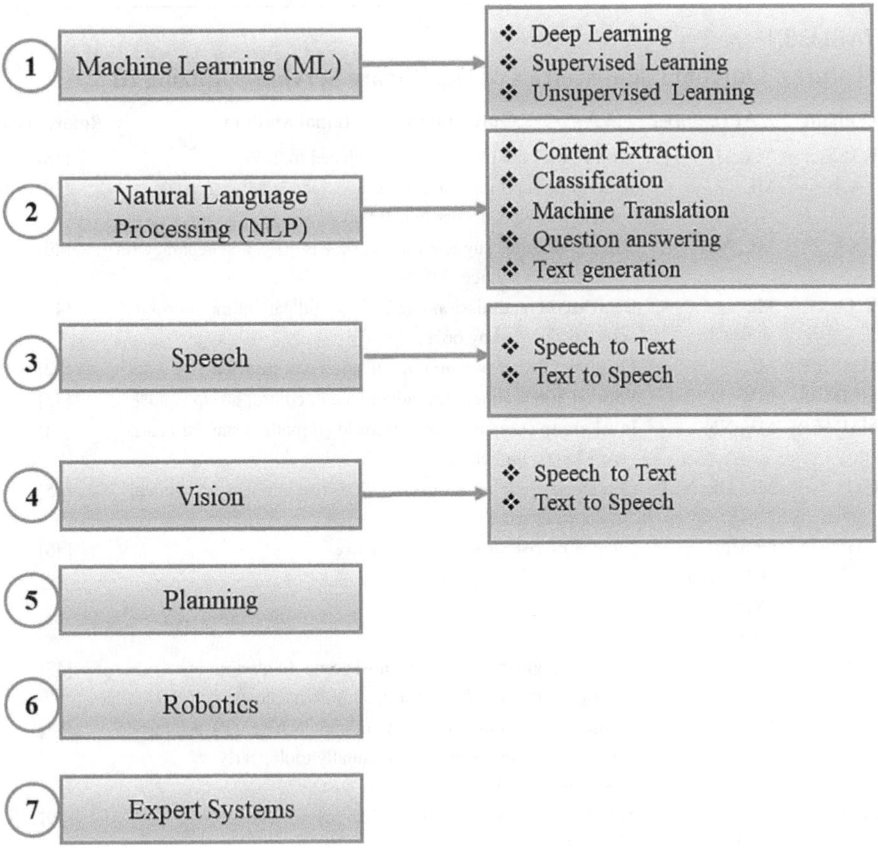

FIGURE 8.2 Different AI methods and their use [19].

8.2.4 Deep Learning (DL)

DL is a sub-category of ML, which depends on the artificial neural network (ANN). The ANN has inspired by the process of information in the biological system, which has multiple layers. In DL, deep which is nothing but the number of layers through data, can be transmitted. With DL's help, it is possible to allow a computer to perform a task based on the existing data relationship.

8.2.5 Machine Vision/Computer Vision

In this method, the information is extracted from the image automatically. The extracted data from images is helpful for industrial robots, automatic inspection, vehicle guidance, and process guidance. Table 8.1 shows the implementation of AI in the healthcare industry.

TABLE 8.1
Features, Outcome, and Metrics of Healthcare Services Utilizing AI

Domain	AI Features	Outcome vs. Traditional Method	References
RASS	MV, ML	With RASS, the overall cost is reduced to 22%	[38]
VNA	ML	In 20 most common conditions, accuracy is 85%; for safe urgency advice, accuracy is 92.60%	[39]
VNA	ML, NLP	Top 3 conditions suggestion accuracy is 50%, and accuracy for safe urgency advice is 97%	[40]
VNA	ML	Decrease in readmission rate is 75% and patient monitoring costs is reduced by 66%.	[41]
MMMER	ML	Flagging 75% of potential medication inaccuracies	[42]
MMMER	MV	With AI 100% adherence, adherence in control group is 50%	[43]
MMMER	MV, NN	In AI group comparing with traditional method standard care are 17.90% higher.	[44]
CTP	ML, DL, NLP	Compared to earlier procedure, 80% rise in oncology clinical trial enrollment	[45]
CTP	Optical Character Recognition (OCR), ML	24%–50% rise over regular practices	[46]
PDP	ML	Increase in glucose level prognosis time. In alerting of hypoglycemic events, alerting 98.50% accuracy	[47]
PDP	NLP	Compared to manual review by clinicians here 99% accuracy. Fifty-four patient screening manually took nearly 50–70 hours. Saving in time and cost	[48]
PDP	DCNN	Area Under Curve 0.99, 96% accuracy associated to radiologists	[49]
PDP	NLP, ML	Compared to 79% accuracy in traditional diagnosis, 100% accurate detection is obtained	[50]
MID	MV	In identifying 50 common diseases eye problem, the accuracy is 94.50%	[51]
MID	ANN, CNN, DCNN	From 69 trial studies sensitivity and specificity is 79.10% and 88.30% respectively. In 14 readings validity among DL approaches and healthcare professionals pooled sensitivity are 87% 86.70% and specificity is 92.50% 90.50%	[52]
MID	DL, MV	In USA system created 5.70% decrease of false positives and 9.40% decrease of false negatives. In UK 1.20% decrease of false positives and 9.40% decrease in false negatives. With clinical readers comparison takes place	[53]
MID	MV, ML, NN	Here 66% accuracy with dermatologists and 72% accuracy with AI	[54]
PHM	NN	For test population 86.10% diagnosis accuracy	[55]
PHM	ANN, ML	87% sensitivity, 96.10% specificity also 95.40% accuracy	[56]
PHM	ML	Superiority of lifespan enhancement for 31% of patients, 5,2 months lengthier life, dropping 4% in hospitalization in addition to 7% in ER visits	[36]

8.3 ROBOTIC-ASSISTED SURGICAL SYSTEMS (RASS) AND COMPUTER-ASSISTED SURGERY (CAS)

Since 1985, healthcare industries are using RASS [23]. In various fields such as spine surgery, general surgery, cardiac surgery, orthopedic surgery, thoracic surgery, gynecology, transplant surgery, gastrointestinal surgery, and urology, RASS has been implemented successfully [24–31]. The types of RASS are shared control surgery, supervisory controlled surgery, and tele surgical. With the help of RASS, doctors can do precise operations with a camera and tool with a small incision. The RASS can collect the data and details, including video recording during surgery, and it is possible to analyzer this recording later [32].

In the surgical system, CAS is the second option to assist the doctors during surgery. The RASS, which includes robot-assisted technology while CAS uses computer technology, can plan the surgery, and increase guidance during surgery. The CAS generally gives navigation and positioning to surgical instruments [33]. CAS improves the efficiency of surgery. Also, it captures the images from the patients, diagnosis, surgical navigation, and surgical simulation [34]. Compared with open surgery, RASS surgery is better because it reduces surgery complications, blood loss and stays in hospital [35].

The study found that RASS operative time was 20% more overall; it was 22% less costly than conventional open surgery [36]. In recent years, Smart Autonomous Robotic Assistant Surgeon (SARAS) has been developed with AI that can decide on the right time and predict the future surgery situation. In the healthcare industry, with the SARAS system, it is possible for patient's safety and quality of life; hence the efficiency of hospitals is increased [37]. Figure 8.3 shows SARAS-operated arms-operated robot for surgery.

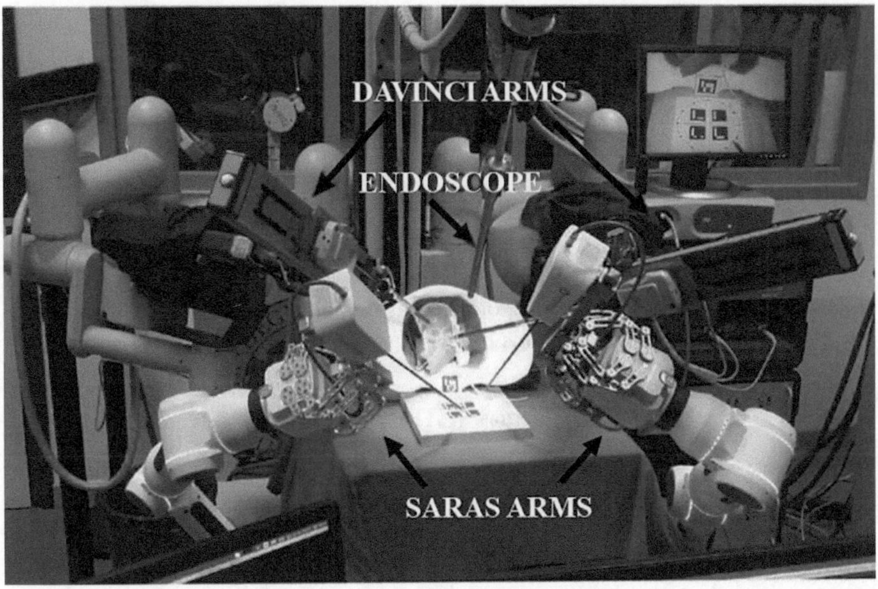

FIGURE 8.3 SARAS-operated arms-operated robot for surgery [36].

8.4 VIRTUAL NURSE ASSISTANTS (VNAs) FOR HEALTHCARE

In today's AI world, various business sectors are using a virtual assistant for easy workflow. Due to VNAs, it is possible to decrease the workload of healthcare-related persons; additionally, it can reduce unexpected appointments to the hospital. The virtual assistant plays a vital role in the healthcare industry because it can listen, talk, and guide. Due to this virtual assistant, it was found that improvements in the healthcare industry have been observed in the last two decades. The recent technologies, which include ML, DL, NLP, MV, and NN, were successfully implemented, and due to this, human conversation is reduced [57,58].

8.4.1 Medication Management and Medication Error Reduction (MMMER)

With the help of MMMER, it is possible to reduce cost expenditure, and its effect is a reduction in unnecessary injuries and deaths. In the United States of America, a total of $528 billion costs have been spent on drug-related medicine therapy. In 2016, the United States of America spent 16% expenditure of healthcare [59]. At the same time, error in the prescription of a drug by healthcare professionals leads to the waste of cost for the healthcare industry. In the United States of America, it was annually 7,000 deaths due to the wrong medicine and drugs [60]. AI is beneficial in the healthcare industry for medication error reduction, reduced expenses, reduced time, reduced drug overdoses, and medication safety.

8.4.2 Improving Medical Safety

Med Aware has built a system for alerting medication errors, and this system is based on the ML approach. This particular system has given a 56.20% valid alert for the presence of medication errors with accuracy. Additionally, 18.80% of medication alerts have given with medium accuracy [42].

8.4.3 Monitoring Medication Non-Adherence

It is one of the significant issues in the healthcare industry regarding cost and the outcome [61]. In a clinical trial, an AI-based NN network algorithm was used for adherence monitoring. This NN algorithm has combined with an MV trial, and its result is 17.90% adherence, which is higher than therapy procedure [44].

8.4.4 Clinical Trial Participation (CTP)

With AI, it is possible to design clinical trials, and it can be used to find patterns from unstructured data such as images, texts, speech. The NLP gives interaction of the information between human and computer from written as well as spoken language. The dataset has been collected, such as data before actual trials, medical literature, and electronic health records. AI needs to use real clinical trials to observe the patients constantly and automatically. Using AI leads to reliability, better assessment, and improved adherence. The IBM had conducted clinical trials based on the background information. In this study, AI was used, and it results in 80% improvement in the clinical trials. With the use of AI for clinical trials, this study increased the efficiency of the clinical trials [45]. One more research was carried out with the CTP to identify whether patients are eligible or not for clinical trials. This study was performed for the lung and breast cancer patients. This study found that only 24%–50% are eligible for clinical trials [46].

8.5 PRELIMINARY DIAGNOSIS AND PREDICTION (PDP)

For many decades, healthcare professionals are using patient's diagnosis data and health history data to offer more precise analysis of patients and more accurate health prediction. AI has been successfully implemented in various healthcare fields, and its result gives a specific medical diagnosis. The use of AI in the healthcare sector for diagnosis purpose and its example has been discussed below.

8.5.1 Diabetes Prediction

In the present world, a lot of people are facing the diabetes problem. While overcoming this problem, AI-based diabetes prediction has been implemented; hence, it provides more precise data for the health analysis of patients. The prognostic study of diabetes can be done with four application types: patient self-management tools, a retinal screening, predictive population risk, and clinical decision support. First, in patients' self-management tools, some devices and sensors have been planted in patients' bodies. Here activity-tracking devices and AI-based glucose sensors have been used. Additionally, the artificial pancreas has been used. Second in the retinal diagnosis, with retina scan, the detection of exudates, maculopathy, and other abnormalities are possible. Third, in the predictive population risk, the prediction study needs to carry out; it will identify the complications, readmissions, and hospitalization. Fourth, in clinical decision support, it is possible to detect and monitor diabetes as well as neuropathy and nephropathy.

The first AI-based glucose-monitoring system has been implemented by Medtronic's guardian connect, which can envisage the deviations in blood glucose level with the aid of ML. With the help of a sensor, it is possible to predict the changes 60 minutes earlier. Also, in 5-minute interval time, the sensor can monitor the body glucose level [47].

In this research, diabetes prediction is carried out with different ML algorithms. Different ML algorithms such as decision tree, support vector machine, and Naïve Bayes algorithms were used to detect diabetes. The highest classification accuracy, i.e., 76.30% is obtained with the Naïve Bayes algorithm compared with other algorithms. The validation of this study was carried out with Receiver Operating Characteristic curve [62]. Figure 8.4 shows implemented model for diabetes prediction.

8.5.2 Cancer Prediction

The US-based researchers have developed a system that can interpret mammography outcomes with the help of pathology and radiology patients. This interpretation has been carried out with the NLP-based application, and the accuracy of this application is 99%. This accuracy is more than the manual trials of the clinical process.

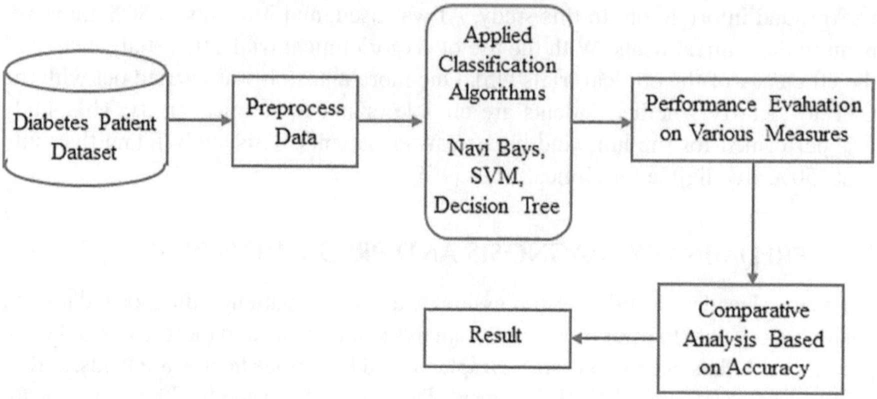

FIGURE 8.4 Implemented model for diabetes prediction [62].

Also, this system reduces the time required compared with manual clinical trials, which generally take 50–70 hours [48].

8.5.3 Tuberculosis Diagnosis

In tuberculosis diagnosis, the tests have been performed with the help of a deep convolution NN-based Google Net and Alex Net, which have the ability to learn the negative and positive X-rays. The tuberculosis test was performed on 150 patients. The 96% accuracy is obtained by combining both Google Net and Alex Net. The accuracy with the DL approach is increased as compared with the ML approach. In upcoming years the DL-based diagnosis of tuberculosis plays a significant role [49].

8.5.4 Psychiatric Diagnosis

The US-based researcher has developed an AI-based system that includes ML and NLP approaches for psychiatric diagnosis and gives 100% accuracy in diagnosis. In the conventional method, the accuracy was 79% which is less than the newly developed system. Due to AI, this system is more precise and gives better results than human diagnosis, which has errors. The system developed by these researchers analyzed the speech of psychiatric patients, and it differentiates from other patients. A psychologist misses a moment in psychiatric patients, but this computer-based system notices than minor moment; hence, this system is more suitable than the conventional one [50].

8.6 MEDICAL IMAGING AND IMAGE DIAGNOSTICS (MID)

Medical imaging is one of the best solutions to acquire patient data which have more complex. A traditional approach that includes image scans is more skilled, and it needs a lot of years of training. In recent AI-based method has been implemented for medical imaging. The DL result of these medical images increases accuracy and speed and reduces the cost compared with the conventional way. MID has been used in the healthcare industry to screen cancer, recognize the neurological disease, detect cardiovascular abnormalities, identify thoracic complications, etc. The combination of AI and medical image diagnosis can increase in the year 2026 to $256 billion, and the current value is $18 billion in the United States of America [63].

In one of the research convolutional neural network (CNN) is used to detect the predominantly compromised teeth. Here the peripheral radiographic images were acquired for the analysis point of view. After the diagnosis of teeth with this CNN, it is clear that the diagnostic accuracy was 81% and 76.7%, respectively. From the result, it can understand that it is one of the best methods for diagnosis [64]. Figure 8.5 shows the detection of predominantly compromised teeth using CNN.

8.6.1 Medical Imaging with Deep Learning

In medical imaging, DL has been used for diagnosis purposes. Deep CNNs, ANNs, and CNNs are the types of DL algorithms. The healthcare professional has 90.50%, while DL-based medical imaging has more accuracy, i.e., 92.50%. With DL for medical imaging, the complication has reduced, which is better than the conventional approach [21].

8.6.2 Image Diagnosis for Oncology

Many people are losing their lives in the entire world due to breast and lung cancers; hence, it is necessary to detect this cancer in the early stage [53]. The conventional method for breast cancer detection is digital mammography. But to understand and read these mammography images is a hard task for the radiologist because it may be false positive or false negative. Hence, there are many inaccuracies present in this method for detecting breast cancer; there is also anxiety for patients and more work for radiologists.

Some researchers have developed the system with AI and the DL model to predict breast cancer. This system is more superior to human experts. In the United States of America and United Kingdom, studies were conducted for breast cancer prediction. In the United States of America, false-negative and false-positive rates are reduced by 9.40% and 5.70%, respectively. Similarly, in the United Kingdom, false-negative and false-positive rates are reduced by 9.40% and 1.20%, respectively. The comparative study has been carried out with a conventional approach, i.e., clinical trials and AI systems [53].

The research was conducted at Stanford University with a NN algorithm. Here, cancer patient images were taken, and NNs have to identify cancer and automatically carry out the diagnosis. The NN-based prediction gives 72% accuracy, which is better than the healthcare professional with a specialty [55].

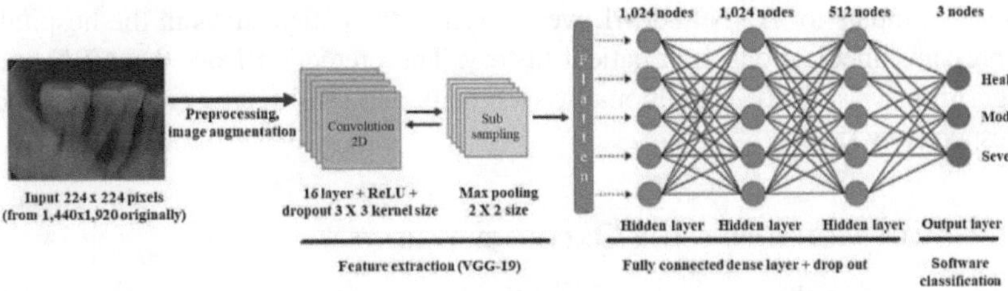

FIGURE 8.5 Detection of predominantly compromised teeth using CNN [64].

8.6.3 Optical Coherence Tomography (OCT) Diagnosis

In 2014 at the United States of America, 5.4 million OCT has been scanned. The AI-based DL approach is used to detect eye problems. The eye problem includes age-based macular degeneration, diabetic retinopathy, and glaucoma. The AI is a properly recognized eye problem with 94.50% accuracy through OCT scans [56]. To treat congenital cataracts, the Chinese researcher has developed a CNN-based AI system to identify, estimate, and advise treatment for this disease. This system provides the best results for diagnosing congenital cataract disease, but sometimes its misdiagnosis [64].

8.7 PATIENT HEALTH MONITORING (PHM)

Due to patients' continuous health monitoring, it is possible to reduce the staying time of patients in the hospital, and it increases the recovery time. The PHM generally provides the data which contains instructions, reminders, alerts, which is helpful for preventive action. In this study, researchers have used AI for continuous and periodic health monitoring of patients. Also, with AI, symptoms have been checked continuously by healthcare professionals. The AI-based health monitoring can be used in the home as well as in a hospital. This AI-based PHM system is the second-best method for 15 evaluated services in the healthcare sector [65].

8.7.1 Heart Failure Monitoring

Some researchers from Italy have developed the AI-based system, computer-based assistance for diagnosis, which is helpful for heart disease treatment. This system is more beneficial for the nurses and general practitioners because these healthcare professionals are not experts in cardiology. For nurses and general practitioners, while deciding on heart disease, this system is more suitable. This system generally collects heart disease information and predicts the current health of patients. While predicting heart disease, historical data is more important. The AI-based NN gives the best result from all group data, such as mild, moderate, and severe. The NN correctly classified 98 patients' data from out of 100 patients while training. During testing, 31 patient's data were correctly classified out of 36 patient's data. The overall accuracy was 86.10%, which best for heart disease monitoring [54].

8.7.2 Health Monitoring After Surgery

The ANN and ML are used by one research to envisage the death of patients in the hospital. The inputs for ANN and ML were given as the patient stays in the hospital, blood pressure, medications, and patient history. The prediction from the ANN and ML algorithms was 95.40%; additionally, specificity and sensitivity were 96.10% and 87%, respectively [55].

8.7.3 Health Monitoring for Oncology Patients

For improvement in the survival rate and the cancer patients' superiority, Finnish Firm Kaiku health has established the health intervention platform for symptom checking. This Kaiku firm took 766 cancer patient's data for the evaluation of symptom checking. The symptom monitoring during the chemotherapy gives them 150 days longer lifetime, which is shown by Kaiku health firm. Also, from this study, 31% of patients have enhanced their superiority of life, and 4% of patients have reduced hospitalization time. A 7% reduction in the emergency room visits was observed [56].

8.8 ADDITIONAL QUANTITATIVE METHODS USED IN BIOMEDICAL APPLICATION

8.8.1 Neural Network-Based ECG Anomaly Detection

The NN-based analysis was used for the ECG anomaly recognition on FPGA and trade-off analysis. The features reduction had carried out with the principle component analysis. In this work, a total of 181 samples were taken for prediction purposes. The multilayer perceptron algorithm was used for prediction; the training and testing prediction was carried out. With different activation, function mean squared error plots have been carried out. Different hidden and input layer classification accuracy was calculated, it was found that accuracy varies from 98% to 100%. In the result part, it was observed that 100% accuracy was obtained with 8 hidden layers [66].

8.8.2 A Fuzzy Neural Network Model for Post-surgery Risk Prediction

The post-surgery risk is a major problem in the healthcare industry. Hence this problem can be reduced with fuzzy NN model-based prediction. In this work, 120 patient's data were collected for the clinical trial. It was observed that 63 patients have successfully functioned. In addition, 30 patients have complications after surgery, and 28 patients need additional treatment after surgery. A fuzzy NN-based model predicts accurate results for post-surgery risks [67].

8.8.3 Heart Stroke Prediction with GUI Using Artificial Intelligence

The GUI using AI was made for heart stroke prediction in this work. Different inputs such as age, gender, married, work type, heart disease, BMI, smoking status, hypertension, glucose level, and stroke were considered. Each input parameter contains 10 records. Different algorithms such as decision tree, Naïve Bayes, and random forest were used to predict heart stroke. The classification accuracy for the decision tree, Naïve Bayes, and random forest algorithm was 95.9%, 96.4, and 98.2%, respectively. Hence, GUI was made with a random forest classifier for heart stroke prediction [68].

8.9 KEY ELEMENTS FOR SUCCESSFUL IMPLEMENTATION OF AI-BASED SERVICES IN HEALTHCARE

In this study, we emphasized the healthcare services and areas in which AI has been implemented. From this study, a lot of AI methods are available in the healthcare industry for improvement purposes. These AI-based systems have been observed by healthcare professionals.

Based on this study, the AI methods have been implemented in healthcare industry improvement and enhance the quality of workflow in the healthcare industry. Due to implementation of AI in the healthcare industry, it can:

- Accurate diagnosis of disease
- Monitor the health of patient condition and predict the disease
- Improvement in the quality of treatment
- Reduction in complication present in the surgery
- Save the time consumed in healthcare
- Control the misuse of medication
- Give the help while deciding on the clinical process

To successfully implement AI methods in the healthcare industry, one needs to consider the following factors.

- Certification for the medical devices
- In AI-based diagnosis need to give correct and proper inputs
- A dataset after proper clinical validation is needed in large amount
- The AI system developer and healthcare professional should validate the data together

8.10 OPPORTUNITIES AND CHALLENGES

AI-based diagnosis in the healthcare industry plays a crucial role in improvement in this industry. Hence compared with the conventional approach, this approach is more suitable. With the use of AI in healthcare, there are essential opportunities which are listed below:

- Improvement in operational efficiency and reduce medical cost
- Improvement in disease diagnosis
- Reduction in medical error and improvement in service quality
- Enhanced productivity and creation of new jobs
- Improvement in patient engagement
- Reduction in healthcare cost

The application of AI in the healthcare industry improves people's quality, but at the same time, new challenges are coming. Because of AI, sometimes human lives are in danger. The present difficulties in the AI-based healthcare industry are as follows:

- Cyber security because of the security and privacy
- Loss of jobs
- Need of training
- Managerial control loss
- Transformation pain
- Accountability of system usage

8.11 CONCLUSION AND FUTURE WORK

As the conclusion of this study, AI has many opportunities in the healthcare industry, including reducing healthcare costs, reducing healthcare professionals' workload, and easily giving more accurate diagnoses for various diseases. Due to the increase in the cost of the healthcare sector, AI is necessary and plays a vital role in cost reduction. Due to the shortage of nurses and healthcare professionals, so need to adopt an AI-based system for disease diagnosis. In developing countries, most of the population is poor, so they cannot take expensive treatment; in these countries, AI has overcome this problem. If a full implementation of AI in the healthcare sector occurs simultaneously, many costs will save and improve the quality of life. In developing countries, AI-based systems can implement patient health monitoring, primary diagnosis, VNA, and preventive health monitoring. The healthcare industry's AI-based solution saves $150 billion by 2026 in global healthcare [3]. This study clearly shows that the AI-based system can impact the healthcare industry; hence, healthcare-related companies and research institutes should adopt AI fully.

AI has been widely used in many areas of the biomedical field. AI can solve the complex problem of the biomedical field; hence AI is more suitable in this field for diagnosis purposes. Also, AI provides better decisions with models hence overcoming the human error problem, so AI has huge demand in this field. Due to the use of AI in this field, drastic changes are coming with better results; hence, it increases people's lives. The decision-making in a clinical trial has been increased accuracy with AI, reducing the clinical trial error and time. AI overcomes the major problem of the healthcare industry, which increases the quality of this industry. Also, the safety of this industry increased.

Hence, it attracted more researchers to make the healthcare industry's societal transformation using AI.

In upcoming years, i.e., it can be possible to design a model for AI-based healthcare service providers and healthcare service developers. From this present study, it is clear that cloud-based, easily combined platform and open-access technologies with AI can be implemented in the future. Also, future AI-based systems can be fast, accurate, and fully automated compared with today's system. It is also possible to use hospital information, national databases for health, and imaging databases for learning purposes.

ACKNOWLEDGMENT

The authors acknowledge RMD Sinhgad School of Engineering Pune for continuous support.

REFERENCES

1. Frost and Sullivan. (2016) Transforming healthcare through artificial intelligence systems. http://ww2.frost.com/news/press-release/600-m-6-billion-artificial-intelligence-systems-poised-dramatic-market-expansion-healthcare.
2. Accenture Technology Vision (2019). Full report, Accenture. Retrieved 19.12.2019. https://www.accenture.com/_acnmedia/PDF-94/Accenture-TechVision-2019-Tech-Trends-Report.pdf.

3. Artificial Intelligence. (2017). Healthcare's new nervous system. Accenture Consulting https://www.accenture.com/_acnmedia/pdf-49/accenture-health-artificial-intelligence.pdf.
4. Lee S, Lee D. (2020). Healthcare wearable devices: An analysis of key factors for continuous use intention. *Serv. Bus.* 14, 503–531.
5. Yoon S, Lee D. (2019). Artificial intelligence and robots in healthcare: What are the success factors for technology-based service encounters, *Int. J. Healthc. Manag.* 12, 218–225.
6. Ramesh A, Kambhampati C, Monson J, Drew, P. (2004) Artificial intelligence in medicine. *Ann. R. Coll. Surg. Engl.* 86, 334–338.
7. Safavi K, Kalis B. (2019). How AI can change the future of health care. *Harv. Bus. Rev.* Available online: https://hbr.org/webinar/2019/02/how-ai-can-change-the-future-of-health-care (accessed on 15 June 2020).
8. Mesko B. (2016). Artificial intelligence will redesign healthcare. Available online: https://www.linkedin.com/pulse/artificialintelligence-redesign-healthcare-bertalan-mesk%C3%B3-md-phd (accessed on 10 May 2020).
9. Liang H, Tsui B, Ni H, Valentim C, Baxter S, Liu G. (2019). Evaluation and accurate diagnoses of pediatric diseases using artificial intelligence. *Nat. Med.* 25, 433–438.
10. Amato F, López A, Peña-Méndez E, Vanhara P, Hampl A, Havel J. (2013). Artificial neural networks in medical diagnosis. *J. Appl. Biomed.* 11, 47–58.
11. Bennett C, Hauser K. (2013). Artificial intelligence framework for simulating clinical decision-making: A Markov decision process approach. *Artif. Intell. Med.* 57, 9–19.
12. Dilsizian S, Siegel E. (2014). Artificial intelligence in medicine and cardiac imaging: Harnessing big data and advanced computing to provide personalized medical diagnosis and treatment. *Curr. Cardiol. Rep.* 16, 441.
13. Esteva A, Kuprel B, Novoa R, Ko J, Swetter S, Blau H, Thrun S. (2017). Dermatologist-level classification of skin cancer with deep neural networks. *Nature* 542, 115–118.
14. Rigby M. (2019). Ethical dimensions of using artificial intelligence in healthcare. *AMA J. Ethics* 21, E121–E124.
15. Luxton D. (2014). Artificial intelligence in psychological practice: Current and future applications and implications. *Prof. Psychol. Res. Pract.* 45, 332–339.
16. Miyashita M, Brady M. (2019). The health care benefits of combining wearables and AI. *Harv. Bus. Rev.* Available online: https://hbr.org/2019/05/the-health-care-benefits-of-combining-wearables-and-ai (accessed on 18 June 2020).
17 Abomhara M, Køien G. (2015). Cyber security and the internet of things: Vulnerabilities, threats, intruders and attacks. *J. Cyber Secur.* 4, 65–88.
18. Rao DAS, Verweij G. (2017). *What's the Real Value of AI for Your Business and How Can You Capitalise.* PwC Publication, PwC, 2017, https://www.pwc.com/gx/en/issues/analytics/assets/pwc-ai-analysis-sizing-the-prize-report.pdf
19 Vaananen A, Haataja K, Julkunen K, Toivanen P. (2021). AI in healthcare: A narrative review, F1 research. https://doi.org/10.12688/f1000research.26997.1.
20. Russell SJ, Norvig P. (2010). *Artificial Intelligence: A Modern Approach* (3rd ed.). Prentice Hall 2010, https://zoo.cs.yale.edu/classes/cs470/materials/aima2010.pdf
21. Jordan MI, Bishop CM. (2004.). Neural networks. In Allen B. Tucker (ed.). *Computer Science Handbook.* 2nd ed., (Section VII: Intelligent Systems). Boca Raton, FL: Chapman & Hall/CRC Press LLC.
22. Chowdhury GG. (2003). Natural language processing. *Ann. Rev. Info. Sci. Technol.* 37(1), 51–89.
23. Kwoh YS, Hou J, Jonckheere EA, et al. (1988). A robot with improved absolute positioning accuracy for CT guided stereotactic brain surgery. *IEEE Trans. Biomed. Eng.* 35(2), 153–160.

24. Kypson AP, Chitwood WR. Jr (2004). Robotic applications in cardiac surgery. *Int. J. Adv. Robot Syst.* 1(2), 87–92.
25 Melfi FM, Menconi GF, Mariani AM, et al. (2002). Early experience with robotic technology for thoracoscopic surgery. *Eur. J. Cardiothorac. Surg.* 21(5), 864–868.
26 Hyun MH, Lee CH, Kim HJ, et al. (2013). Systematic review and meta-analysis of robotic surgery compared with conventional laparoscopic and open resections for gastric carcinoma. *Br J Surg.* 100(12), 1566–1578.
27. Herron DM, Marohn M. (2008). SAGES-MIRA robotic surgery consensus group: A consensus document on robotic surgery. *Surg Endosc.* 22(2), 313–325, discussion 311-2.
28. DiGioia AM, Jaramaz B, Picard F, et al. (2004). *Computer and Robotic Assisted Hip and Knee Surgery.* Oxford University Press. 127–156., https://books.google.co.in/books/about/Computer_and_Robotic_Assisted_Hip_and_Kn.html?id=z8Q8mAhPtoC&redir_esc=y.
29. Shweikeh F, Amadio JP, Arnell M, et al. (2014). Robotics and the spine: A review of current and ongoing applications. *Neurosurg. Focus.* 36(3), E10.
30. Hameed AM, Yao J, Allen RD, et al. (2018). The evolution of kidney transplantation surgery into the robotic era and its prospects for obese recipients. *Transplantation* 102(10), 1650–1665.
31. Lee DI. (2009). Robotic prostatectomy: What we have learned and where we are going. *Yonsei Med. J.* 50(2), 177–181.
32. Invester presentation Q4 (2019). Intuitive surgical. Retrieved December 13, 2019, https://isrg.gcs-web.com/static-files/0fc01a59-8d32-481f-9872-b262fd1f87b2.
33. Jenny JY. (2006). The history and development of computer assisted orthopaedic surgery. *Orthopade* 35(10), 1038–1042.
34. Kenngott HG, Wagner M, Nickel F, et al. (2015). Computer-assisted abdominal surgery: New technologies. *Langenbecks Arch Surg.* 400(3), 273–281.
35. Ho C, Tsakonas E, Tran K, et al. (2011). Robot-assisted surgery compared with open surgery and laparoscopic surgery: Clinical effectiveness and economic analyses, Canadian Agency for Drugs and Technologies in Health, Ottawa (ON); 2011. PMID: 24175355.
36. Basch E, Deal AM, Dueck AC, et al. (2017). Overall survival results of a trial assessing patient-reported outcomes for symptom monitoring during routine cancer treatment. *JAMA.* 318(2), 197–198.
37. Sham JG, Richards MK, Seo YD, et al. (2016). Efficacy and cost of robotic hepatectomy is the robot cost-prohibitive, *J. Robot. Surg.* 10(4), 307–313, doi: 10.1007/s11701-016-0598-4.
38. Kristensen SE, Mosgaard BJ, Rosendahl M, et al. (2017). Robot-assisted surgery in gynecological oncology: Current status and controversies on patient benefits, cost and surgeon conditions - a systematic review. *Acta Obstet. Gynecol. Scand.* 96(3), 274–285.
39. Carr-Brown J, Berlucchi M. (2016). Pre-primary care: An untapped global health opportunity. https://www.wiltonpark.org.uk/wp-content/uploads/Pre-Primary-Care-An-Untapped-Global-Health-Opportunity.pdf
40. Chambers D, Cantrell AJ, Johnson M, et al. (2019). Digital and online symptom checkers and health assessment/triage services for urgent health problems: Systematic review. *BMJ Open.* 9(8), e027743.
41. Sensely. (2019). An integrated payer/provider wanted to intervene in a timelier manner with its Chronic Heart Failure (CHF) patients.
42. Schiff GD, Volk LA, Volodarskaya M, et al. (2017). Screening for medication errors using an outlier detection system. *J. Am. Med. Inform. Assoc.* 24(2), 281–287.
43. Labovitz DL, Shafner L, Reyes GM, et al. (2017). Using artificial intelligence to reduce the risk of nonadherence in patients on anticoagulation therapy. *Stroke.* 48(5), 1416–1419, doi:10.1161/STROKEAHA.116.016281. National Library of Medicine, HHS Public Access.

44. Bain EE, Shafner L, Walling DP, et al. (2017). Use of a novel artificial intelligence platform on mobile devices to assess dosing compliance in a phase 2 clinical trial in subjects with schizophrenia. *JMIR Mhealth Uhealth*. 5(2), e18.
45. Haddad T, Helgeson J, Pomerteau K, et al. (2018). Impact of a cognitive computing clinical trial matching system in an ambulatory oncology practice. In *Abstract Presented at American Society of Clinical Oncology (ASCO), Annual meeting,* 36(15), 6550.
46. Calaprice-Whitty D, Galil K, Salloum W, et al. (2020). Improving clinical trial participant prescreening with artificial intelligence (AI): A comparison of the results of AI-assisted vs standard methods in 3 oncology trials. *Ther. Innov. Regul. Sci.* 54(1), 69–74, https://doi.org/10.1007/s43441-019-00030-4.
47. Christiansen MP, Garg SK, Brazg R, et al. (2017). Accuracy of a fourth-generation subcutaneous continuous glucose sensor. *Diabetes Technol. Ther.* 19(8), 446–456.
48. Patel TA, Puppala M, Ogunti RO, et al. (2017). Correlating mammographic and pathologic findings in clinical decision support using natural language processing and data mining methods. *Cancer* 123(1), 114–121.
49. Lakhani P, Sundaram B. (2017). Deep learning at chest radiography: Automated classification of pulmonary tuberculosis by using convolutional neural networks. *Radiology* 284(2), 574–582.
50. Corcoran CM, Carrillo F, Fernández-Slezak D, et al. (2018). Prediction of psychosis across protocols and risk cohorts using automated language analysis. *World Psyc.* 17(1), 67–75.
51. De Fauw J, Ledsam JR, Romera-Paredes B, et al. (2018). Clinically applicable deep learning for diagnosis and referral in retinal disease. *Nat Med.* 24(9), 1342–1350.
52. Liu X, Faes L, Kale AU, et al. (2019). A comparison of deep learning performance against health-care professionals in detecting diseases from medical imaging: A systematic review and meta-analysis. *Lancet Digit. Health.* 1(6), e271–e297.
53. McKinney SM, Sieniek M, Godbole V, et al. (2020). International evaluation of an AI system for breast cancer screening. *Nature.* 577(7788), 89–94.
54. Esteva A, Kuprel B, Novoa RA, et al. (2017). Dermatologist-level classification of skin cancer with deep neural networks. *Nature.* 542(7639), 115–118.
55. Guidi G, Iadanza E, Pettenati MC, et al. (2012). Heart failure artificial intelligence-based computer aided diagnosis telecare system. In *International Conference on Smart Homes and Health Telematics*, Springer, Berlin, Heidelberg, 278–281.
56. Monsalve-Torra A, Ruiz-Fernandez D, Marin-Alonso O, et al. (2016). Using machine learning methods for predicting in hospital mortality in patients undergoing open repair of abdominal aortic aneurysm. *J. Biomed. Inform.* 62, 195–201.
57. Samuel D, Cuzzolin F. (2021). Unsupervised anomaly detection for smart autonomous robotic assistant surgeon (SARAS) using a deep residual autoencoder. *A Preprint*, https://saras-project.eu/.
58. Laranjo L, Dunn AG, Tong HL, et al. (2018). Conversational agents in healthcare: A systematic review. *J. Am. Med. Inform. Assoc.* 25(9), 1248–1258.
59. Radziwill NM, Benton MC. (2017). Evaluating quality of chatbots and intelligent conversational agents. arXiv preprint arXiv: 1704.04579.
60. Watanabe JH, McInnis T, Hirsch JD. (2018). Cost of prescription drug-related morbidity and mortality. *Ann. Pharmacother.* 52(9), 829–837.
61. Williams DJP. (2007). Medical errors. *J. R. Coll. Physicians, Edinb.* 37(4), 343.
62. Sisodia D, Sisodia D. (2018). Prediction of diabetes using classification algorithms. In *International Conference on Computational Intelligence and Data Science*, 132, 1578–1585.
63. Iuga AO, McGuire MJ. (2014). Adherence and health care costs. *Risk Manag. Healthc. Policy* 7, 35–44.
64. Long E, Lin H, Liu Z, et al. (2017). An artificial intelligence platform for the multihospital collaborative management of congenital cataracts. *Nat. Biomed. Eng.* 1(2), 0024.

65. Väänänen A, Haataja K, Toivanen P. (2020). Survey to healthcare professionals on the practicality of AI services for healthcare [version 1; peer review: 1 approved with reservations]. *F1000Res.* 9(760), 760.
66. Wess M, Sai Manoj PD, Jantsch A. (2017). Neural network based ECG anomaly detection on FPGA and trade off analysis. In *IEEE International Conference on Symposium on Circuits and Systems.* Doi: 10.1109/ISCAS.2017.8050805.
67. Shatalova O, Filist S, Korenevskiy N, et al. (2021). Application of fuzzy neural network model and current voltage analysis of biologically active points for prediction post-surgery risks. *Comput. Meth. Biomech. Biomed. Eng.* Doi: 10.1080/10255842.2021.1895128.
68. Babu D, Karunakaran V, Gopinath S, et al. (2021). GUI based prediction of heart stroke using artificial intelligence. In *Material Today Proceedings*, Doi: 10.1016/j.matpr.2021.03.696.

9 Role of Artificial Intelligence in Transforming Agriculture

K. Kanagaraj
MEPCO Schlenk Engg. College

K. Nalini
Ayya Nadar Janaki Ammal College

CONTENTS

9.1 INTRODUCTION

Environmental elements such as rainfall, temperature, and humidity play a vital role in the agriculture life cycle in current farming systems. On the other hand, growing trends in deforestation and increasing pollution result in environmental changes. Due to these factors, farmers find it difficult to take decisions to prepare the soil, sow

DOI: 10.1201/9781003220176-9

seeds, and harvest. Each crop needs a specific type of nutrition from the soil. There are three most important nutrients to be present in soil for the growth of plants: nitrogen (N), phosphorus (P), and potassium (K) [1]. The deficiency of any of the nutrients can result in poor quality of crops. Apart from the above factors, weed protection also plays an important role in crop cultivation [2]. If not controlled in a timely manner, it can lead to an increase in production costs as well as the absorption of nutrients from the soil, resulting in nutritional deficit.

Government of India and IBM have initiated a pilot study for using artificial intelligence (AI) in Indian agricultural farms at Madhya Pradesh, Maharashtra, and Gujarat and provide information about soil fertility and climatic conditions of farms. The study has a focus on improving farming practices using AI, satellite data, and Internet of Things (IoT) sensors. Pradhan Mantri Fasal Bima Yojana is a crop insurance scheme that uses AI, remote sensing images, and modeling tools for settling the claims of farmers within a short duration of time. Government of Karnataka in association with Microsoft assists farmers in using AI-based solutions for agricultural issues like sowing area, yielding time, weather forecast, and production time. To evaluate the lands, the Maha Agri Tech project employs satellite pictures and data analytics from the Maharashtra Remote Sensing Application Centre and the National Remote Sensing Centre. This is carried out for different crops based on the soil fertility, moisture content, and rainfall. Figure 9.1 shows the different stages in the life cycle of crop cultivation.

Industries are focusing on AI technologies to generate more yields by growing healthier crops, controlling pests, monitoring soil fertility, and developing growing conditions to improve a wide range of agriculture-related tasks in the entire food supply chain. Figure 9.1 shows the life cycle of crop cultivation. With the wide range

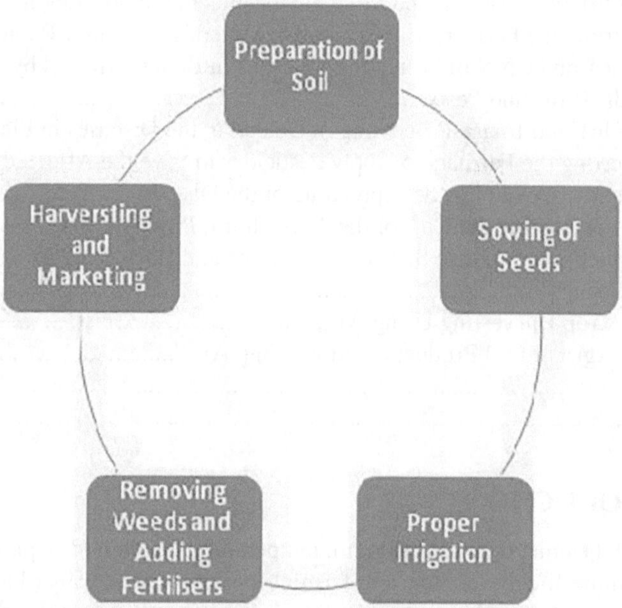

FIGURE 9.1 Life cycle of crop cultivation.

of AI techniques, it is possible to help the farmers with smart farming to maximize their yield as well as the profit [3].

9.2 ROLE OF AI IN DETERMINING THE NATURE OF THE SOIL AND RECOMMENDING SUITABLE PLANTS

Indian agriculture is encountering a lot of problems due to the lack of using/applying appropriate technologies. Our nation has plenty of agricultural land, and majority of them receive enough seasonal rains mostly during appropriate time. Several good practices are being followed by our farmers for several decades. Though this is a good practice, they are resistant of overcome some of the practices, particularly using previous knowledge of rainfall, yield, and understanding the changing nature of soil. Predicting the current nature of soil and selecting crops that can grow well in the soil will provide better results [4]. This helps them to cultivate crops that they believe cannot be grown in that soil.

Several new initiatives are being taken by the government to test the soil type and recommend suitable crops [5]. However, they are not easily accessible to them, and only a minimum number of them adopt them due to the fear of losing their income. Soil type identification will be helpful for recommending crop type using the following process:

- Collect soil sample as per the requirements
- Test for soil type using manual/automatic process
- Apply suitable machine learning technique for classifying the soil
- Recommend a suitable type of crop for cultivation
- Recommend suitable construction methods and practices for that soil

As shown in Figure 9.2, once the soil type is tested, AI-based techniques can be used to forecast different types of crops that can be cultivated. Also farmers will be informed about the different factors to select a suitable type of crop for cultivation.

IoT-based sensors can be utilized to identify the soil organic matter (SOM) substance, which is the key marker for deciding the richness of the soil. It is additionally a marker for the physicochemical properties and quality of soil. Among the wide extent of methods, Obvious (VIS, 400–780 nm) and near-infrared (NIR, 780–2,526 nm) spectroscopy are the better and effective procedures for checking the SOM [4], since unearthly reflectance of soil is adversely related with the SOM substance, and the SOM substance can be gotten from measured soil reflectance range [1,6].

Soil nano-biosensors are kept beneath the soil and followed utilizing GPS. These biosensors upgrade the wellbeing status of the soil and send the report of soil pH, mineral substance (nitrogen, phosphorus, and potassium), moisture level, and pesticide residue. The information is upgraded and put away in a database at customary interims, which makes a difference to require choice of the trim determination, seed, and fertilizer amount to the agriculturists. Intel has developed nanochips possessing all over a vineyard in Oregon, USA. The sensors record temperature values every minute [3]. Researchers have successfully developed wireless nanosensors consisting

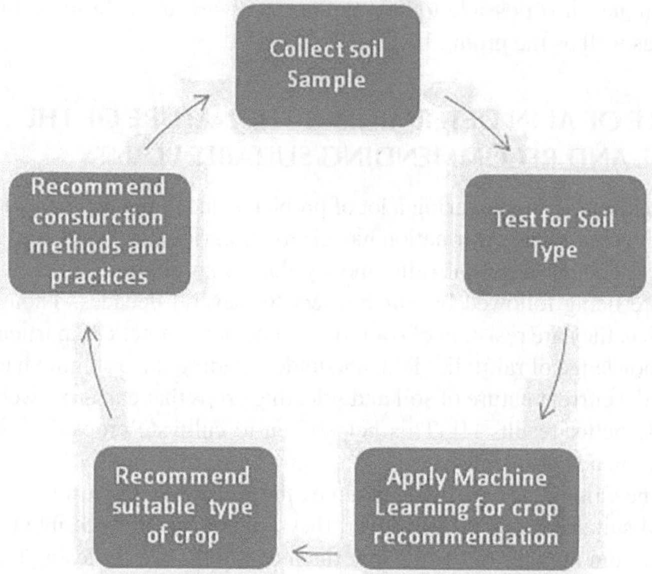

FIGURE 9.2 Role of AI in soil type identification and crop recommendation.

of micromachined microelectromechanical system cantilever beams coated with water-sensitive nanopolymer for moisture optimization.

Deep learning (DP) has developed quickly in recent years. This strategy permits for a computational demonstration comprising numerous handling layers to memorize information representations with different levels of deliberation. After information preprocessing by vital components analysis, the SVM gave the finest expectation coming from the SOM substance (RPD = 2.28). Also, Ji et al. [7] utilized 441 soil tests (400–2,450 nm) to foresee the SOM content, for which the SVM had an RPD = 2.16, but the most excellent precision came from PLSR-back propagation (BP) with an RPD = 2.36. The inversion precision of models such as PLSR, BPN, and SVM is for the most part higher than that of LR, and the foremost commonly utilized modeling strategy [8] utilizes the BPN to upgrade the moment factors created from an arbitrary timberland to supply a variable choice procedure. Concurring to earlier ponders, the current approach to anticipate the SOM substance by VIS-NIR spectroscopy finds the highlight range and after that builds up a forecast demonstration. Most of these thoughts have centered on the preprocessing of soil spectral data and the screening of useful element spectra. However, we still require a high-performance modeling method that eases the preprocessing requirements of spectral data, which is also vital for ensuring exact predictions.

By understanding the structure of the deep nonlinear network approach, the BP algorithm is used to approximate complex functions. The results obtained show how the DP machine should change its internal parameters to discover the complex structure of a larger data set, demonstrating the powerful ability to learn the essential characteristics of the data set from a smaller sample [9]. In addition, convolutional neural networks (CNNs) have been applied to image recognition research. CNN uses convolution and grouping operations to extract abstract feature maps layer by layer from the data, thereby learning the structural features and basic relationships in the spectral data [10]. In this way, the spectral curve can be regarded as a wavelength · 1 grayscale image, and the same filling ability can be used for convolution operations on a deeper lattice. Therefore, it should be feasible to directly use the original or converted spectral data for SOM inversion. With the improvement of computing power and the rapid development of DP, it becomes increasingly necessary to explore how to apply DP to predict SOM content from the VIS-NIR wavelength.

As this method can be trained to understand the important features in a data set, comparing the results based on the full spectrum with the selected feature spectrum will get better results. In addition, the best performing model can be used to fit spectral reflectance data and test whether certain data preprocessing steps can be eliminated. It is very useful to apply machine learning to detect the properties of soil. Accurate and absolute identification of soil characteristics and soil quality classification can be accomplished through visual and manual techniques and laboratory identification tests.

Detailed laboratory soil tests and investigations provide accurate and basic information about soil strength and behavior characteristics as needed. Apply the information obtained previously and the principles and concepts of soil mechanics to analyze the problems involved in a given situation and make design and construction decisions.

9.3 ROLE OF AI IN ESTIMATING THE WATER REQUIREMENT FOR THE CROPS AND THE DETERMINING THE AVAILABILITY OF WATER IN WATER BODIES AND THE EXPECTED AMOUNT OF RAIN

Using AI, farmers can examine the water availability by using a suitable weather forecasting technique and other factors that help them to prepare the type of crop to be grown and when to sow the seeds. The water availability for cultivation can be had from three important sources (Figure 9.3):

1. Rainfall, seasonal and unexpected
2. Rivers and canals
3. Water bodies like dams, lakes, and wells

Sowing crops at the right time is a great challenge in agriculture. This problem can be rectified by AI. In association with International Crops Research Institute, Microsoft had collected the data of climate change in semi-arid tropics since 1986 in a specific area of Andhra Pradesh, which is analyzed using AI and moisture adequacy index was calculated, which helped to assess the degree of rainfall and soil moisture. Crop sowing apps help the farmers to choose the right time for sowing. Proper estimation of the rainfall and water availability in water bodies will be good information for the famers to select the suitable type of crop to be cultivated depending on availability of water. The existing weather forecasting techniques help the farmers to estimate the

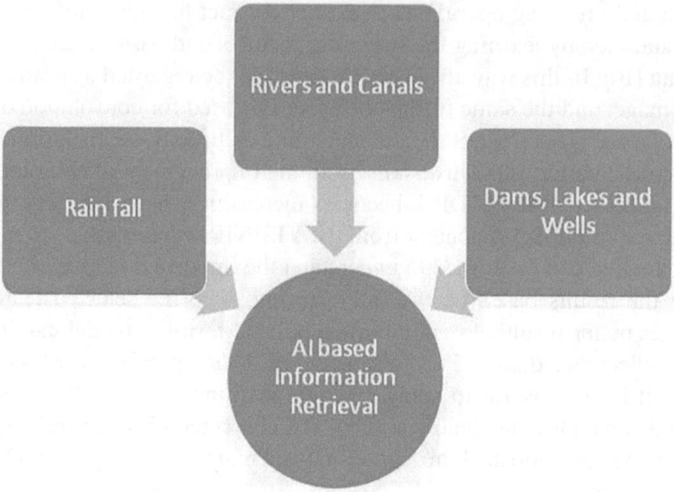

FIGURE 9.3 Use of AI in estimating water availability and recommending suitable type of crop.

rainfall to some extent; however, estimating the water availability in the water bodies requires the use of modern data-processing and AI-related tools (Figure 9.3).

Increasing demand for food with good quality and large quantity has created the need for smart and optimal farming practices. The research reports from different agencies state that within the next 30 years, there is a need to feed 2 billion more people without an increase of cultivatable land. They have also predicted that by 2060, the agriculture sector has to generate more food than it has produced during the previous 8,000 years. While various intelligent solutions are available currently to increase the yield, many barriers have prevented farmers in adopting digital transformation. Some of the key issues include requiring too much physical labor to make the solution work or relying on remote internet connectivity. Huge amounts of agricultural data, on the other hand, is generated but seldom utilized.

Accurate weather monitoring and providing timely inputs based on the water availability and requirement can drive several useful farming activities that help the farmers to decide on the suitable time and crop for cultivation. The satellite weather data provided by several government agencies in association with the local weather stations provides a very precise and accurate data for the farmer and enable them to make day-to-day decisions. If the prediction is for rainfall to occur, then the farmer can save water by not watering the farm and refining the irrigation schedule. The amount of rainfall becomes crucial in deciding the correct time to spray fertilizer and avoid them from being washed out. It is important for a farmer to know when it will rain. If AI can predict the rain accurately and how much and when that will happen, farmers can make meaningful decisions.

Several firms have recently applied AI, machine learning, and advanced analytics to farm data in order to extract important insights and automatically generate suggestions for better decisions. Farmers can visualize data and alerts linked to essential aspects including weather forecasts, soil conditions, evaporation and transpiration rates, and crop stress using a uniform dashboard. In addition, AI data enables insights for irrigation, planting, and yield timing.

A project was recently implemented in Karnataka to use IBM's Watson platform to predict the price of tomatoes and corn. IBM has developed a price forecasting system for the Karnataka Agricultural Price Council, which can predict market price trends and tomato production patterns for at least 15 days. The dashboard uses IBM's Watson assessment value and combines satellite imagery and weather data to assess the ground surface and monitor crop health in real time. It can detect pests and diseases, estimate tomato yields and yields, and predict prices. Previously, production estimates were mainly based on planted area data. Other data, such as prices in major markets in neighboring countries, are also used for price forecasts. The solution is applicable to the three main tomato growing areas of Kolar, Chikkaballapur, and Belgavi and the two main corn growing areas of Davangere and Haveri. Through more ultra-local data entry, we can improve the accuracy of the weather forecast.

9.4 ROLE OF IoT IN RETRIEVING THE MINERAL CONTENTS IN THE SOIL REGULARLY AND ALERTING THE FARMERS TO ADD SUITABLE MINERALS WHENEVER REQUIRED

The growing information technology can support to do mineral exploration with reduced cost. Geochemistry and geophysics are the conventional methods used in mineral exploration. Nowadays, satellite imaging and remote sensing are majorly used to find the minerals through hyperspectral images. Remote sensing and satellite imaging are also slightly complex to use. Based on recent survey, the drones are also used to capture the land images that are very clear with good resolution. All along, the current technologies like IoT and machine learning can be used for mineral exploration with reduced cost than the state-of-the-art techniques and other latest technologies. IoT and AI already exist in the geosciences and exploration.

In land, soil or rocks can be sensed with sensors to read its characteristics. The soil surface in particular can be used to find the minerals availability on that land area. Based on the IoT technology, the characteristics and other features can be sensed with the help of sensors. Those values can be taken as input and processed with the help of machine learning technologies to develop the classification and exploration model that helps to find the mineral deposits in the soil. For instance, the natural radioactivity of the soil can be sensed with the gamma ray sensor. The gamma photons on soil surface or rock surface can be read using this sensor. The gamma photons will help to find the minerals. Hence, the read values can be processed using machine learning technologies to find the availability of such minerals in the soil. In this way, there are various sensors to sense the other properties of the soil/rock to discover the various minerals in the particular land.

Around the world, many countries are implementing latest information technologies in finding the mineral contents from the soil. In Finland, the minerals are explored using IoT technology. Human operators in northern Sweden keep an eye on what's happening on and plot their attack on lucrative iron ore seams. But, unlike in the past, these mining experts are sitting in an office thousands of kilometers away, relying on IoT sensors, cameras, subsurface long-term evolution networks, and a sophisticated private cloud to supervise their blasting and tunneling in reclaimed land. The Bulgarian gold mine in Europe uses sensors in conveyor belts and lighting and RFID tags on helmets of their employees for enhanced asset tracking. As Indian economy is globalized, it is now becoming essential to find the natural deposits in cost effective and eco-friendly ways to assimilate the global developments in science and technology.

Exploration of many mineral is done in geophysical and geochemical methods using magnetic, electrical, radioactivity, and gravitational properties. Now these physical methods can be applied with the help of IoT and machine learning technologies. Many geophysical sensors are available to read the soil surface parameters that can be applied with machine learning technology for classification and exploration of minerals. The IoT device with geophysical sensor will be used to collect the soil/rock surface parameters such as magnetic, electromagnetic, induced polarization, radioactivity, gravitation, conductivity, and density. Data from surveying devices gets uploaded using wireless technology to a locally available database and finally passed off to a cloud service for further processing in machine learning for characterization of the explored minerals. The classification model developed using machine learning technology will get trained by ingesting the data read through IoT device. The trained model can be scalable for '*N*' number of samples. The classification model can produce an accurate and more efficient characterization of minerals from the sample soil. Using these, farmers can choose to cultivate the crops suitable for their soil to get high yield and produce healthy crops.

9.5 USE OF IoT AND CNN IN PROTECTING CROPS FROM BEING AFFECTED BY ANIMALS, BIRDS AND PESTS

Agriculture is the primary sector in India. But recently, the growth of the agricultural sector is getting diminished due to various reasons. Damaged crops are also a major reason to attain bad crop yield. The agricultural field near the forest/mountain area is affected by wild animals/birds/insects [11]. A lot of ways are getting implemented to safeguard the agricultural field from animals/birds/insects. Those methods are sometimes more harmful to wildlife animals/insects/birds and also to the environment. Nowadays, technologies are getting implemented in various fields, even in agricultural field. But most of the technologies focus on finding the plant diseases, plant irrigation, etc. There is little research focus in the area of securing the agricultural land from intruders. Moreover, the farmers are not getting any knowledge related to the demand for the crop that they cultivate. Some research is going in forecasting demand for crop. The farmers can have interaction with a mobile application with visual and audio mode [12]. Animal monitoring system was developed using radio tracing and wireless sensor technology [13]. A lot of researches were done using wireless sensor network for detecting the wildlife animals crossing the road. The animal monitoring system is developed using Zigbee-based wireless sensor network. The wireless sensor network is used to identify the elephant's existence on railroads [14]. But no system is available to give suggestions to farmers for selecting right crop based on the demand. It is urged to develop wildlife friendly safeguarding systems along with the crop selection systems.

Pest is a drastic problem for the farmers, which reduce the growth, yield, and price of the crop products. Lots of visual-based recognition methods linked with geolocation are used to detect pests when they make initial attacks on the crops [15]. Asian citrus psyllid pest is detected in Florida using machine vision and DP-CNN [16]. It reduced the time for detecting the pest and thus saving citrus plants. The traditional system for securing agricultural field is not animal and environment friendly. A lot of systems are developed using information technology. There is no system available that is environment friendly as well as increases the crop yield. The proposed system will provide better security without harm to wild animals as well as to environment. The farmers are getting very low price for their crop in the market as those crops are not in demand. A part of the proposed system will give awareness to farmers before they cultivate the crops whether they have demand in future or not. Although a lot of techniques are available for image classification, the DP is playing vital role and bringing more accuracy of classification than any other state of art technologies. Even for prediction of time series, data can attained through DP models. Most of the existing systems are available only for disease detection in the plant and plant irrigation. But the major reason is intrusion, which affects at the end of cultivation time leading to a lot of losses for the farmer but not addressed by any readily available systems.

The protection of agriculture field using technology has started long ago. The automatic monitoring of the growth of the crop is done by the IoT solutions to protect crop from animals. A lot of research on crop protection from animals and monitoring of plant diseases is going on. DP models are created to detect plant diseases. Laser technologies are used for protecting crops from birds. The crops are protected from insects using repellants or pesticides. In DP, CNN model is developed to identify the image. Many improved CNN models are developed to identify the images as human or not. However, in spite of lots of research on crop protection, there is no complete system to protect crop from animals/insects/birds. We identified the gap in the research that no complete system for crop protection and demand-based selection crops is available [16]. Hence, we propose complete automated protection system and demand-based crop selection system. In Himachal Pradesh, solar fencing is used for protecting crop from animals.

Growth in agricultural sector is very important at all the time. Although agricultural sector is improving, the farmers are not getting rational income. The crop yield is getting affected by various factors such as rainfall and temperature. There are a lot of researches going on to resolve those issues. But there is no research regarding protection of crop and selection of crop based on demand prediction. Though many technologies are used for plant irrigation, prediction of plant diseases, etc., when the crop is ready for harvesting, if any threats/damages occur for crops, farmers will be completely get fed up with their job. The importance given for growing crops should the same for their protection. The protection using traditional method is not effective for all kinds of threats to crops. According to Parliamentary Standing Committee (2012–2013), in our country, pests and diseases are eating 20%–30% of food per year, which costs around Rs. 45,000 crores, produced through agricultural workers.

The report says about damages by pests and diseases. The crop damages are high in Maharashtra, and scaring techniques with common sound are not worth to distract the birds. All India Network Project on Vertebrate Pest Management has conducted a survey, and based on that, it was found that not all the birds and animals are damaging the crops. Only few birds and animals are damaging the crops. So, animal- and bird-specific scaring sounds will give better solutions. The income of the farmers is affected due to crop damage done by snow leopards and wolves in Spiti region of the Indian Trans-Himalaya. IBM is developing an integrated pest management system, which will regulate the pesticides to the crops. Agronomy experts have identified few strategies to protect crop from wild animals. Low-cost animal friendly crop protection system is proposed in myGov Innovations. But no system is proposed that includes all the intruders like animal, birds, and insects. Most of the researches are in disease monitoring and detection. Artificial neural network is used to develop predictive model. Most of the machine learning techniques with IoT are also used to predict accurately of time series data. The demand data set is also time series data, which may be used to predict with machine learning techniques. Still, DP techniques are showing better accuracy than any state-of-the-art techniques. The wireless network with better energy throughput is also used for developing protection system [17]. Some of the interesting research projects that use machine learning in agriculture are as follows:

- International Plant Protection Convention – Food and Agriculture Organization of the United Nations having standards for plant protection.
- Asia and Pacific Plant Protection Commission is ensuring the plant protection by having pest management and Invasive/Migratory Species Management.
- North American Plant Protection Organization is providing the forum to work in partnership for the development of science-based standards planned to protect farm lands.
- Republic of Tajikistan provides laws to protect plant.
- Spain is offering a crop protection industry, which can be attributed to the rising trend for organic farming in the country.
- Taiwan is having BASF crop protection innovation pipeline.

The crop damages are high in all the countries, especially the countries like Italy and Nepal. Wild boar, monkeys, porcupine, goral, deer, and bear are damaging crops from 0.23% to 46.48%. The Danish National Advanced Technology Foundation has invested a lot of money to project where to improve the efficiency of the machines in the field in account of wildlife and surroundings and also the key focus of the project is to develop algorithms to detect and recognize obstacles using wireless sensor networks. The movable objects are captured with the help of wireless sensor network using passive infrared sensors to identify the objects in motion in Korea. The alarm system using MCS and GSM is developed in China and New Zealand with infrared monitoring system for quick actions. The crops are protected from elephants through chili fencing in Tanzania for not harming elephants. Protection of plants from diseases and pests is the science of crop protection. Decision support system was developed for dynamic management of diseases. Solar panel fencing is used to scare birds by creating parametric sonic shield in Virginia.

9.6 ROLE OF IoT AND IMAGE PROCESSING IN DETECTING THE DISEASES IN PLANTS AND ALERTING THE FARMERS TO APPLY PESTICIDES TO SAVE THE AFFECTED PLANTS AND TO AVOID FURTHER SPREADING OF THE DISEASE

The existing methods for detecting diseases in plants are done simply using naked eye observation by farmers and experts. To carry out this, a large human resource is required and not feasible for large farms with multiple acres of land [18]. However, farmers in several countries do not possess basic facilities or even knowledge that they can get in touch with experts. Also sometimes consulting experts costs even higher [7,19]. In such conditions, detection of the diseases automatically by just looking from the symptoms in the leave of the plants is the cheap and better solution. This also aids the support of machine vision to provide image-based automatic process control, inspection along with the guidance of robots [6].

Detection of diseases based on their visualized clinical symptoms is a concept based on image recognition techniques. AI analyzes the crop images, and the data input can make a clear distinction between the diseases with same symptoms [2]. The X-FIDO program utilizes a DP-CNN and a data fusion algorithm at the abstract level to detect the symptoms about olive quick decline syndrome on leaves of *Olea europaea* L. infected by *Xylella fastidiosa*. Virtual digicam or comparable devices are used to seize the snap shots of leaves of various types and identify the leaves laid low with ailment (Figures 9.4 and 9.5). Then distinctive forms of photograph-processing strategies are applied on them, to system those pictures, to get unique and beneficial features needed for the purpose of reading leaves. The set of rules written beneath illustrate the step-by-step technique for photograph reputation and segmentation procedures:

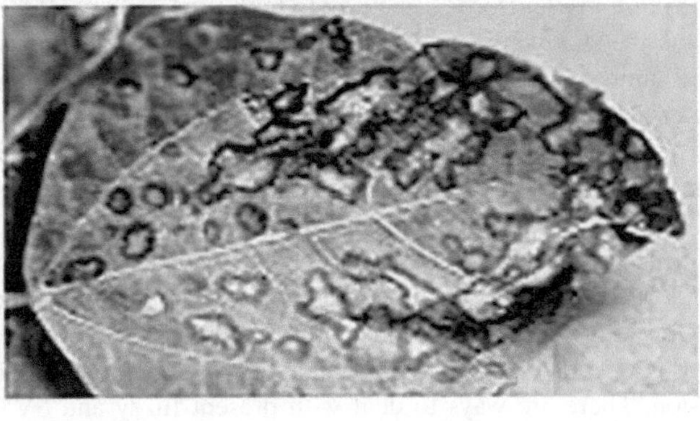

FIGURE 9.4 A leaf classified as affected by disease using machine learning.

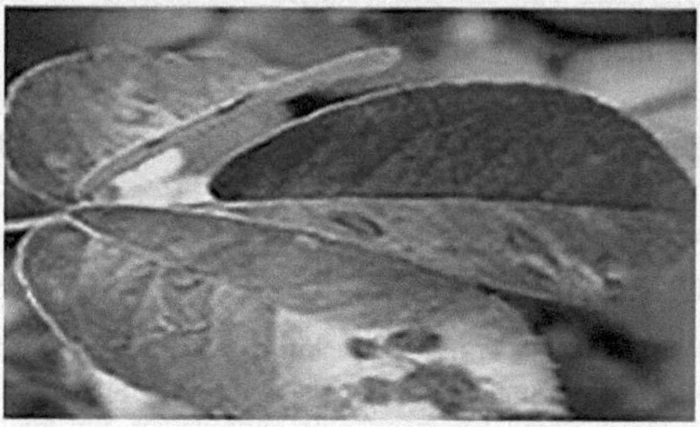

FIGURE 9.5 A group of leaves classified as affected by disease using machine learning.

1. Capturing of picture is the initial step, which acquires an image using a digital camera.
2. The quality of the image is enhanced using preprocessing, and the affected area is clipped for further processing. Also suitable filters are used to smoothen the image, and image enhancement is performed if necessary.
3. As most of the leaves are green in color, masking is done to separate the correct image from the affected image based on a threshold value. Within the infected clusters, the masked cells will be removed.
4. The required portions are finally segmented to classify the leaves affected with diseases using genetic algorithm.

9.7 ML IN FORECASTING THE COST OF THE AGRICULTURAL PRODUCTS AND RECOMMENDING SUITABLE SEASON FOR PLANTING AND HARVESTING TO MAKE BETTER PROFITS

Real-time data analytics from different sources can be combined and analyzed using AI and assist in building a good supply chain based on the demand and price ratio. Gobasco uses real-time analytical data from corner to corner of the country using automated pipelines based on AI, which gives the information of current supply chain status of agriculture [5]. It affords best prize to both the farmers and the consumers.

To overcome the shortcomings of traditional autoregressive methods with regard to limiting the number of simultaneous predictions, a method based on matrix factorization, unsupervised scalable representation learning, and DeepGLO combined with global matrix factorization and local time network can be used [20]. These methods explore a huge training data space made up of larger variables and durations. Regarding the prediction of prices of agricultural products at the national level, we must make predictions based on a limited number of sample data; these data lack part of the construction period and/or product items, and these methods cannot be applied directly.

There are few ways to deal with agriculture item cost by AI, introducing basic recurrent neural network (RNN) in agrarian item value gauging now and then works on the precision. There are ways to deal with present fuzzy and SVM and Naïve Bayesian algorithm, where the assessment strategies are restricted, and the benefits against regular techniques are muddled. The methodologies empower the choice of fundamental estimating system, for example, basic RNN, LSTM, and GRU, as indicated by the predefined rules, for example, time-point precision, worldwide exactness, and measurable attributes of created future time series. For the determination, it explains the viability and impediment of controlling the impact between a particular past time point and the current utilizing the door (for example, neglect, peephole) component in LSTM and GRU.

The initial step toward the prediction mechanism is to predict the longer term price value only through the past price history. The essential idea is to prepare the essential predictor p (n, k), which receives n steps from the past price sequence and returns the price after k steps in the future. The basic predictors are often applied to any of the parts of n steps of the past sequence. The anticipated value in '*k* step' after the past sequence is obtained depending on the predictor π (*n*, *k*), which is expressed as π (*n*, *k*). Each π (*n*, *k*) is an ANN that's trained to predict '*k* step after' from the previous n step sequence using LSTM, GRU, and straightforward RNN (SRNN) to get the network. The prediction duration of *k* ranges from 1 to *k*_max. Since previous data is out there hebdomadally, the duration of the anticipated time step is going to be 1 week.

The fundamental forecasters were trained by the info of the past 20 weeks because the cost of the agricultural products keeps changing every year. We can also get the predicted price at the end the year using the available data. This depends on the power of forecasting mechanism within the case where the info is out there for less duration than the longest cycle like 20 weeks of less. The second reason is for data availability. Because there are missing periods and empty values within the original data set and not possible insert the interpolated values, we'd like to pick a comparatively shorter period for data preparation.

RNNs (LSTM, GRU, and simple RNN) were used to estimate the price value at '*k*-week after' where *k* is an integer value in RNN with one middle layer of 50 nodes was trained with 90% of stochastically picked training data from the historical price series, and the remaining 10% was utilized for validation. There were 312 weeks in all in the original sequences, and continuous sections were chosen for training and validation data. For all types of RNNs, back-propagation learning converged to stability after 200–300 epochs. We trained for up to 10,000 epochs and discovered that GRU and SRNN overfit after 200 and 300 epochs, respectively. Training was given to 10,000 epochs and identified that GRU overfitted after 200 epochs and SRNN overfitted after 300 epochs. LSTM showed improvement during the training up to 10,000 epochs and obtains the maximum performance in the latter part of the epochs.

LSTM, with a gate mechanism connecting the past and the present at a certain time, successfully found the sharp convex surface (peak) corresponding to the sudden rise or fall of the value change in the time series, even for the long forecast term after 15 weeks, he reported. However, the GRU cannot capture such abrupt changes embedded in the original time series. It describes TATP-based forecasts and their results in periods of less than 1 year. In this article, TATP and DFTS and their results are studied in a short period of 1 year. The precision of the validation data for all the basic predictors is less than 0.01, which means that they show high performance on 10% of the validation data in the original past series. In this way, we can provide farmers with accurate price forecasts of agricultural products.

9.7.1 Crop Harvesting Using AI

There is a huge investment for maintaining labor in agriculture for sowing, weed clearance, fertilizer and pesticide application, and harvesting. AI robots can be used as an alternative for labors, which reduces the time, and the crops can also be sorted based on their quality. This technique reduces the labor demand for agricultural practices. In Japan, AI robots have been tested in tomato farms, and it is found that it reduced labor cost by 20% [9].

9.7.2 Agricultural Product Grading Using AI

By analyzing the color, shape, and size images of the vegetables, fruits, or grains taken by the farmers, the agricultural products can be categorized based on their quality and can be graded in the market [10]. Intello Labs in Bengaluru utilizes DP for image analysis and grades the agricultural products.

9.8 CONCLUSION

With increasing population throughout the world, demand for food is increasing exponentially. Agricultural production has to double, and moreover, the environmental resources are now becoming a tough for the farmers. They can't predict the sudden climate change that produces a drastic effect on crop yield. This demand can be met by introduction of modernized techniques in agriculture that improve the yield even in unfavorable environmental conditions. AI applications in agriculture have developed applications and tools that help farmers in a controlled farming by providing them proper guidance about water management, soil fertility, moisture content, temperature in the environment, crop rotation, type of crop to be grown, optimum planting, pest control, work load, organize data for farmers to improve different tasks and nutrition management. While using the machine learning algorithms in connection with images captured by satellites and drones, AI-enabled technologies predict weather conditions, analyze crop sustainability, and evaluate farms for the presence of diseases or pests and poor plant nutrition on farms with data like temperature, precipitation, wind speed, and solar radiation.

Farmers without a network connection can now take advantage of the benefits of AI through simple tools like SMS-enabled mobile phones and planting applications. At the same time, farmers with Wi-Fi access can use AI applications to get customized ongoing AI plans for their land. With these solutions powered by IoT and AI, farmers can meet the global demand for more food and sustainably increase production and income without consuming valuable natural resources. Some of the AI-enabled technologies are introduced on real-time study in certain small and large agricultural firms are reported as follows:

1. Plantix can recognize the deficiency of nutrients in the soil with the help of image recognition technology. It was developed by a German concern PEAT.
2. Agricultural robots are used to detect weeds and pests in the agricultural land, and it can be even removed at regular periods.

3. Sky Squirrel Technologies and Trace Genomics use machine learning methods for detecting the health status of the crop. This helps to find the disease at early stage before it gets it invasion of many crop plants.
4. Biosensor solutions dealing with AI in agriculture has started DRAGON (data-driven precision agricultural services and skill acquisition) project with fund supported by European Union horizon 2020 research and is aimed in implementing low-cost precision farming techniques.

In the future, AI will help farmers become experts in agricultural technology and will use data to optimize the performance of single-row plants. AI in agriculture can not only help farmers to realize agricultural automation, but can also switch to precision planting when there are few resources to improve crop yield and quality. Companies involved in improving machine learning or AI-based products or services (such as agricultural training data, drones, and automated machine manufacturing) will gain technological advancements in the future, providing the industry with more useful applications and helping the world to solve the growing food production problems.

REFERENCES

1. Ammulu, K. (2021). The impact of artificial intelligence in agriculture. *International Journal of Advanced Research in Science, Communication and Technology*, 4(1): 157–161.
2. Dandawate, Y. and Kokare, R. (2015). An automated approach for classification of plant diseases towards development of futuristic decision support system in Indian perspective. In *International Conference on Advances in Computing. Communications and Informatics, ICACCI 2015*, pp. 794–799. IEEE, Kochi, India.
3. Yiannis, A. (2018, December, 19). Applications of artificial intelligence for precision agriculture. Retrieved from https://edis.ifas.ufl.edu/publication/AE529.
4. Abdulridha, J., Ampatzidis, Y., Ehsani, R. and Castro, A. D. (2018). Evaluating the performance of spectral features and multivariate analysis tools to detect laurel wilt disease and nutritional deficiency in avocado. *Computers and Electronics in Agriculture*, 155: 203–2011.
5. Curchoe, C. (2020, December 17). Artificial-intelligence is intertwined in precision-agriculture. Retrieved from https://www.agdaily.com/technology/artificial-intelligence-is-intertwined-in-precision-agriculture.
6. Ampatzidis, Y. and Cruz, A. C. (2018). Plant disease detection utilizing artificial intelligence and remote sensing. In *International Congress of Plant Pathology (ICPP) 2018: Plant Health in a Global Economy*, July 29–August 3, Boston, MA.
7. Das, S. (2019, December, 27). Retrieved from https://analyticsindiamag.com/top-ai-powered- projects-in-indian-agriculture-sector.
8. Dominguez-Morales, M. et al., A. (2011). Technical viability study for behavioral monitoring of wildlife animals in Donana: An 802.15.4 coverage study in a Natural Park. In *Proceedings of the International Conference on Data Communication Networking* (DCNET), pp. 1–4, 18–21.
9. Panasonic (2018, May 23). Introducing AI-backed tomato harvesting robots, Tokyo, Japan. Retrieved from https://news.panasonic.com/global/stories/2018/57801.html.
10. SanjayChaudharyl, M. B. (2015). Agro advisory system for cotton crop. *AGRINETS Workshop, COMSNETS. IEEE.*

11. Giordano, S., Seitanidis, I., Ojo, M., Adami, D. and Vignoli, F. (2018). IoT solutions for crop protection against wild animal attacks. 1–5. Doi: 10.1109/EE1.2018.8385275.
12. Liqiang, Z., Shouyi, Y., Leibo, L., Zhang, Z. and Shaojun, W. (2011). A crop monitoring system based on wireless sensor network. Springer Nature, Switzerland AG, *Procedia Environmental Sciences*, 11: 558–565. Doi: 10.1016/j.proenv.2011.12.088.
13. Huircán, J. I. et al., (2010). Zigbee-based wireless sensor network localization for cattle monitoring in grazing fields. *Computers and Electronics in Agriculture*, 74(2): 258–264.
14. Kanagaraj, K., Swamynathan, S. and Karthikeyan, A. (2019). *Cloud Enabled Intrusion Detector and Alerter Using Improved Deep Learning Technique*. Springer Nature Singapore Pvt Ltd, ICIIT 2018, CCIS 941, pp. 17–29.
15. Zhang, J., Khan, S. A., Hasse, C., Ruf, S., Heckel, D. G. and Bock, R. (2015). Pest control. Full crop protection from an insect pest by expression of long double-stranded RNAs in plastids. *Science (New York, N.Y.)*, 347(6225): 991–994. Doi: 10.1126/science.1261680.
16. Bhardwaj, A. K. et al., Solar Power fencing for crop protection from monkey menace and domestic/wild animals in Himachal Pradesh Retrieved from https://www.nabard.org/demo/auth/writereaddata/ModelBankProject/1302170923merged_document_9.pdf.
17. Bruggers, R. L. and Ruelle, P. (1982). Efficacy of nets and fibres for protecting crops from grain-eating birds in Africa. *Crop Protection*, 1(1): 55–65.
18. Brahimi, M., Boukhalfa, K. and Moussaoui, A. (2017). Deep learning for tomato diseases: Classification and symptoms visualization. *Applied Artificial Intelligence*, 31(4): 1–17, Ahmedabad, India.
19. Konstantinos, P. and Ferentinos. (2018). Deep learning models for plant disease detection and diagnosis. *Computers and Electronics in Agriculture*, 145: 311–318.
20. Ojha, T., Misra, S. and Raghuwanshi, N. (2015). Wireless sensor networks for agriculture: the state-of- the-art in practice and future challenges. *Computers and Electronics in Agriculture*, 118: 66–84.

10 Internet of Things (IoT) and Artificial Intelligence for Smart Communications

Dnyaneshwar S. Mantri
SIT Lonavala

Pranav M. Pawar
BITS Pillani

Nandkumar Kulkarni
SKNCOE Pune

Ramjee Prasad
Arhus University

CONTENTS

10.1 INTRODUCTION

The Internet of Things (IoT) is the prime need of control applications, and it not only supports controlling the heterogeneous devices but also provides the means of analysing the data. In Industry 4.0 Revolution, it is used to connect the devices over internet and communicate with higher data rates. In nutshell, IoT can be expressed as 'ecosystem of connected physical objects that are accessible though internet [1].' The variety of heterogeneous devices 'connected' over the internet are 'smart' and can be remotely monitored and controlled. In short, IoT can be considered as the

DOI: 10.1201/9781003220176-10

FIGURE 10.1 Attributes of IoT (Fig. [1,2]).

combination of data, technology and medium integrated and support to 5G objectives. The attributes of IoT can be defined in different ways as shown in Figure 10.1.

The survey included more than twenty different definitions – important words including devices, system, connected, internet, data, virtual, objects, and networks. The IoT is a 'seamless blend of embedded intelligence, pervasive connectivity, and deep analytical insights that produces unique and disruptive value for enterprises, individuals, and societies,' according to many definitions.

Also, IoT can be defined as 'a network of interconnected physical things that may be accessed over the internet.' This definition is more important because it conveys the different functional parameters of IoT as it is not only used for point-to-point communication but can connect any type of physical device over the internet in multi-point way with precise monitoring and control [1,2]. Hence, the components of new generation IoT may be represented using the integration of data, technology and medium as shown in Figure 10.2.

The IoT and Artificial Intelligence (AI) are communication services that connect things and humans. It's more intricate and dynamic than the internet itself. IoT refers to readable and recognised 'things,' whereas AI refers to intelligent machines like computers. As technology advances at a faster rate, it is becoming increasingly difficult to keep up.

In this context, intelligence is defined as the ability to solve problems, while AI is defined as the combination of human intelligence interacting (simulated) with machines to provide accurate results. The basic elements of AI in relation to IoT data include cognition, expertise, strategy, learning, and communication with perception. In short, according to John McCarthy AI is 'The Science and Engineering making INTELLIGENT machines'.

In the smart communication system, it is of utmost importance to have control over networks to transmit the analysed data. The controlling action is done by the IoT principles and analysis of data by the AI algorithms. The redundant data generated by the multiple sensors are reduced at the source itself by the use of source

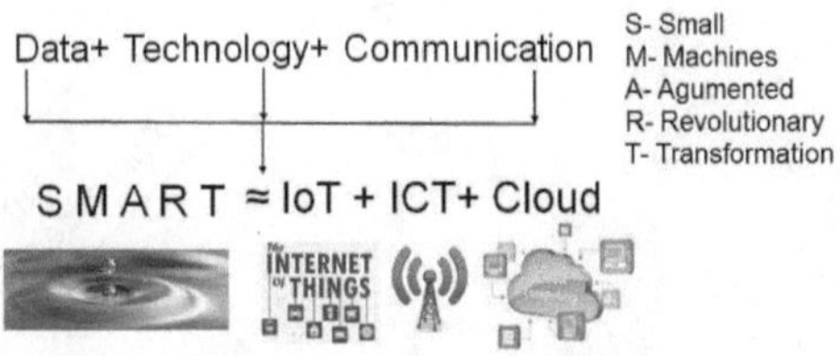

FIGURE 10.2 Components of new generation IoT.

coding algorithms to increase the efficiency, while the reliability is increased by the channel coding algorithm by adding controlled redundancy. The IoT is used to have control over the different communications. AI research has been specifically used for problem-solving precisely and accurately [3]. The chapter aims to discuss the application scenarios for IoT and AI, with enabling technologies for making smart communication using IoT and AI, and challenges to achieve smart communication using AI-based IoT networks.

The remaining chapter is organised into six sections. Section 10.2 discusses application scenarios of IoT and AI. Section 10.3 discusses related work in the area of making smart communication using AI-IoT. In continuation with it, Section 10.4 focuses on the IoT road map and service model, and Section 10.5 gives highlights of different enabling technologies for AI-IoT. Section 10.6 concentrates on the proposal for enhancement of AI-IoT and challenges to achieve it. Section 10.7 concludes the chapter.

10.2 APPLICATION SCENARIOS OF IoT AND AI

IoT, updated technological developments and AI are integral parts of daily lives. IoT-enabled devices are used in the automation process, home security, healthcare applications and so on, i.e. anything, everything and anywhere. If we look from the business point of view, companies may be small or big; everyone is relying on the combination of IoT and AI to meet customer satisfaction. At the same time, if the application of IoT is considered as smart city, then it demands happy and secure life for human; it includes, healthcare, transportation, waste management, education, planning, connected cars, home and environment with services, and so on, means small things are important and all must happen with minimum delay, quick access, utilise minimum energy and go green concept. The application use case of IoT and AI is shown in Figure 10.3.

The AI is used to solve complicated and complex problems by trial and error methods to optimise the performance of robots and other machine-controlled devices, the results are repeated and used to compare the performances according to the stored results. The major components used by AI in applications are learning, reasoning, problem-solving, perceptions and language understanding. These all are helpful specifically in the M2M, D2D, D2M, D2P and other types of communications. The applications are common in the IoT and AI, which modifies the analysed data using AI/ML algorithms.

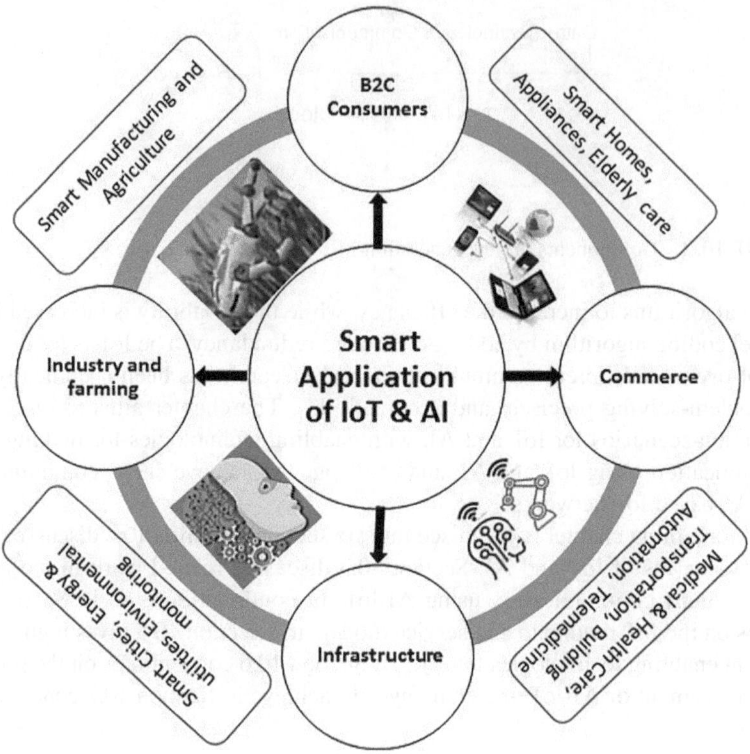

FIGURE 10.3 Application use case of IoT and AI.

Figure 10.3 shows that the application is categorised into four parts – business to consumers, e-commerce, infrastructure, industry and farming. It covers almost all the applications related to IoT and AI [1–4].

Both IoT and AI are smart communication services between things to things and humans to things. AI is used to provide the solution to complex problems with support of humans. IoT is referred to as the things to be recognised and it is concerned with smart embedded machines. The common and most important applications of IoT and AI are healthcare, robotics, smart cities, education, business, entertainment, gaming, and agriculture, both have an infinite number of applications.

10.3 RELATED WORK

This section examines the current state-of-the-art work in the area of IoT and AI [1,2]. The author presents a detailed review of numerous data fusion strategies, mathematical foundations as well as research proposals for study in the field of IoT, AI, and cloud computing. Ref. [3] defines AI-IoT as the integration of AI with IoT based on AI components such as Learning, Prospective, Reasoning, and Behaving.The survey also opens many domains of research in AI-IoT. More efforts are required to improve the edge intelligence. The deep learning approach used in evaluation provides the control accuracy, and multimodal interactions [4] express that IoT and AI can be used to improve the classroom performance with minimum errors as an application scenario. The optimal algorithm developed allocated the classroom on the basis of attendance and overcomes the overflows. According to Ref. [5], as the number of IoT devices grows, a large amount of data is generated, necessitating the use of big data technology to assess it alongside AI. As a result, the authors suggest the Cyber IoT device, which includes numerous tools integrated as AI-talk. It avoids the involvement of real IoT devices. In [6], the author proposes the fusion of different technologies used in healthcare IoT (H-IoT) considering the two steps, as sensors are used to generate the real-time data and processors are used to compute the values with big data technology. The involvement of AI completely transformed the way of analysis bridging the gap between computing powers and deployed networks. The software-defined radio (SDR), internet of nano things (N-IoT), and tactile internet (TI) are some of the techniques used in innovative H-IoT. Ref. [7] discusses the development of the online evaluation framework based on knowledge of machine learning algorithms named RANK Algorithm. It provides the online assessment in real time for industrial environment with cyber-physical systems and distributed cooperative control mode for security. Ref. [8] provides a comprehensive survey and remedies in the communication using AI. It categorises the system architecture on the basis of data and edge calculations. Ref. [9] considers the use cases with the help of medical data analysis using AI cloud computing in the Covid-19. Ref. [10] proposed the future AI wireless communication network on the basis of five themes, Sensor AI, Network Device AI, Access AI, User Device AI and Data Provenance AI. Refs. [11,12] propose the emerging technology 6G which is combination of IoT, 5G and AI, which are key elements in the development of less latency network in communication system. Ref. [13] says that security is an important aspect of any new technological development with large data; at the same time, feature extraction needs to be obtained using minimum information, and it is done by reinforcement learning algorithm (RL). Ref. [14] provides the detail survey and proposes probabilistic solutions for various problems in design and development of network with IoT and AI. Ref. [15] provides the general ubiquitous IoT architecture for context aware learning. The constrained application protocol (CoAP) uses various variants for computation and is resource efficient. At last, it is important to have IoT architecture as for any AI/ML to obtain high computing performances.

10.4 IoT ROAD MAP AND SERVICE MODEL

The things in the IoT are connected with various modes and are used for the data collection and transmission. The roadmap of IoT is shown in Figure 10.4. In the initial stage of IoT developments around 2020, RFID tags were used to identify the devices and used for courting applications where Industry 1.0 will be compared and is for supply chain helpers for process initialisation. The RFID tags were seen as prerequisites for IoT. As the revolutions are charged in a decade span technological developments are growing vertically. The evolutions are used to meet the demands of user not only in one area such as logistic movements, surveillance, security, health care and safety but many more. In the third revolution, internet completely changed the scenario and distant devices were able to connect and communicate for the data

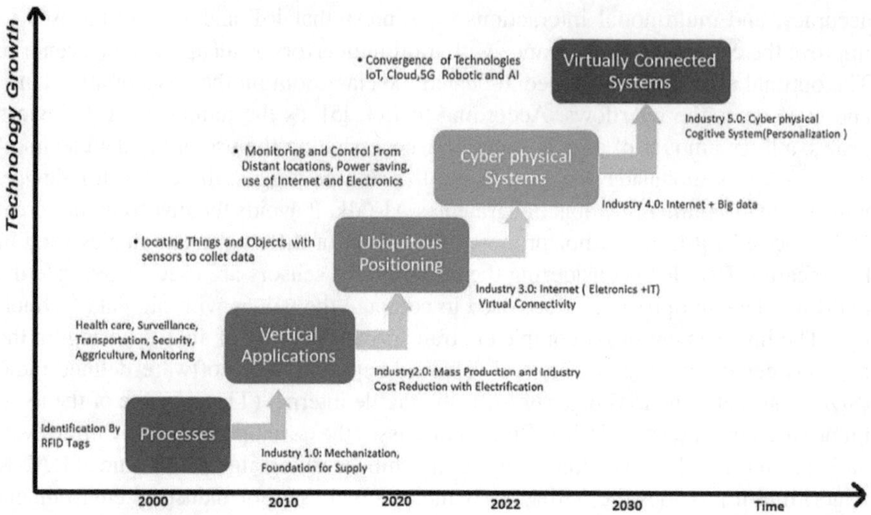

FIGURE 10.4 IoT roadmap.

collections. In Revolution 4.0, monitoring and control were actually introduced with big data analysis technique.

The machines, things and devices securely communicate over internet and wireless media and termed as cyber-physical systems for connecting the physical to virtual world. Now the updates are knocking the doors with convergence of technology for seamless connectivity, virtual storage of real-time data with precise accurate algorithms using AI and machine learning (AIandML) leading to Industry 5.0 Revolution in 2030, i.e. cognition of virtual and imagination.

The future world of things is not the concern of a single innovative technology; rather, a number of complementing technological breakthroughs give skills that, when combined, help to bridge the gap between the physical, virtual, and imaginary worlds, i.e. Human Bond Communication (HBC).

With wave of technology developments in the IoT and AI, it is important to know the service model used for the efficient utilisation in industry and business modelling. The basic service model is shown in Figure 10.5. The service model is divided in to three layers, the outer layer has things which are heterogeneous and base is sensor in both IoT and AI techniques used to collect the real-time data from field. The complete system decides the policies for data converting and communication and utilisation along with values and impact on technological developments. At last, all things are recognised by the industry, i.e. outer layer has things, policies and standards, value and impact with industries.

The second layer has four components

a. Sensors and Devices: They are used to collect a variety of data and processes to reduce the redundancy. Many devices have inbuilt sensors with all processing capabilities as AI is working on the machine for computational intelligence. Precise and accurate data processing is the important property.

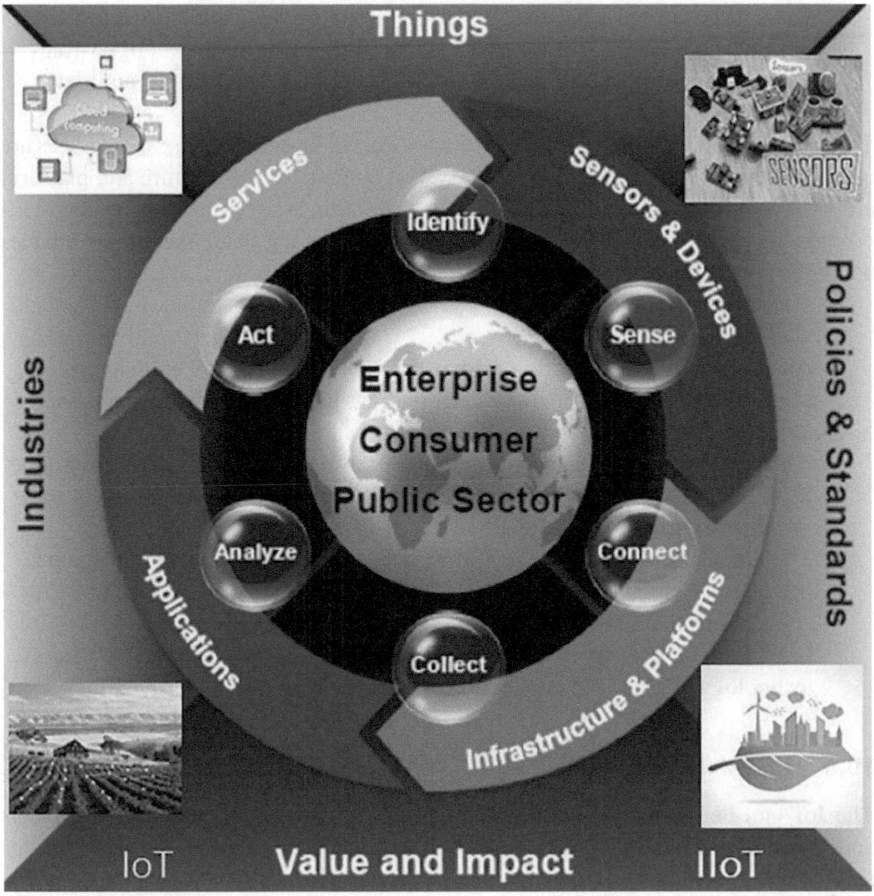

FIGURE 10.5 Internet of Things service model.

b. Infrastructure and Platforms: It is used to provide the infrastructure which support the flexibility along with technological revolution, and they need to provide the connectivity, communication media, and computational facility for reliable transmission of data over internet. As billions of things are connected over internet, they generate rapidly increasing data rates and IoT will require network that must handle the billion more devices.
c. Applications: Integration of IoT, AI and 5G techniques has an infinite number of applications; you name it, and you will have it. These are not limited to control and monitoring but also in analysis of data. The applications include military, agriculture, disaster minoring, robotics, healthcare, drones, AR/VR, cloud computing and so on.
d. Services: Mostly cloud-based services are preferred which use the virtual data for analysis and communication using any one service platform: IaaS – Infrastructure as a Service: virtualised environment,

PaaS – Platform as a Service: multi-tier architecture, DaaS – Desktop as a Service, and SaaS – Storage as a Service. It provides high level of abstraction of computation and storage model in cooperative communication.

The inner layer of the service model is integral part of layer 2. Sensors and devices are used for identification and sensing the physical quantity. Infrastructure and platform are used for collection and connectivity between things over internet. Applications provide the base for the analysis of constantly rising data. Finally, the service enables action for future developments.

In a nutshell,

- An IoT ecosystem is made up of web-enabled smart devices that gather, send, and act on data from their surroundings using embedded systems such as CPUs, sensors, and communication hardware.
- Although individuals can interface with the devices to set them up, give them instructions, or access the data, the machines do the majority of the job without human interaction.
- The connectivity, cooperation, networking, and communication protocols that these web-enabled devices use are primarily determined by the IoT applications that are installed.
- Integration of 5G, cloud, AI and machine learning can be used by future technology to make data collection processes easier and more dynamic.

10.5 IoT AND AI ENABLING TECHNOLOGIES

The IoT can be thought of as a massive network made up of networks of objects and computers linked by a variety of intermediary technologies, with various technologies such as RFIDs and wireless connections serving as enablers of this interconnectivity. AI works at the top of IoT platform for data analysis and precise problem-solving capability where involvement of human can be avoided. The functions that are enabled using IoT are as follows:

- Tagging Things: Important for location addressability and tracing the objects.
- Feeling Things: Sensors, the important component of IoT and AI, used to get the redundancy free data from the environment with mounting on objects.
- Shrinking Things: Technological developments used to connect and interact with the things or smart devices.
- Thinking Things: The intelligent control over the things and devices can be achieved by embedded technology of device and sensors.

Figure 10.6 represents the technology enablers for IoT and AI with top use cases, i.e. connecting physical world to virtual

Networked: Connected Cars & Autonomous Vehicle: V2X, a new way of communication between the user, cars, and infrastructure, permits to envision numerous applications focusing on Trusteed Cooperative communication in

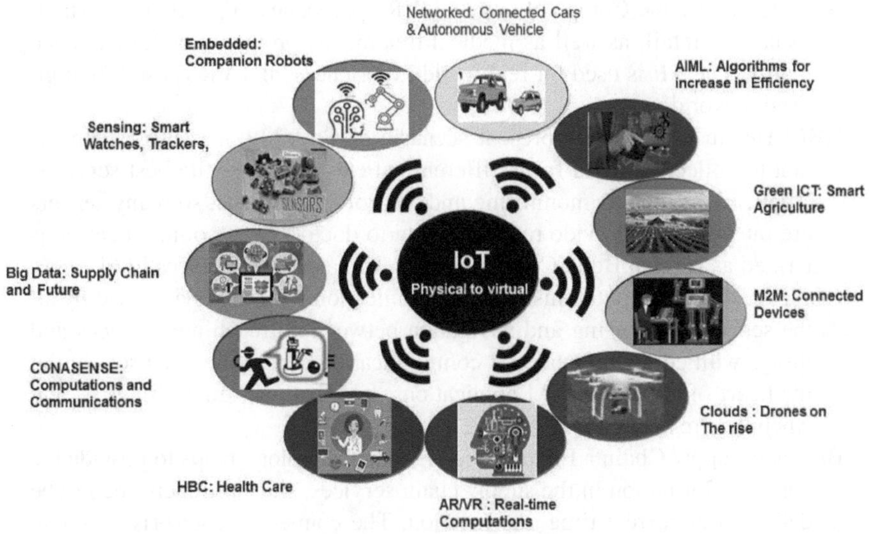

FIGURE 10.6 IoT and AI enabling technologies.

VANET as heterogeneous communications. On-board sensors are used to safety measures, and ad-hock network is to communicate with each other. It is one kind of communication named as Vehicle to everything over internet.

AIML: Algorithms for Increase in Efficiency: Billions of devices are connected over internet and generate rapidly increasing number of heterogeneous data which will be analysed and used for future comparison.

Green ICT: Smart Agriculture: Green ICT refers to technology that helps the environment reduce carbon emissions, and smart agriculture is one such example. The smart sensors are used to test the quality of land and products like fruits and grains. The connected technology with IoT and AI helps to improve the productivity and efficiency. Smart sensors gives real-time data of temperature, Soil moisture, water consumptions in the green house and irrigation systems.

M2M: Connected Devices: With waves of industrial revolutions, many industries are automated and machines are equipped with sensors which communicates over internet to manager regarding the input, output and if any maintenance is required. M2M devices are embedded with advanced features of IoT and AI, and have the capacity to communicate with each other in the network.

Clouds: Drones on the Rise: The better services are provided by the cloud computing in the high-speed networks like 5G, which has very less latency of the order of microseconds. Drones are an example of a device that may be utilised to help teams in distress by providing precise positions. Further, software and infrastructure can be employed as a communication platform. In dangerous and hazardous situations, it requires high performance and storage which could be provided by cloud computing techniques with virtual storage of data.

AR/VR: Real-Time Computations: AR/VR approaches in gaming, entertainment, and retail, as well as medical treatments, provide a better extended reality. The AR is used for real-world experiences, and VR is used for digitisation world.

HBC: Health Care: In the present scenario of Covid-19, it is utmost important to collect the data from different patients and offer the best services. Applications include monitoring and control applications, so many sensors are integrated to provide real-time data to doctors. The communication is termed as Human Bond Communication with sensor interfaces [2,6].

CONASENSE: Computations and Communications: It is a service used by all the sensors for sensing and navigation between all machines, devices and things with computational and communication capabilities. Sensors are at the heart of all IoT and AI applications, and they can track things as and when they're needed.

Big data: Supply Chain a Future: The big data technology helps to provide the correct information in the supply chain services, since products need to be delivered at correct time and location. The connected enterprises system developed in companies takes the support of IoT and AI/ML algorithms for monitoring applications and plans the route for delivery.

Sensing: Smart City: The important and promising use case of IoT is smart city for making human life mode safe and more secure, not just in terms of communication, connection, and information, but in every way [1].

Embedded: Companion Robots: In a pandemic situation where human interaction is rare, a companion robot is a welcome acquaintance. With Industrial Revolution 5.0, these technologies have the bright prospects where robots and humans work together, which is named as COBAT.

10.6 PROPOSALS FOR ENHANCEMENT OF AI-IoT WITH CHALLENGES

IoT is used for controlling and monitoring various items that are virtually connected to each other and may transmit data in a genuine sense, but AI is utilised for supervisory control with the human mind involved in managing the machines.

Figure 10.7 represents the journey from IoT to PIoT. In the beginning, just a few devices were linked to the internet to govern the process of collecting data, analysing it, and maintaining control. Industrial IoT, or IIoT, is a new revolution that has emerged as a result of technological advancements, industry engagement, and communication between networks and objects. According to application demand, the rise of 5G network provides the seamless connectivity between devices, machines, and sensors as Anything, Anywhere, Anytime, and Anybody, i.e. Internet of Everything (IoET). Up to 2030, the integration of numerous technologies such as robotics, AI/ML, ICT, Cloud, IoT, Human Bond Communication, Deep Learning, and various ML algorithms, as well as the participation of people, may lead to the Personalized Internet of Things (IoET) (PIoT).

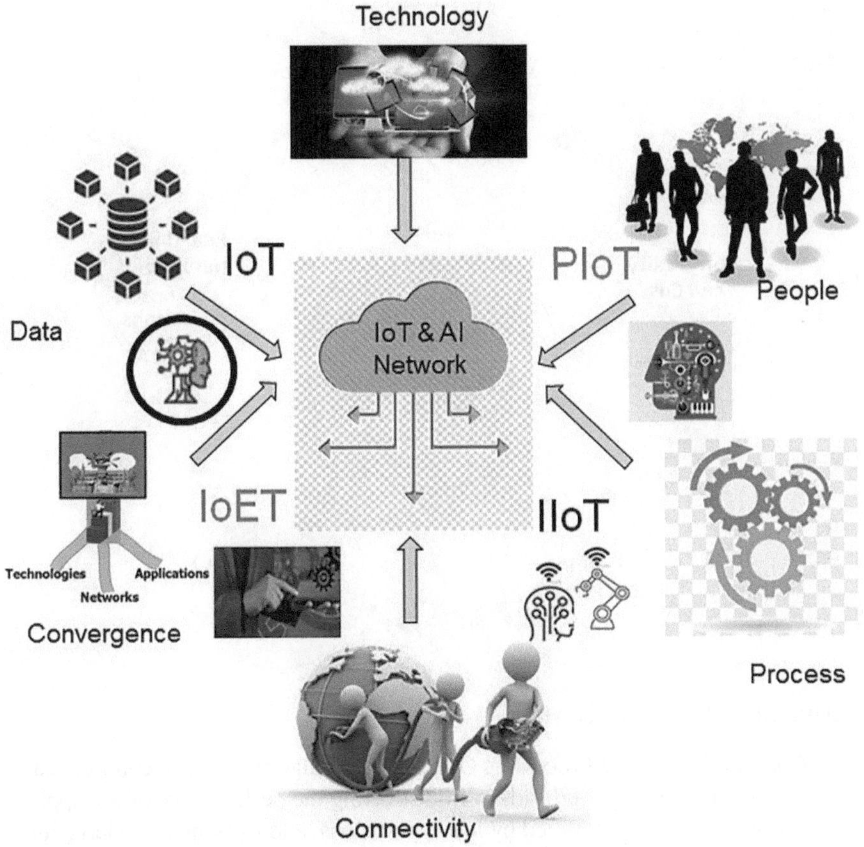

FIGURE 10.7 Journey of IoT to PIoT.

Using AI/ML supervisory and control algorithms, humans and machine collaborate to increase communication and throughput (Industry 5.0). The convergence of technology, applications, networks, people, and machine is the future.

IoT is a heterogeneous, resource-constrained and complex network, and the application of AI in IoT to manage computing, storage, connection, bandwidth, access method, security, and other aspects faces the following main obstacles. (as shown in Figure 10.8) for making AI-IoT-based smart communication in reality.

- Communication and processing overheads
- Managing real-time nature of IoT applications
- Routing and traffic shaping
- Security and privacy
- Complexity and cost of AI-IoT networks

FIGURE 10.8 Challenges for AI-IoT.

- Communication and Processing Overheads: AI algorithms raise communication and processing overheads in IoT, affecting the performance of IoT applications, the services offered by IoT applications, and working of underlaying functionality of IoT network. The communication and processing overheads are increasing as most of AI-based algorithms learn from environment and take decisions accordingly. The learning model is depending on data, which is gathered using a large amount of information exchange between IoT devices and finally, IoT devices may learn from the model and make the best decisions possible. The storage of required data and high amount of computation are not possible at resource-constrained IoT devices, which impose extra overheads on IoT network. The solution to reduce the overhead is to transfer the highly computational tasks to cloud or edge devices [16].
- Managing Real-Time Nature of IoT Applications: IoT applications will be required in the future to make real-time judgments in the same way that humans do in the shortest amount of time. The processing of AI algorithms takes longer time on IoT devices because of the limited processing and storage capabilities of IoT devices. Additionally, doing required computations on a cloud or edge device is not a feasible option since it introduces considerable delays in sending decisions to devices. Therefore, it is necessary to derive the feasible technique which can work by keeping real-time nature of IoT applications.

- Routing and Traffic Shaping: One of the most crucial issues in IoT networks is traffic routing and shaping in order to reduce communication overheads. Many AI-enabled algorithms are useful to solve this issue, but such algorithms are not sufficiently efficient in terms of scalability. The dynamic nature of network also induces significant delays in routing and traffic shaping, as every time new data is available in network and AI model needs to be trained from it. One viable solution is software-defined network (SDN), where maximum part of network is managed by using software than hardware. SDN is easier for managing dynamicity of network and updating the network according to learning model.
- Security and Privacy: The AI-based algorithms have given promising solutions for detecting and preventing the different attacks on IoT network by analysing the pattern of the attack. The majority of AI-based solutions for security and privacy are based on building model using a large amount of available attack information. In such situation, it is difficult for IoT application to detect any new attack, as system is not trained for it. The reinforcement learning-based solutions can be a possible choice in such situations where system learns from rewards, but it may increase computation complexity of system. Therefore, it is necessary to develop the AI-based security and privacy algorithms for IoT which should react in real time to existing attacks and any new attack on system.

Complexity and Cost of AI-IoT Networks: The implementation of AI-IoT network requires efficient IoT devices which should be feasible to perform partial or full AI computation tasks. To perform heavy AI computations,the AI-IoT network should utilize cloud or edge devices, a high-end base station (BS), efficient policies and a high-speed network. The above-mentioned requirements increase the complexity and cost of such network. Therefore, a very important challenge is reducing the complexity and cost of such network to increase the feasibility of it in real-time scenarios for smart communication.

10.7 CONCLUSIONS

The 21st century presents the omnipresent technologies such as wireless access, RFID, network applications, and man-machine interaction. The integration of IoT and AI altogether drives new network with heterogeneous ways of device communication, analysis of data and control. It can achieve person-to-person, person-to-object, and object-to-object communication anytime and anywhere. IoT provides control over data communication, whereas AI focuses on excellence to increase the likelihood of success. The future of IoT and AI is truly unclear and logically based on the speed of changing technology, increases in data rates, and the medium adopted for communication.

REFERENCES

1. Furqan Alam Rashid Mehmood, Lyad Katib, Nasser N. Albogami, and Aiiad Albeshri (2017), 'Data fusion and IoT for smart ubiquitous environments: A survey', *IEEE Access, Special Section on Trends and Advances for Ambient Intelligence with Internet of Things (IoT) Systems*, vol. 5, pp. 533–9554, Doi: 10.1109/ACCESS.2017.2697839.
2. Fatima Alshehri, and Ghulam Muhammad (2021), 'A comprehensive survey of the Internet of Things (IoT) and AI-based smart healthcare', *IEEE Access*, vol. 9, pp. 3660–3678.
3. Jing Zhang, and Dacheng Tao (2021, May 15), 'Empowering things with intelligence: A survey of the progress, challenges, and opportunities in artificial intelligence of things', *IEEE Internet of Things Journal*, vol. 8, No. 10, pp. 7789–7817.
4. Thanchanok Sutjarittham, Hassan Habibi Gharakheili, Salil S. Kanhere, and Vijay Sivaraman (2019, October), 'Experiences with IoT and AI in a smart campus for optimizing classroom usage', *IEEE Internet of Things Journal*, vol. 6, No. 5, pp. 7595–7607.
5. Yun-Wei Lin, Yi-Bing Lin, Chun-You Liu, Jiun-Yi Lin, and Yu-Lin Shih (2020, April), 'Implementing AI as cyber IoT devices: The house valuation example', *IEEE Transactions on Industrial Informatics*, vol. 16, No. 4, pp. 2612–2620.
6. Yazdan Ahmad Qadri, Ali Nauman, Yousaf Bin Zikria, Athanasios V. Vasilakos, and Sung Won Kim (2020, Second Quarter), 'The future of healthcare Internet of Things: A survey of emerging technologies', *IEEE Communications Surveys & Tutorials*, vol. 22, No. 2, pp. 1121–1167.
7. Zhihan Lv, Yang Han, Amit Kumar Singh, Gunasekaran Manogaran, and Haibin Lv (2021, February), 'Trustworthiness in industrial IoT systems based on artificial intelligence', *IEEE Transactions on Industrial Informatics*, vol. 17, No. 2, pp. 1496–1504.
8. Yuanming Shi, Kai Yang, Tao Jiang, Jun Zhang, and Khaled B. Letaief (2020, Fourth Quarter), 'Communication-efficient edge AI: Algorithms and systems', *IEEE Communications Surveys and Tutorials*, vol. 22, No. 4, pp. 2167–2199.
9. Adedoyin Ahmed Hussain, Ouns Bouachir, Fadi Al-Turjman, and Moayad Aloqaily (2020), 'AI techniques for COVID-19', *IEEE Access*, vol. 8, pp. 128776–128795, Doi: 10.1109/ACCESS.2020.3007939.
10. Dinh C. Nguyen, Peng Cheng, Ming Ding, David Lopez-Perez, Pubudu N. Pathirana, JunLi Aruna Seneviratne, Yonghui Li, and H. Vincent Poor (2021, First Quarter), 'Enabling AI in future wireless networks: A data life cycle perspective', *IEEE Communications Surveys & Tutorials*, vol. 23, No. 1, pp. 553–595.
11. Anutusha Dogra, Rakesh Kumar Jha, and Shubha Jain (2021), 'A survey on beyond 5G network with the advent of 6G: Architecture and emerging technologies', *IEEE Access*, vol. 9, pp. 67512–67547.
12. Xiuquan Qiao, Schahram Dustdar, Yakun Huang, and Junliang Chen (2020, September), '6G vision: An AI-driven decentralized network and service architecture', *Department: Internet of Things, People, and Processes, IEEE Computer Society*, pp. 33–40, Doi: 10.1109/MIC.2020.2987738.
13. Aashma Uprety and Danda B. Rawat (2021, June 1), 'Reinforcement learning for IoT security: A comprehensive survey', *IEEE Internet of Things Journal*, vol. 8, No. 11, pp. 8693–8706.
14. Antonio Marcos Alberti, Mateus A. S. Santos, Hirley Dayan Lourenço Da Silva, Ricardo Souza, Jorge Roberto Carneiro, Vitor Alexandre Campos Figueiredo and Joel J.P.C. Rodrigues (2019), 'Platforms for smart environments and future internet design: A survey', *IEEE Access*, vol. 7, pp. 165748–165778.
15. Salsabeel Y. Shapsough, and Imran A. Zualkernan (2020, July-September), 'A generic IoT architecture for ubiquitous context-aware learning', *IEEE Transactions on Learning Methodologies*, vol. 13, No. 3, pp. 449–464, Doi; 10.1109/TLT.2020.3007708.

16. Ijaz Ahmad, Shahriar Shahabuddin, Thilo Sauter, Erkki Harjula, Tanesh Kumar, Marcus Meisel, Markku Juntti, and Mika Ylianttila. (2021, March), 'The challenges of artificial intelligence in wireless networks for the internet of things: Exploring opportunities for growth,' *IEEE Industrial Electronics Magazine*, vol. 15, no. 1, pp. 16–29, Doi: 10.1109/MIE.2020.2979272.

11 Cyber-Security in the Internet of Things

Snehal A. Bhosale
RMD Sinhgad School of Engg

S. S. Sonavane
Symbiosis Skills and Professional University

CONTENTS

11.1 INTRODUCTION

The Internet of Things (IoT) has brought huge changes in day-to-day lives of end-users. IoT is a revolution and an emerging technology. Because of IoT, there are huge improvements in products as well as production efficiency of the companies. Hence, most of the companies are working in online mode and bringing new models or products using IoT technology.

The IoT network of machines and things is formed which is capable of communicating with each other for smart applications. PCs, laptops, smartphones and many other devices are connected using the internet. The applications of IoT are enormous,

DOI: 10.1201/9781003220176-11

for example smart environments like smart homes, smart cities, banking, healthcare, automation, industrial manufacturing, etc. [1]. Due to huge development in the telecommunication sector, IoT collaborates with things and networks anywhere, anytime and in any form.

The IoT industry is proliferating because of applications like wearables, smart cities, smart grids, industrial internet, connected car, large companies, banks, data centers and healthcare providers. The study on the growth of connected devices says that more than 75 billion devices will be connected using IoT by 2025. The growth of connected devices from 2015 to 2025 is shown in Figure 11.1. According to Business Insider, investment in IoT could top $8:30 billion in the coming future. Most of the applications mentioned above contain valuable personal or financial data that must be protected.

In many applications of IoT, devices, sensors, actuators, circuits, processors, gateways and routers connect using wired or wireless technologies. Many telecommunication technologies that connect IoT devices wirelessly are Bluetooth, Zigbee, Wi-Fi, WiMAX, NFC and RFID. For wired connection, co-axial cable, Ethernet cable and routers are used for IoT network formation. Also, a variety of protocols like COAP, MQTT, AMQP, RESTful, XMPP, IPv4/IPv6, Routing Protocol for Low-Power and Lossy Networks (RPL), IPv6 over Low-Power Wireless Personal Area Networks (6LOWPAN), HTTP, etc. are used for communication between IoT devices [2].

In IoT ordinary devices with sensors are allowed to communicate without human intervention. Example: smart speakers, fitness trackers, cars, light bulbs, etc. connect to the internet and communicate to other devices.

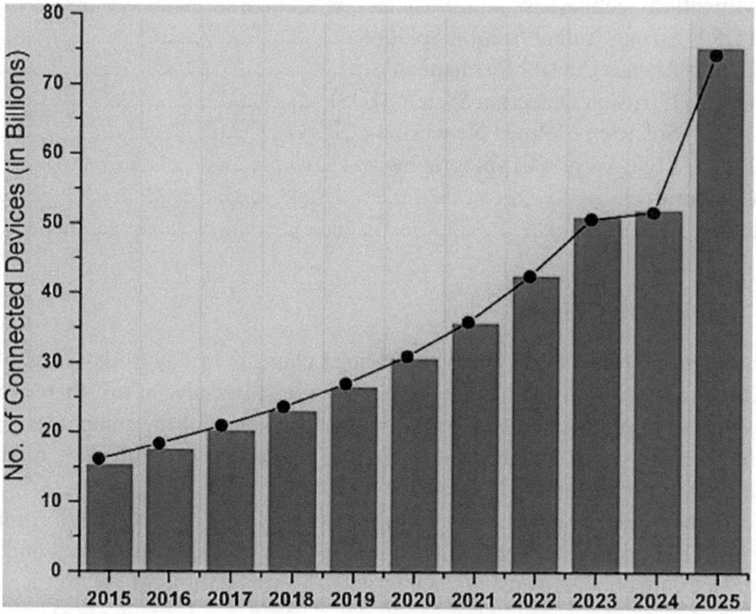

FIGURE 11.1 IoT-connected devices from 2015 to 2025.

11.1.1 Cyber Threats in IoT

Now it is a known fact that sudden growth in IoT networks has introduced many challenges in its functionality. One of the biggest challenges is cyber-attacks in IoT networks. A rapid growth in various IoT applications like health monitoring, smart grids, smart city, smart agriculture, etc. has caused a phenomenal increase in cyber attacking the IoT networks as shown in Figure 11.2. It is challenging to manage security in IoT because of many reasons and characteristics of IoT devices [3–5].

A recent study concludes that IoT-related threats will be more widespread and impactful; hence, cyber risk management needs to work harder to get optimum solutions. It is expected that the worldwide IoT security market will grow by around 35% from 2018 to 2023 because of the increase in the number of cyber-attacks in IoT networks [6–8].

The current risk management system in networking is not suitable for IoT networks because of its dynamic and heterogeneous properties. The IoT system is not developed with standard platform which results in increase in cyber risks. Recently, many cyber-security techniques are emerging that provide many challenges and opportunities to cyber-security management.

If the cyber-security issue is not properly managed, there are many chances that hackers will take advantage of the limitations or weaknesses of the devices and disturb the entire IoT network. Traditional security solutions do not apply to the IoT device network because of the interoperability, heterogeneity and resource-constrained nature of the IoT network. Hence, heavyweight security solutions are not supported by IoT devices. It is a need of time to develop new security standards which accommodate the IoT requirements like privacy, security and reliability [9,10].

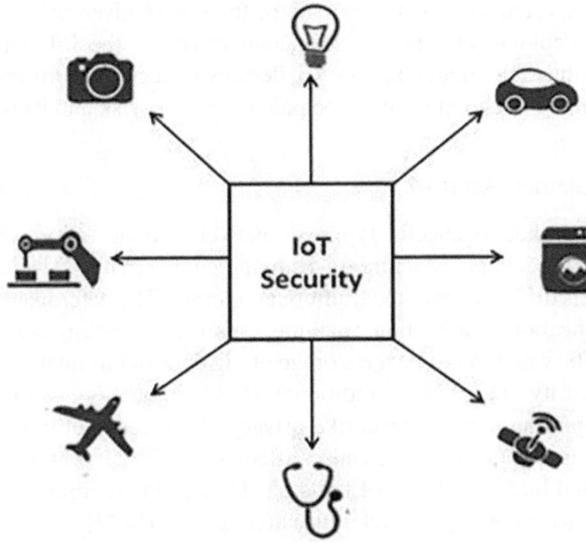

FIGURE 11.2 Cyber-attacks affecting IoT devices.

The heterogeneity characteristic of IoT networks where a variety of devices are connected with different types of communication standards makes IoT networks venerable to many security attacks. More is the variety of devices and their communication standards; more is the risk of cyber-attacks. The heterogeneity characteristic of IoT networks can be well understood with a smart home example. In smart homes, devices like smart fridges, smart washing machines, smart bulbs, smart showers, etc. are communicating with others with different communication standards like Zigbee, Bluetooth, NFC, BLE, etc. Thus we can see that it is difficult to accommodate all communication standards in one security solution. IoT network device scale is very vast. It includes smart cities to smart agriculture, smart grid to smart healthcare and so on [11,12].

Due to huge advancements in wireless technology and internet standards variety of smart devices and things, sensor network technologies, IP protocols, industry 4.0, social media are part of the IoT cyber-security [11–13]. It is cited by Institutions and Nations that regulations and standards are necessary to achieve high levels of cyber-security in IoT. Without a strong cyber-security infrastructure, cyber-attacks will cause huge problems and losses to the global IoT network. Because of the diversified benefits of the internet, IoT has become an integral part of our lives; hence, it is a need of time to have a global cyber-security solution that will enhance the efficiency of the IoT network.

Currently, the IoT of research is going on in the field of IoT cyber-security. Many standard journals like IEEE, INSPEC, Web of Science, etc. are providing IoT of solutions in cyber-security.

11.2 SECURITY ISSUES IN IoT

It has been observed by AT&T cyber-security that 85% of enterprises are in the process to deploy IoT devices but only 10% of them have given priority to security in IoT network deployment. The wireless connectivity in the IoT network makes devices more vulnerable to attacks. As IoT devices collect more information about one's daily life more attention needs to be paid to security risk and its solution.

11.2.1 IoT Generic Architecture

The IoT supports heterogeneous types of network connectivity which include Machine-to-Machine and Machine-to-Human connectivity. The connectivity approach in IoT is different from the traditional internet. The heterogeneous applications like identification, localization, tracking, controlling, sensing etc. lead to three tremendous traffic which requires the storage of big data and its analytics along with maintaining security [14,15]. The traditional TCP/IP model is not suitable for the needs of IoT regarding many aspects like privacy, the security of information, data communication, network etc. which also affect scalability, reliability, Quality of Service (QoS) and interoperability of the data. Though IoT requires a new protocol stack, it must be based on internet reference architecture [16,17].

IoT reference model consists of three layers: Application Layer, Network Layer and Perception Layer. Details of layers and respective threats at each layer are shown

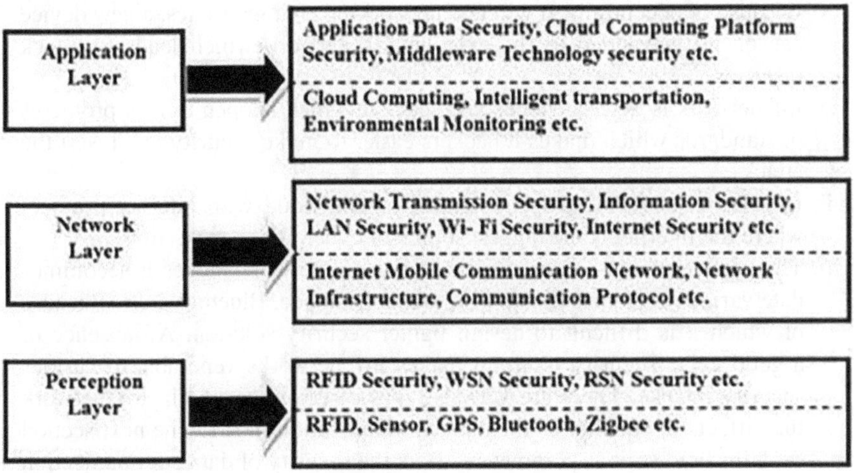

FIGURE 11.3 IoT architecture and layerwise attack.

in Figure 11.3 [18–20]. Out of many issues present in the IoT network, security is one of the most critical issues which need to be addressed at the highest priority. The next topic discusses different security issues in IoT and how IoT network differs from the traditional network in terms of security.

11.2.2 Reasons for Cyber-Attacks in IoT Network

IoT is the reason why the world has become smarter. However, intelligent applications are increased, the more intruders have got the chance to insert the security threats in various forms. The hackers attack the privacy and security of IoT networks by changing and manipulating generated, stored and shared data. IoT networks will be successfully implemented when user data will be protected with safety and privacy. The smart device's data may be exploited by attackers to misuse it which may severely affect the end-user's life. Attackers or hackers may use industrial, personal or governmental data for the wrong purpose. Following are the reasons why the security of IoT networks can be easily broken:

i. IoT devices are resource-constrained in terms of battery power, processing power, memory size, transmission capacity, bandwidth, etc. This characteristic doesn't support traditional network security solutions which require better resources.
ii. These networks are highly distributed and heterogeneous.
iii. The huge data is generated and floated by IoT devices in wireless media with limited bandwidth. Hackers can easily modify, destroy the data packets and interrupt the ongoing transmission.
iv. In many IoT applications, IoT devices are placed in remote areas or physically insecure areas where the attacker can destroy the IoT devices or reprogram them or change the device battery.

v. Because of decentralized wireless networking characteristics, a new device can easily get added to the existing IoT network which leads to attack insertion.
vi. IoT network is heterogeneous and scalable which is open to new protocols or standards which makes it hackers easier to make security and insert the attack.
vii. In IoT, most of the data is uploaded on the cloud with internet protocol where the internet is the biggest source of cyber-attacks insertion.
viii. The heterogeneities characteristic of the IoT network makes it accommodate various protocols like Wi-Fi, Wimax, Zigbee, Bluetooth, etc. Because of which it is difficult to design lighter security solution. An absence of a good cyber-security program makes IoT networks venerable to various security attacks. There are various types of errors caused in IoT network that affects the quality of the data communication in IoT. The next section explains how various parameters affect the security of data communication in the IoT network [21,22].

11.3 POTENTIAL CYBER-ATTACKS IN IoT

There are many attacks present in the wired or wireless networking for long years. However, since the IoT has become popular technology, the scale and variety of attacks present in this type of network have increased tremendously. Today in IoT, millions of devices that are connected are the victims of these cyber-attacks with limited mitigation techniques. The devices in IoT, for example a vehicle connected to your mobile wirelessly or refrigerator connected to network or child TV which is part of the IoT network of a smart home may become access points to attackers by increasing privacy and security risks. The severity of cyber-attacks is increased massively in IoT than in the usual network because in IoT, day-to-day services as discussed above smart TV, smart refrigerator, smart vehicles, etc. are involved through which hackers can get close insights into our life and misuse the data against us. Following are the major attacks:

i. Denial of Service (DoS) Attack:
In a DoS attack, the attacker makes allocated service unavailable to us. This affects severely the companies which are acting as service providers to end-users. DoS attack may cause loss of reputation of the company by denying the use of services to the user. This may affect the business of that company in long run.
ii. Botnet Attack:
Botnets are mainly used by criminals where control is remotely taken by an attacker to steal private information introducing DoS attack exploiting online banking information. The botnet can attack smart devices which are connected to the internet and transfer data automatically. They are present in computers, smartphones, laptops, tabs and smart devices. This attack also crashes the system by sending enormous mails to different devices

which are part of the botnet. By receiving thousands of emails, target device crashes by unable to cope up with these enormous requests through mails.

iii. Data and Identify Theft:
Millions of devices of our daily life, for example mobile phones, smart-watches, ipads, etc. are connected using the internet forming IoT network. This type of network makes it easy for hackers to steal our important data that is combined with smart-watches, fitness trackers, social media and also from the smart fridge, smart meters, etc. These all smart devices along with social media information give an idea about one's personal information and whereabouts. This affects severely on privacy and security of one's personal life and belongings.

iv. Man-in-the-Middle (MITM) Attack:
In a MITM attack, the attacker listens to all the communication between the smart things, modifies it as per its convenience, manipulates data by letting both the devices think that they are getting data from legitimate nodes. This type of attack is dangerous in IoT networks where smart things like the smart vehicle may give wrong information to the smart garage to open it through hackers.

v. Social Engineering Attack:
Through social engineering attacks, the intruder may get information about bank account numbers, passwords, etc. Criminals target individuals to get such information. This can be achieved by sending mails to our email account or SMS to smartphones. Or they can get access to online banking credentials through our laptops or smartphones. In this type of attack, intruders also collect our personal information and use those to blackmail us. Using social sites information, the attacker can easily get to know our whereabouts.

Many other cyber-attacks like Encryption Attacks, Firmware Hijacking, Ransomware, Eavesdropping, Brute Force Password Attack, etc. will affect the IoT technology rigorously. Hence, it is essential to pay attention to the cyber-security of IoT networking to make IoT further popular [23–26].

11.4 NEED OF CYBER-SECURITY IN IoT

The following are the areas of IoT where cyber-security affects severely:

i. Cloud Service Security:
Cloud computing is an area that provides huge information storage and data analytics in IoT. Cloud computing provides the facility to analyze massive data with less cost and provides efficient solutions with storage. However, the cloud computing platform is an open platform; hence, there are huge security risks in its analysis process.

As discussed earlier, cloud computing is an open service offered for user convenience where IoT devices are accessed at any time, and from anywhere without any constraints. Smart IoT devices get access to the cloud either with a wireless connectivity system using the internet or through Wi-Fi. Sensor data of various applications of IoT are stored on the cloud using Database-as-a-Service or Storage-as-a-Service (SaaS) facility [27,28].

Cloud computing technology doesn't give any assurance of complete security of end-user data. Once data is saved or uploaded to the cloud, end-users will not have any control over their data. It will get analyzed and processed by the cloud computing platform. To do this, cloud computing uses a virtual platform. If the inscription of any piece of data isn't completed then a complete batch of that particular data may get exploited by illegal users. There are many means by which security in trusted cloud computing of IoT databases can be achieved. Those are network security, physical security, system security and database security. By using these techniques, trusted cloud computing platforms guarantee end-user's data security and privacy from cyber threats and unauthorized access. In near future, it is necessary to focus on strong encryption technologies, authorization and authentication processes, and potential cyber-security threats to have trusted cloud computing platforms for data security in IoT.

ii. 5G:

The 5G technology mainly has two strong pillars in its development: wireless technology and network technology. In upcoming connectivity technology where smart devices are integrated 5G technology enhances the scalability, reliability, ubiquity, and also cost-effectiveness of the IoT system. IPv4 will be completely replaced by Ipv6 because it supported more IP identifiable objects. There are massive technological changes with 5G technology which includes solutions to more traffic issues and delays by providing better bandwidth. Furthermore, IIoT will also evolve rapidly with IoT. The development of smartphones and mobile devices will allow users to use exponential data flow.

However, though there will be a positive impact on the quality of wireless technology in IoT, there are a lot of risks that will be introduced by attackers by identifying vulnerable targets and launch the different attacks on the IoT network [29]. So when we are considering 5G technology's benefits, we should not forget to address cyber-security issues like security protocols, data privacy, data integrity, information transmission management etc. It is because IoT-related services will grow exponentially with 5G technology. This is one of the big challenges for 5G technology to reduce the cyber-security risks in IoT device networks [30].

iii. QoS Design in Cyber-Security

QoS-based cyber-security is responsible for the improvement in the functionality of the entire IoT network. QoS management can have a better solution for cyber-security in IoT. QoS in IoT cyber-security field has a lot of research potential. There are many fields in the area which are unexplored [31], for example resource-constrained nature of IoT network, scalability and data privacy which are discussed below:

A. Resource Constraint Nature of IoT Network:
The mechanism developed in QoS-based IoT cyber-security must consider the resource constraint nature of IoT networks where bandwidth, memory, battery power, processing power, etc. are limited.
B. Scalability:
The basic characteristic of any IoT network is scalability in which the size of the network is expanded without disturbing its original functionality. QoS-based IoT cyber-security should support this property. It should not degrade the quality of IoT network when it will get expanded.
C. Data Privacy:
Data privacy is one of the most critical issues of IoT security. The QoS-based IoT cyber-security must consider it while giving the appropriate solutions.

Apart from the above-discussed advanced technologies, other trends plays important role in IoT cyber-security. These trends are blockchain-embedded cyber-security design, IoT Forensics, self-management, fault tolerance mechanism, etc. [32–35]. From the above discussion, it is understood that paying attention to cyber-security in IoT is a must.

11.4.1 Need of Standardization

When the IoT concept was designed, security issues have not addressed with the topmost priority. But as the IoT network has become an integral part of our day-to-day activities, cyber-security issues in IoT must be attended with the topmost level of priority.

Because of the heterogeneity of IoT networks, it is challenging to design standardized solutions for IoT cyber-security. The protocols and communication standards need to be modified so that they will support all heterogeneous things. It is quite challenging due to the complex structure of IoT. A standardized IoT architecture accommodates different interfaces, data models and protocols and supports a wider range of devices, programming languages and operating systems by making cyber-security difficult to achieve [36–48]. IETF and IEEE have designed the new security protocols that will play an important function in protecting worldwide IoT networks [49,50].

IEEE 802.15.4 standards have finalized the rules for lower-level communication, which will ultimately affect the standardization of higher-level IoT communication protocols. The Routing Over Low-Power and Lossy Networks has proposed RPL protocol which mainly works in routing issues of IoT communication and works at the network layer. 6LOWPAN which is a compressed version of IPV6 protocol supports fragmentation and re-assembling functions of packets at the adaptation layer at protocols stack. At the communication layer, the COAP protocol designed by the CORE group of RETF supports communication of IoT network [51–57].

Despite huge work in the standardization of IoT protocols going on, there is no standardized framework present that will integrate various standards with IoT services, applications and protocols in today's scenario. Because of extreme standards, it is a need of time to build detailed and generalized infrastructure which will accomplish integrity and interoperability of IoT services. Due to the extent variety of technologies, protocols, devices and services, it seems difficult to achieve standardization in IoT architecture. This affects hugely cyber-security solutions in IoT.

11.4.2 Data Issues

In IoT, various applications like medical insurance records, user's personal information, account transaction data, business-related information, health-related information, etc. generate a huge amount of data. If any information from this sensitive data is leaked, then it will have a huge adverse impact on a person's or company's performance and work. The IoT cyber-security infrastructure must focus on issues like data privacy, confidentiality and data integrity.

In IoT cyber-security, data confidentiality is one of the major issues. To address this issue a well-configured system can be installed which will allow access only to authorized entities that can further process the data by denying access to unauthorized persons. To achieve this two mechanisms are used which are access control and authentication process [57].

11.5 MITIGATION TECHNIQUES

So far, various threats and security attacks in IoT have been discussed in this chapter. The conventional mitigation techniques are not applicable for IoT networking. IoT security demands customization of traditional solutions as per applications. The security solutions for IoT networks must incorporate adaptability of various types of devices along with scalability characteristics of IoT networks. The following are few security solutions proposed for cyber-security attacks in IoT networks.

11.5.1 Strong Authentication Solutions

Many IoT devices have the facility of the password authentication process to access their services. Weak passwords give chances to attackers to misuse the resources of the IoT network. Hence, it is highly recommended to use a strong password. Two methods have received a strong recommendation for the same. Those are as follows:

i. Biometric Authentication: It involves fingerprinting, face recognition, eye or retina recognition, etc. Though biometric authentication is a robust authentication solution, it has issues like complexity in the algorithm and high cost [58].
ii. Multifactor Authentication: It supports a multilayer authentication process. It has 2–3 steps which include a combination of passwords, authentication cards and biometric authentication (fingerprint or retina identification). This kind of combination makes a robust authentication process. Though this includes complex and costly solutions, these methods are highly recommended to avoid cyber-security attacks in IoT [59].

11.5.2 Access Control Mechanism

Like robust authentication solutions, biometric and multifactor authentication methods are used for attack avoidance in access control mechanisms. In biometric access, many biological attributes like fingerprinting, face recognition and eye or retina recognition are considered in multilayer format. It is used when any resource needs to be accessed. This robust access control method protects the IoT devices from cyberattacks to a great extent [60]. To reduce higher costs and complexity in processing many researchers have proposed methods that include Physical Unclonable Function (PUF) and hardware obfuscation. These methods reduce the complexity in access control mechanisms by avoiding key storage still with a guarantee of security in IoT networks.

11.5.3 Intrusion Detection System (IDS)

Intrusion Detection System (IDS) is a method that detects any abnormal or suspicious activity happening in the network. IDS detects various attacks with a combination of algorithms fed to it. Depending on the algorithms, IDS will identify the attack as well as the attacker present in the network. Few IDS can remove these attacks and attackers [61,62].

Intrusion detection is a security mechanism that depends on the analysis of data collected in the network to identify any abnormal activity symptoms to discover the attack and trigger an alarm. To design IDS for IoT network, one must consider its characteristics. The IoT networks are infrastructure-less, ad-hoc and heterogeneous networks. Also, they are formed by devices with resource-constrained characteristics in terms of memory, processor, bandwidth, storage capacity and battery. Hence, suitable IDS for any attack detection in IoT network should be the one that consumes fewer resources.

These IDS are efficient when attack patterns are similar, but if the patterns are changed then the effectiveness of IDS gets reduced. The solution for this limitation is to use adaptable methods using Artificial Intelligence (AI). Machine learning and deep learning are the two methods on which researchers are focusing nowadays.

i. Machine Learning IDS:

Researchers of paper [63] have proposed host-based attack detection and removal techniques using open flow protocol for security. For smart home applications, they have used Security-as-a-Service that monitors and takes necessary action if any suspicious activity happens. Though the proposed method has many advantages it suffers from two limitations: It doesn't support scalability that means if a new device is added to the network, the database needs to be updated manually. Also, this system monitors only specific devices. It doesn't cover all devices of the smart home. For known attacks, this system works perfectly however unknown attacks cannot be detected.

In [64], the authors proposed a method that collects raw data and changes it into the standard normalized format. Decision tree, Support Vector Machine (SVM) and Bayesian network are used to validate the classification scheme.

The machine learning approach categorizes whether the attack is present or not. This approach has given a good True Positive Rate and less False Positive Rate. This method detects new attacks also. However, few complex attacks are not detected using this methodology.

The method proposed by the researcher [65] detects the anomalies in the network by classifying them into fraudulent and benevolent nodes. The mechanism in the proposed system creates the antibodies to mitigate the attack similar to the human immune system. There is another method proposed by [66] to detect and remove the attack in the IoT network. For classification, they have used Filter and Wrapper Methods using Correlation Coefficient algorithm, decision trees and cuttlefish algorithm.

ii. Deep Learning:

When a dataset is large, machine algorithms are not sufficient to detect the attack effectively. The authors in [67] have suggested the IDS which is implemented in two steps: They have used Least Square SVM (LS-SVM) technique. In the first step, the entire dataset is divided into clusters that heredities the characteristics of whole datasets. In the second step, intrusions are determined using the LS-SVM technique. This method supports static as well as incremental data.

11.5.4 Software-Defined Networking (SDN)

In conventional networking routers and switches are used to transfer the data. They use a control plane and data plane for data transmission. The control plane decides the path between source and destination and the data plane transfers the data using the prescribed path. Unlike conventional networks, where the control plane and data plane are integrated, in Software-Defined Networking (SDN) both planes are separate. The control plane is a software entity that controls the task of routing. The data plane is executed in hardware [68]. SDN monitors the network traffic and if any malicious activity happens it isolates the suspicious nodes from the network. Though SDN works fine for known attacks but it cannot identify the unknown attacks. The combination of SDN and IDS can identify the newer attacks efficiently [69].

11.5.5 Light-Weight Cryptography

Cryptography is a necessary technique to secure data at rest or in transfer by encapsulating it in unreadable format and decrypt it at another end at the receiver to maintain integrity and confidentiality of data in networking. The traditional cryptographic algorithms which use symmetric and asymmetric algorithms cannot be directly applicable to IoT networks because of their resource-constrained nature. Asymmetric cryptography requires more resources as it requires two different keys for encryption and decryption whereas in symmetric algorithms single key is sufficient for encryption and decryption. Hence, symmetric keys cryptographic algorithms are suitable for IoT networking. When light-weight encryption algorithms are to be designed, the size, power consumption, throughput and delay of IoT devices and networks must be considered. Nowadays, PUF is used to avoid physical layer attacks and end-point security by inserting security hardware layer in IoT protocol stack. This method requires lesser resources than the conventional encryption method. One method which uses light-weight symmetric key algorithms is developed by the researcher in the paper [70]. The authors have provided light-weight cryptographic algorithm which provides the effective and secure transfer of data within IoT devices.

11.6 CONCLUSION

This chapter covers the cyber-security issues in today's IoT technology. IoT has become a very popular network nowadays where millions of devices are connected with wireless technologies and the internet. Many applications in our day-to-day life are part of IoT. Devices like smart refrigerators, smart cars, smart fridges, smart fitness trackers, etc. at smart homes communicate with each other and also send the information to mobiles and laptops. This is where hackers take charge of one's personal information and steal important data as well as money. In this chapter, we have discussed different cyber-attacks in IoT networks along with the reasons for the presence of attacks. When IoT was under development process, security issues have given low priority while designing it. However as the popularity of IoT networks has increased, security has become one of the most important issues. The cyber-security solutions of computer networks do not apply to the IoT network because of its resource constraints characteristics. In this chapter, we have also covered the mitigation techniques to overcome cyber-security attacks. We have discussed Robust Authentication, Robust Access Control, IDS, Software-Defined Network and Light-Weight Cryptography. Machine learning and deep learning fields of AI are the emerging fields for cyber-security IDS. There is a lot of research potential in the cyber-security area as it is the most phenomenon area nowadays.

REFERENCES

1. L. Xu, W. He, and S. Li, Internet of Things in industries: A survey, *IEEE Trans. Ind. Informat.*, vol. 10, no. 4, pp. 2233–2243, 2014.
2. R. Roman, J. Zhou, and J. Lopez, On the features and challenges of security and privacy in distributed internet of things, *Comput. Netw.*, vol. 57, no. 10, pp. 2266–2279, 2013.
3. L. Atzori, A. Iera, and G. Morabito, The internet of things: A survey, *Comput. Netw.*, vol. 54, pp. 2787–2805, 2010.
4. J.R.C. Nurse, S. Creese, and D. de Roure, Security risk assessment in Internet of Things systems, *IT Prof.*, vol. 19, pp. 20–26, 2017.
5. V. Malik, and S. Singh, Security risk management in IoT environment, *J. Discret. Math. Sci. Cryptogr.*, vol. 22, pp. 697–709, 2019.
6. Markets and Markets. IoT security Market worth $35.2 billion by 2023. 2019. Available online: https: //www.marketsandmarkets.com/PressReleases/iot-security.asp (10 February 2021).
7. PwC. Managing emerging risks from the Internet of Things. 2016. Available online: https://www.pwc.com/us/en/services/consulting/cybersecurity/library/broader-perspectives/managing-iot-risks.html (accessed on 10 February 2021).

8. Irdeto. New 2019 global survey: IoT-focused cyberattacks are the new normal. 2019. Available online: https://resources.irdeto.com/global-connected-industries-cybersecurity-survey/new-2019-globalsurvey-iot-focused-cyberattacks-are-the-new-normal (accessed on 12 March 2021).
9. L. Atzori, A. Iera, and G. Morabito, The internet of things: A survey, *Comput. Netw.*, vol. 54, no. 15, pp. 2787–2805, 2010.
10. D. Bandyopadhyay, and J. Sen, Internet of things: Applications and challenges in technology and standardization, *Wireless Pers. Commun.*, vol. 58, no. 1, pp. 49–69, 2011.
11. H. Suo, J. Wan, C. Zou, and J. Liu, Security in the internet of things: A review. In *Proceedings of the Computer Science and Electronics Engineering (ICCSEE),* vol. 3. Hangzhou, China, IEEE, 2012, pp. 648–651.
12. M. Covington, and R. Carskadden, Threat implications of the Internet of Things. In *Proceedings of the 5th International Conference on Cyber Conflict (CyCon).* IEEE, Tallinn, 2013, pp. 1–12.
13. Y. Lu, Industry 4.0: A survey on technologies, applications and open research issues, *J. Ind. Inform. Integ.*, vol. 6, pp. 1–10, 2017.
14. O. Said, and M. Masud, Towards internet of things: Survey and future vision, *Int. J. Comput. Networks*, vol. 5, no. 1, pp. 1–17, 2013.
15. M. Ge, H. Bangui, and B. Buhnova, Big data for internet of things: A survey, *Future Generat. Comput. Syst. J.*, Doi: 10.1016/j.future.2018.04.053, homepage: www.elsevier.com/locate/fgcs.
16. R. Khan, S.U. Khan, R. Zaheer, and S. Khan, Future internet: The internet of things architecture, possible applications and key challenges. In *Proceedings of the 2012, 10th International Conference on Frontiers of Information Technology*, Islamabad, India, 17–19 December 2012.
17. M. Weyrich, and C. Ebert, Reference architectures for the internet of things, *IEEE Softw.*, vol. 33, pp. 112–116, 2016.
18. M. Bauer, M. Boussard, N. Bui, J.D. Loof, C. Magerkurth, S. Meissner, A. Nettsträter, J. Stefa, M. Thoma, and J.W. Walewski, *IoT Reference Architecture. In Enabling Things to Talk.* Springer: Berlin/Heidelberg, Germany, 2013; pp. 163–211.
19. I. Mashal, O. Alsaryrah, T.Y. Chung, C.Z. Yang, W.H. Kuo, and D.P. Agrawal, Choices for interaction with things on Internet and underlying issues, *Ad Hoc Netw.*, 2015, vol. 28, pp. 68–90.144.
20. M. Wu, T.J. Lu, F.Y. Ling, J., Sun, and H.Y. Du, Research on the architecture of Internet of Things. In *Proceedings of the 2010 3rd International Conference on Advanced Computer Theory and Engineering (ICACTE),* Chengdu, China, 20–22 August 2010.
21. M. El-hajj, M. Chamoun, A. Fadlallah, and A. Serhrouchni, Analysis of authentication techniques in Internet of Things (IoT), In *Proceedings of the 2017 1st Cyber Security in Networking Conference (CSNet),* Rio de Janeiro, Brazil, 18– 20 October 2017, pp. 1–3.
22. W. Shang, Y. Yu, R. Droms, and L. Zhang, *Challenges in IoT Networking via TCP/IP Architecture*; Technical Report 04, NDN, Technical Report NDN-0038; Named Data Networking. Available online: http://nameddata.net/techreports.html.
23. S. Rathore, and J.H. Park, Semi-supervised learning based distributed attack detection framework for IoT, *Appl. Soft Comput.*, vol. 72, pp. 79–89, 2018.
24. E.J. Cho, J.H. Kim, and C.S. Hong, *Attack Model and Detection Scheme for Botnet on 6LoWPAN*; Springer: Berlin/Heidelberg, Germany, 2009; pp. 515–518.
25. N. Moustafa, B. Turnbull, and K.R. Choo, An ensemble intrusion detection technique based on proposed statistical flow features for protecting network traffic of internet of things, *IEEE Internet Things J.* vol. 6 (3), pp. 4815–4830, 2019.
26. https://www.forbes.com/sites/chuckbrooks/2021/02/07/cybersecurity-threats-the-daunting-challenge-of-securing-the-internet-of-things/?sh=2e7116735d50. (Article published on 07 Feb 2021).

27. J. Gubbi, R. Buyya, S. Marusic, and M. Palaniswami, Internet of things (IoT): A vision, architectural elements, and future directions, *Future gener. Comput. Syst.*, vol. 29, no. 7, pp. 1645–1660, 2013.
28. A. Whitmore, A. Agarwal, and L.D. Xu, The internet of things—a survey of topics and trends, *Inf. Syst. Front.*, vol. 17, no. 2, pp. 261–274, 2015.
29. X. Duan and X. Wang, Authentication handover and privacy protection in 5G hetnets using software-defined networking, *IEEE Commun. Mag.*, vol. 53, no. 4, pp. 28–35, Sep. 2015.
30. W.H. Chin, F. Zhong, and R. Haines, Emerging technologies and research challenges for 5G wireless networks, *IEEE Wireless Commun.*, vol. 21, no. 2, pp. 106–112, Apr. 2014.
31. F. Al-Turjman, E. Ever, and H. Zahmatkesh, Small cells in the forthcoming 5g/IoT: traffic modelling and deployment overview, *IEEE Commun. Surv. Tut.*, 2018. Doi: 10.1109/COMST.2018.2864779.
32. F. Al-Turjman and S. Alturjman, Context-sensitive access in industrial internet of things (IIoT) healthcare applications, *IEEE Trans. Ind. Informat.*, 2018. Doi: 10.1109/TII.2018.2808190.
33. L. Li, S. Li, and S. Zhao, QoS-aware scheduling of services-oriented internet of things, *IEEE Trans. Ind. Informat.*, 10(2), pp. 1497–1505, 2014.
34. Y. Lu, Blockchain: a survey on functions, applications and open issues, *J. Ind. Inform. Manag.*, 2018, online published, Doi: 10.1142/S242486221850015X.
35. . Lu, Blockchain and the related issues: a review of current research topics, *J. Manage. Analy.*, 2018, online published, Doi: 10.1080/23270012.2018.1516523.
36. R. Mahmoud, T. Yousuf, F. Aloul, and I. Zualkernan, Internet of things (IoT) security: Current status, challenges and prospective measures. In *Internet Technology and Secured Transactions (ICITST), 2015 10th International Conference*, IEEE, London, UK, 2015, pp. 336–341.
37. X. Jia, O. Feng, T. Fan, and Q. Lei, RFID technology and its applications in internet of things (IoT). In *Proceedings of the 2nd IEEE International Conference on Consumer Electronics, Communications and Networks (CECNet),* Yichang, China, Apr. 2012, pp.1282–1285.
38. M.C. Domingo, An overview of the internet of things for people with disabilities, *J. Netw. Comput. Appl.*, vol. 35, no. 2, pp. 584–596, 2012.
39. F. Alsubaei, A. Abuhussein, and S. Shiva. Security and privacy in the internet of medical things: Taxonomy and risk assessment, local computer networks workshops (LCN Workshops). *42nd Conference on. IEEE*, Singapore, 2017, pp. 112–120, Doi: 10.1109/LCN.Workshops.2017.72.
40. S.A. Alabady, F. Al-Turjman, and S. Din, A novel security model for cooperative virtual networks in the IoT era, *Springer Int. J. Paral. Prog.*, 2018, Doi: 10.1007/s10766-018-0580-z.
41. C. Sun, Application of RFID technology for logistics on internet of things, *AASRI Procedia*, vol. 1, pp. 106–111, 2012.
42. S. Babar, P. Mahalle, A. Stango, N. Prasad, and R. Prasad, Proposed security model and threat taxonomy for the internet of things. In *Proceedings of the Recent Trends in Network Security and Applications*, Springer, Berlin Heidelberg, 2010, pp. 420–429.
43. P.N. Mahalle, B. Anggorojati, N.R. Prasad, and R. Prasad, Identity authentication and capability based access control (IACAC) for the internet of things, *J. Cyber Secur. Mobil.*, vol. 1, no. 4, pp. 309–348, 2013.
44. A.R. Sadeghi, C. Wachsmann, and M. Waidner. Security and privacy challenges in industrial internet of things. In *Annual Design Automation Conference*, ACM, San Francisco, CA, USA, 2015, pp. 54.

45. A. Belapurkar, A. Chakrabarti, H. Ponnapalli, N. Varadarajan, S. Padmanabhuni, and S. Sundarrajan, *Distributed Systems Security: Issues, Processes and Solutions*. Wiley Publishing: Chichester, 2009.
46. M. Farooq, M. Waseem, A. Khairi, and S. Mazhar, A critical analysis on the security concerns of internet of things (IoT), *Perception*, vol. 111, no. 7, pp. 1–6, 2015.
47. S. Sicari, A. Rizzardi, L. Grieco, and A. Coen-Porisini, Security, privacy and trust in internet of things: The road ahead, *Comput. Networ.*, vol. 76, pp. 146–164, 2015. [Online]. Available: http://www.sciencedirect.com /science/article/pii/ S13891286140 03971.
48. R. Roman, P. Najera, and J. Lopez, Securing the internet of things, *Computer*, vol. 44, no. 9, pp. 51–58, 2011.
49. J. Granjal, E. Monteiro, and J.S. Silva, Security for the internet of things: A survey of existing protocols and open research issues, *IEEE Commun. Survey Tuts.*, vol. 17, no. 3, pp. 1294–1312, 3rd Quart., 2015.
50. K.T. Nguyen, M. Laurent, and N. Oualha, Survey on secure communication protocols for the Internet of Things, *Ad Hoc Netw.*, vol. 32, pp. 17–31, Sep. 2015.
51. IEEE standard for local and metropolitan area networks—Part 15.4: Low-rate wireless personal area networks (LR-WPANs), in IEEE Std. 802.15.4-2011 (Revision of IEEE Std. 802.15.4–2006), 1–314, 2011. DOI: 10.1109/IEEESTD.2011.6012487
52. IEEE standard for local and metropolitan area networks—Part 15.4: Low-rate wireless personal area networks (LR-WPANs) amendment 1: MAC sublayer, IEEE Std. 802.15.4e-2012 (Amendment to IEEE Std. 802.15.4–2011), 1–225, 2012, DOI: 10.1109/IEEESTD.2012.6185525
53. N. Kushalnagar, G. Montenegro, and C. Schumacher, IPv6 over Low power wireless personal area networks (6LoWPANs): Overview, assumptions, problem statement, goals, RFC 4919, 2007, [Online]. Available: https://www.rfc-editor.org/rfc/pdfrfc/rfc4919.txt.pdf.
54. G. Montenegro, N. Kushalnagar, J. Hui, and D. Culler, Transmission of IPv6 packets over IEEE 802.15.4 networks, RFC 4944, 2007, [Online]. Available: https://www.rfc-editor.org/rfc/pdfrfc/rfc4944.txt.pdf.
55. J. Hui and P. Thubert, Compression format for IPv6 datagrams over IEEE 802.15.4-based networks, RFC 6282, 2011, [Online]. Available: https://www.rfc-editor.org/rfc/pdfrfc/rfc6282.txt.pdf.
56. T. Winter, et al., RPL: IPv6 routing protocol for low-power and lossy networks, RFC 6550, 2012, [Online]. Available: https://www.rfceditor.org/rfc/pdfrfc /rfc6550.txt.pdf.
57. C. Bormann, A. Cast,ellani, and Z. Shelby, CoAP: An application protocol for billions of tiny internet nodes, *IEEE Internet Comput.*, vol. 1, no. 2, pp. 62–67, Mar./Apr. 2012.
58. R. Amin, N. Kumar, G.P. Biswasb, R. Iqbal, and V. Chang, A light weight authentication protocol for IoT-enabled devices in distributed cloud computing environment, *Fut. Generation Comput. Syst.*, vol. 78, pp. 1005–1019, 2018.
59. A. Ometov, V. Petrov, S. Bezzateev, S. Andreev, Y. Koucheryavy, and M. Gerla, Challenges of multi-factor authentication for securing advanced IoT applications, *IEEE Network*, vol. 33, no. 2, pp. 82–88, 2019.
60. B. Ali, and A. Awad, Cyber and physical security vulnerability assessment for IoT-based smart homes, *Sensors*, vol. 18, no. 3, p. 817, 2018.
61. P. Mell, Understanding intrusion detection systems. In *IS Management Handbook*, 8th Edition, Imprint Auerbach Publications, 409–418, 2003, E-book ISBN: 9780429117374.
62. B. Morin, L. Me, H. Debar, and M. Ducasse, M2D2: A formal data model for IDS alert correlation, In *International Workshop on Recent Advances in Intrusion Detection*, Verlag Berlin Heidelberg, 115–137, 2002.

63. M. Nobakht, V. Sivaraman, and R. Boreli, A host-based intrusion detection and mitigation framework for smart home IoT using OpenFlow, In *11th International Conference on Availability, Reliability and Security (ARES)*, Salzburg, Austria, IEEE 2016.
64. K. Stroeh, E.R.M. Madeira, and S.K. Goldenstein, An approach to the correlation of security events based on machine learning techniques, *J. Internet Serv. Appl.*, vol. 4, no. 1, pp. 7, 2013.
65. H. Rathore, and S. Jha, Bio-inspired machine learning based wireless sensor network security, In *2013 World Congress on Nature and Biologically Inspired Computing*, Fargo, ND, USA, IEEE, 2013.
66. S. Mohammadi, H. Mirvaziri, M. Ghazizadeh-Ahsaee, and H. Karimipour, Cyber intrusion detection by combined feature selection algorithm, author, *J. Info. Secur. Appl.*, vol. 44, pp. 80–88, 2019.
67. E. Kabir, J. Hu, H. Wang, and G. Zhuo, A novel statistical technique for intrusion detection systems, *Fut. Generation Comput. Syst.*, vol. 79 (1), pp. 303–318, 2018.
68. N. McKeown, T. Anderson, H. Balakrishnan, G. Parulkar, L. Peterson, J. Rexford, S. Shenker, and J. Turner, OpenFlow: Enabling innovation in campus networks, *ACM SIGCOMM Comput. Commun. Rev.* vol. 38, no. 2, pp. 69–74, 2008.
69. D. Papamartzivanos, F.G. Marmol, and G. Kambourakis, *Introducing Deep Learning Self-Adaptive Misuse Network Intrusion Detection Systems,* in IEEE, vol. 7, pp. 13546–13560, 2019.
70. S. Rajesh, V. Paul, V.G. Menon, and M.R. Khosravi, A secure and efficient lightweight symmetric encryption scheme for transfer of text files between embedded IoT devices, *Symmetry*, vol. 11, p. 293, 2019 Doi: 10.3390/sym11020293.

12 Smart Materials for Electrochemical Water Oxidation

Shital B. Kale, Dhanaji B. Malavekar and Chandrakant D. Lokhande
D.Y. Patil Education Society

CONTENTS

12.1 INTRODUCTION (*IS THERE ANY ALTERNATIVE TO FOSSIL FUELS?*)

About 1.5 million years ago, a spark lead to the first human-made fire and ignited a great revolution for species, a revolution of energy. Since then, energy is one of the main drivers of the economic and social development of humankind. Our ancestors were producing energy by burning biomass. With the special ability to think, human civilization evolved through the bronze and iron ages using this source of energy. In 300 BCE, humans started to harness energy from various sources and also started to search for new ones. In the 1800s, mankind had the second energy revolution, the use of fossil fuels. By burning fossil fuels, the world has transformed with industrial machinery and communication. World's current primary energy sources, such as coal, oil and natural gas, are hydrocarbons formed over the course of millions of

DOI: 10.1201/9781003220176-12

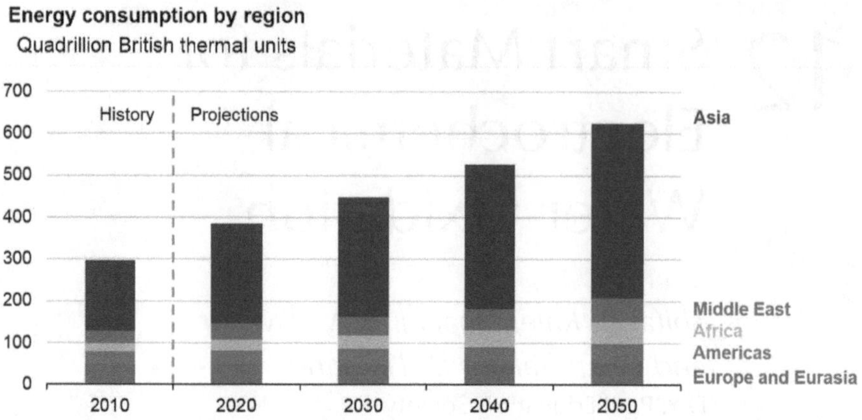

FIGURE 12.1 Energy consumption for selected regions from 2018 with projection to 2050. (U.S. EIA, 2019.)

years in the earth's crust. To fulfill the energy need, we dig or drill this stuff out for a long time. Along with the increasing world population, we are becoming a more and more digitalized civilization due to which the energy demand is also increasing. According to the report of the Energy Information Administration (EIA), between 2018 and 2050, there will be a 50% increment in the world's energy consumption (Figure 12.1 shows energy consumption for selected regions) [1].

Right now, fossil fuels supply 80% of that energy, but we are aware that the stock of fossil fuels under the earth's crust is limited and going to be exhausted soon. Because of their origins, fossil fuels have high carbon content, which causes the greenhouse effect. Due to this, it is widely agreed that fossil fuels cannot be the major energy supplier; this motivates the search and use of clean and sustainable alternative energy sources [2].

The answer to the question, 'Is there any alternative to fossil fuels?' is yes – hydrogen fuel (H_2). Hydrogen is possibly the main carrier for the new wave of a renewable form of energy and has been termed as the hydrogen economy or more recently transitioning to a hydrogen society [3,4]. Hydrogen is zero-emission fuel burned with oxygen and it has high energy content than gasoline. Due to high energy output and carbon-free combustion, hydrogen is considered an ideal alternative to fossil fuels [5]. As pure hydrogen does not occur naturally, steam reforming, coal gasification and water electrolysis processes are used for hydrogen production. Steam reforming is the major contributing process in hydrogen production. Natural gas such as methane is mixed with water vapor under high pressure in a reformer vessel. The hydrogen is produced via a strongly endothermic reaction $CH_4 + 2H_2O \rightleftharpoons CO_2 + 4H_2$. To produce hydrogen from coal, coal gasification is used.

The drawback of both above-mentioned methods is the requirement of fossil fuels for hydrogen production. Also, a large amount of CO_2 is produced during these processes, which is harmful to the environment. Water electrolysis is the only method to produce hydrogen without the use of fossil fuels, in which decomposition of water takes place into molecular oxygen (O_2) and hydrogen (H_2) by passing an electric current. It is also a carbon-free way of hydrogen generation. Also, if the electricity required for water splitting is generated from any renewable source such as wind or solar farms, it will be an indirect approach to use renewable energy sources.

12.2 ELECTROCHEMICAL WATER SPLITTING

Electrochemical water splitting is the most feasible hydrogen production method at the time. Electrolysis is the process by which a compound is broken down into its constituent molecules using electricity. The reaction of electrochemical water splitting can be written as

$$2H_2O(l) \rightarrow 2H_2(g) + O_2(g) \tag{12.1}$$

This overall water-splitting reaction consists of two half-reactions, cathodic hydrogen evolution reaction (HER) and anodic oxygen evolution reaction (OER). Figure 12.2a shows the schematic of electrochemical water splitting.

During OER at the anode, oxidation of water takes place as follows:

$$2H_2O\,(l) \rightarrow O_2(g) + 4H^+(aq) + 4e^- \tag{12.2}$$

while during HER, hydrogen ions are reduced to form molecular hydrogen at the cathode as follows:

$$4H^+(aq) + 4e^- \rightarrow 2H_2(g) \tag{12.3}$$

The standard water-splitting potential is 1.23 V versus reversible hydrogen electrode (RHE) at standard temperature and pressure. Hence theoretically, a voltage difference of 1.23 V must be applied between anode and cathode in order to perform water-splitting reaction. However, in practice, some extra potential in addition to the standard one is necessary to overcome kinetic barriers at electrodes, which results in low efficiency of water splitting. This extra potential is called overpotential (η). To minimize energy loss, we need to minimize this overpotential. In this situation, electrocatalyst plays an important role by increasing the rate of electrochemical reaction via participating in it [6]. Even though electrochemical water splitting is known since the 19th century, its practical applications are limited due to the OER. Among the HER and OER, OER is sluggish and complicated [7,8]. Also, it demands high overpotential that affects the overall water-splitting process. For these reasons, OER has been studied with extra efforts for decades to minimize overpotential.

12.3 MECHANISM OF OXYGEN EVOLUTION REACTION (OER) AND EVALUATION PARAMETERS

The possible mechanisms for OER at the electrode surface are proposed by many research groups [9,10]. Typically, all these mechanisms involve the formation of intermediates, for example, E=O, E-OH, and E-OOH (where E represents the electrocatalysts).

OER mechanism:

$$M + H_2O_{(l)} \rightarrow MOH + H^+ + e^- \text{ Step I}$$

$$MOH + OH^- \rightarrow MO + H_2O_{(l)} + e^- \text{ Step II}$$

$$MO + H_2O_{(l)} \rightarrow MOOH + H^+ + e^- \text{ Step III}$$

$$MOOH + H_2O_{(l)} \rightarrow M + O_{2(g)} + H^+ + e^- \text{ Step IV}$$

Hence, to produce one molecule of oxygen, four electrons must be transferred at each step of the multistep OER reaction. Therefore, an efficient electrocatalyst must accelerate the charge-transfer process to get over the energy barrier at the electrode and electrolyte interface [11,12]. Thus, it is well desirable to investigate and develop novel, low-cost and efficient materials as alternative electrocatalysts for OER, which can realize large-scale water electrolysis. Different parameters used to evaluate an electrocatalyst are important.

12.3.1 Overpotential (η)

As seen above, the extra potential in addition to standard potential is called overpotential. Therefore, it is one of the significant parameters to assess the OER performance. The overpotential required by an electrocatalyst to reach any current density, j, can be calculated by Eq. (12.4).

$$\eta_{@j} = E_{RHE} - 1.23 \text{ V} \tag{12.4}$$

However, 10 mA cm^{-2} current density is widely accepted for evaluating the electrochemical OER performance of electrocatalysts. A good OER electrocatalyst should reach a fixed current density with minimum overpotential. The scan rate during linear sweep voltammetry (LSV) measurement makes a great impact on the current response. The current increases with increasing scan rate. Hence, while examining and comparing the LSV of more than two electrodes, the scan rate must be similar.

12.3.2 Tafel Slope (b)

The current in electrochemistry is related exponentially to the overpotential as shown in Eq. (12.5):

$$i = a' e^{\eta/b'} \tag{12.5}$$

where 'i' is current, 'η' is overpotential and 'a'' and 'b'' are constants.

By taking logarithmic scale, the above equation will be

$$\eta = a + b\log(i) \tag{12.6}$$

where 'a' and 'b' are constants. Eq. (12.6) is nothing but the Tafel equation. It relates to the overpotential and rate of electrochemical reaction. The low value of the Tafel slope suggests the high rate of reaction or fast reaction kinetics, while a high value of Tafel slope suggests the low rate of reaction or slow reaction kinetics. To determine the Tafel slope first, the Tafel plots are plotted from LSV curves for each electrode and then their linear part is fitted with the above Tafel equation.

12.3.3 Electrochemical Active Surface Area (ECSA)

The origin of electrocatalytic performance can be better understood by the electrochemical active surface area (ECSA) of electrocatalysts. For this, first double-layer capacitance (C_{dl}) can be measured using either cyclic voltammetry (CV) or electrochemical impedance spectroscopy (EIS). Using Eq. (12.7), ECSA for an electrode can be calculated through C_{dl}.

$$\mathrm{ECSA} = \frac{C_{dl}}{0.04\left(\mathrm{mF\ cm^{-2}}\right)} \tag{12.7}$$

where C_{dl} is double-layer capacitance and 0.04 is the specific capacitance value of a standard 1 cm^2 flat surface.

12.4 ELECTROCATALYSTS FOR OER

The high cost and limited reserve of state-of-the-art electrocatalysts (RuO_2 and IrO_2) restrict their large-scale production, commercialization and widespread application. But after studying various materials over the decades, it is now concluded that the electrocatalysts based on transition metals are suitable substitutes for precious metals. Among them, cobalt, nickel, iron and manganese-based materials are of more interest [13]. Therefore, their oxides, chalcogenides, phosphides and layered double hydroxides (LDH) have been intensively investigated in the last few years for electrochemical water-splitting application. Figure 12.2b illustrates the elements of interest from the periodic table.

12.4.1 Metal Oxides

Due to the 3d shell electronic configuration, cobalt oxide is more widely used for OER than other transition metal-based oxides. Especially, spinel structured cobalt oxide (Co_3O_4) has been the center of attraction due to its good electric conductivity and good stability under high anodic potential [14,15]. Li et al. [16] prepared ultrathin Co_3O_4 nanomeshes for OER application. To obtain 10 mA cm^{-2} current density, it requires 307 mV overpotential with 76 mV dec^{-1} Tafel slope. Acedera et al. [17] controlled the phase composition and morphology of Co_3O_4 nanoparticles and

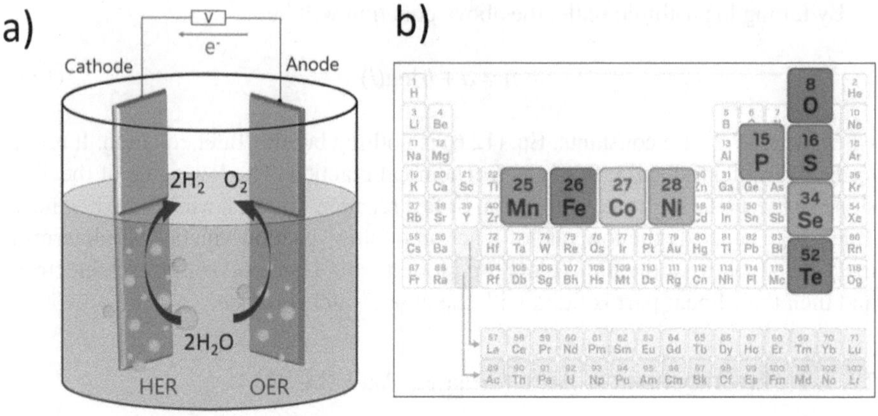

FIGURE 12.2 (a) Schematic of the electrolysis of water and (b) illustration of the elements of interest from the periodic table.

achieved the maximum OER performance of 334 mV overpotential and 61 mV dec^{-1} Tafel slope. Different phases of cobalt oxide other than Co_3O_4 have been investigated for OER. Along with cobalt oxide, nickel oxide is also regarded as a promising electrocatalyst for OER [18–20]. In electrocatalytic OER application, rampant use of Ni foam is done either to support the electrocatalyst powder or to grow thin film. Taking the advantage of readily available metallic nickel in the form of foam, Babar et al. [21] grew porous NiO by thermally oxidizing Ni foam that demands 310 mV overpotential to reach benchmarking current density and shows 54 mV dec^{-1} Tafel slope.

Ferrites and manganese oxides are also studied as OER electrocatalysts [22–25]. Recently, Zhuang et al. [26] have synthesized monodispersed iron oxide nanoparticles of uniform size but various crystallinities. Amorphous FeO_x showed better OER performance than crystalline Fe_3O_4, which demonstrated the importance of crystallinity to determine OER activity. Manganese oxide is a wildly studied material in the field of supercapacitor due to its various oxidation states. A study on the origin of the OER activity of manganese oxide films given by Plate et al. [23] suggests that Mn^{III} and Mn^{IV} are active for OER, while Mn^{II} is inactive. As compared to cobalt and nickel oxides, the reports on iron and manganese oxides are limited. But there are a large number of reports on the binary metal oxides in which Mn or Ni is used as a dopant in spinel cobalt oxides or ferrites.

Béjar et al. [27] have successfully synthesized Co_3O_4 spheres and $NiCo_2O_4$ rosettes by a simple chemical bath. This study shows that the incorporation of Ni to Co_3O_4 has increased the Brunauer-Emmett-Teller (BET) surface area of Co_3O_4 1.7 times, which results in the decrement of overpotential by 155 mV. Similarly, ternary nickel-iron oxide ($NiFe_2O_4$) reported by Dalai et al. [28] shows enhanced OER performance than corresponding NiO and Fe_2O_3. Such enhancement of OER performance of various binary metal oxides than the corresponding single metal oxides is reported by many researchers such as $NiCo_2O_4$ by Gao et al. [29] and Devaguptapu et al. [30], $NiFe_2O_4$ by Liu et al. [31] and Chen et al. [32], $CoFe_2O_4$ by Ou et al. [33], Huang et al. [34]

and Zang et al. [35], $MnFe_2O_4$ by Zhang et al. [36], $MnCo_2O_4$ by Zeng et al. [37], etc. Among all these binary electrocatalysts, $NiCo_2O_4$ and $NiFe_2O_4$ show better performance than others, especially the Mn containing binary electrocatalysts. This may be due to the suppression of Jahn–Teller distortion in spinel structures after the addition of Mn [38]. This suggests that more than one transition metal produces a synergistic effect that effectively regulates the physicochemical properties of oxides and further improves catalytic activity. Taking the advantage of multimetals, Han et al. [39] reported cobalt-iron-nickel oxide (CoFeNi-O), which showed excellent OER performance. Most of the above-mentioned metal oxides show outstanding OER performance; still, their usage is constrained due to the poor intrinsic electronic conductivity. Therefore, research is focused to improve the conductivity of oxides by compositing them with carbon compounds such as graphene and carbon nanotubes (CNTs).

12.4.2 Metal Sulfides

Most of the metal sulfides have strong and highly covalent M-S (M is metal site) and S-S bonds [40]. Different sulfide minerals such as chalcocite (Cu_2S) and pyrite (Fe_2S) exist in nature [41]. Along with these, the most important property of metal sulfides is their intrinsic electronic conductivity. In metal sulfides, the energy gap between metal and sulfide 3p6 level is smaller as compared to the metal and oxygen 2p6 level in the corresponding metal oxide. This gives a higher intrinsic conductivity to metal sulfides than corresponding oxides [42]. Due to all these favorable properties, nanostructured sulfides of transition metals have emerged as a novel class of OER material. Nickel sulfide and cobalt sulfide are widely studied as single metal sulfides for OER application and have also shown good catalytic performance [43–50]. Feng et al. [46] synthesized hollow spheres of Co_9S_8 through a template- and surfactant-free method, which displayed OER performance of 285 mV overpotential at 10 mA cm^{-2}. As discussed in the recent article by Kale et al. [51], OER performance of Co_3S_4 is mainly dependent upon the surface morphology and surface area. They synthesized thin films of Co_3O_4 by Successive Ionic Layer Adsorption and Reaction (SILAR) method with different morphologies. From which the thin film with rough and porous morphology and maximum surface area, shows 275 mV overpotential and 53 mV dec^{-1} Tafel slope. Shang et al. [52] used Ni foam to grow nickel sulfide nanostructure. Similarly, Zou et al. [53] prepared nanosheet arrays of FeS by direct sulfurization of iron foam (IF) and used them as a low-cost preelectrocatalyst for generating highly active electrocatalyst ($Fe\text{-}O_2$) for OER (as shown in Figure 12.3). The OER performance of manganese sulfide is reported by Pujari et al. [54]. They have prepared various phases of manganese sulfide in thin-film form via ion exchange process, out of which α-MnS shows good OER performance than γ and α MnS phases.

As the metal site in sulfides acts as an active site, the multimetal sulfides have more activity than single metal sulfides. Hence, nickel and cobalt, iron and manganese-containing binary or ternary metal sulfides possibly show excellent OER activity. Different compositions of nickel-cobalt sulfides are reported in which spinel structure ($NiCo_2S_4$) has attended more research interest due to the presence of

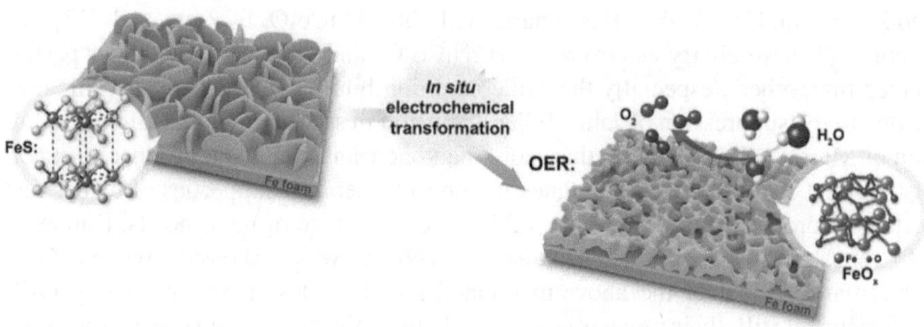

FIGURE 12.3 Schematic illustration of in-situ electrochemical transformations of nanosheet arrays of FeS grown on iron foam (FeS/IF) and in-situ generated Fe-O_2 OER electrocatalysts. (Reproduced with permission from Ref. [53].)

octahedrally coordinated cations. Nearest metal–metal bonds in multimetal sulfides are favorable for the adsorption of reactants and intermediates [55,56]. Manganese-cobalt sulfide is one of the trending materials in energy storage applications [57,58]. Li et al. [59] and Zhang et al. [60] reported binder-free manganese-cobalt sulfide for electrocatalytic application. In both reports, it exhibited excellent activity with high current density. While preparing binary metal sulfides, many researchers follow the root of sulfurization of corresponding oxides. But in our recent article, we have prepared thin films of amorphous manganese-cobalt sulfide via the chemical deposition method [61]. To achieve 10 mA cm^{-2} current density, as-prepared thin films require 243 mV overpotential. Also, it showed that stoichiometric tuning of the composition of metals is very important for optimum performance (Figure 12.4).

Along with the binary metal sulfides, recently some ternary metal sulfides are also reported for electrocatalytic OER application and have shown even better performance than single and binary metal sulfides. Zhao et al. [62] have synthesized Ni-Co-Fe mixed metal sulfides on nickel foam (NiCoFeS/NF) via electrodeposition-solvothermal process. Hill-shaped bulges structured NiCoFeS require 230 mV to drive 100 mA cm^{-2} current density.

12.4.3 Metal Phosphides

Metal phosphide-based electrocatalysts have been studied for OER since the early 1990s. Among these transition metal phosphides of Co, Ni, Fe and Mn have been extensively evaluated. Oh et al. [63] synthesized Co-P foams through electrodeposition and found that it requires 300 mV overpotential to reach benchmarking current density in alkaline electrolyte. In similar approaches, various nanostructures including nanowires, nanorods, and nanotubes have been synthesized and investigated for electrocatalytic performance [64,65]. For instance, the carbon cloth–supported nanoneedle structured Co-P electrocatalyst performed well for overall water splitting [66]. Since the transition metal ions are considered reaction centers for OER, their population on the material surface plays a vital role. Also, the phosphorus to metal ion ratio influences the electronic conductivity of the electrocatalyst, which

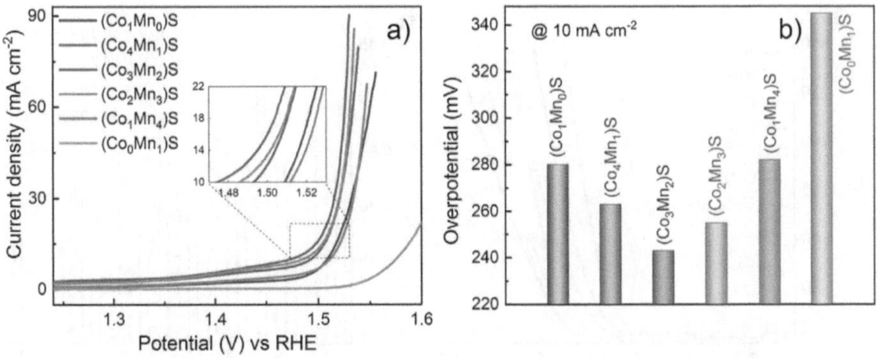

FIGURE 12.4 Electrochemical OER performance: (a) LSV curves with the magnified view shown in inset (iR-corrected), (b) overpotential required to reach 10 mA cm^{-2} current density for (Co$_x$Mn$_y$)S thin-film electrocatalysts with different Co/Mn ratios. (Reproduced with permission from Ref. [61].)

also impacts their catalytic activity. In this view, various studies evaluated the metal phosphide electrodes with varying P:M ratios [67,68].

Many often, the coexistence of binary and ternary metal ions in place of a single metal ion shows synergistic enhancement in the overall efficiency. For this reason, Li et al. [65] developed ternary $Fe_{2-x}Mn_xP$ ($0 \leq x \leq 0.9$) nanorods and found an OER activity dramatically increasing with the incorporation of Mn ($\eta = 440$ mV for 10 mA cm^{-2} for $x = 0.9$). Wang et al. [69] synthesized Fe-doped Ni_2P and observed a higher water-splitting activity for $Ni_{1.85}Fe_{0.15}P$ than Ni_2P. In a similar approach, among a series of Fe-doped Ni_2P developed by Chen and coworkers [70], $(Ni_{0.87}Fe_{0.13})_2P$-Ni exhibited excellent OER activity in alkaline electrolyte. A similar synergism was also observed for other metal phosphides. A positive effect of Fe doping was observed also for CoP in which the Fe-Co-P validated a better activity than the pristine compounds [71,72]. Forwarding this approach with a scalable technique, Liu et al. [73] used Fe-Co metal-organic complex precursor to develop hollow and conductive nanostructures of Fe-Co-P. It displayed outstanding oxygen evolution performance (252 mV overpotential) (Figure 12.5).

A breakthrough phenomenon for outstanding OER activity of metal phosphides is introduced by Yoo et al. [75]. This suggests that in metal phosphides, the surface oxidized sites are the true catalytic site and not the bulk phosphide phase. Prof. Hu et al. [76] found nickel phosphide nanoparticles are Janus catalysts and the in-situ formed Ni_2P/NiO_X core–shell structure gives excellent OER activity. In this aspect, metal phosphides employed for electrochemical water splitting behave as a precatalyst that in-situ transforms into surface oxidized materials including oxides, hydroxides and oxyhydroxides that outperform the ex-situ synthesized similar materials. The surface oxidized metal sites serve as active reaction sites and the phosphide core advances the charge-transfer processes. These phenomena were further validated by reports on CoP [77], Co_2P [78] and even bimetallic and trimetallic phosphides such as MnCoP [79], Co_3FeP [80] and FeCoNiP [74].

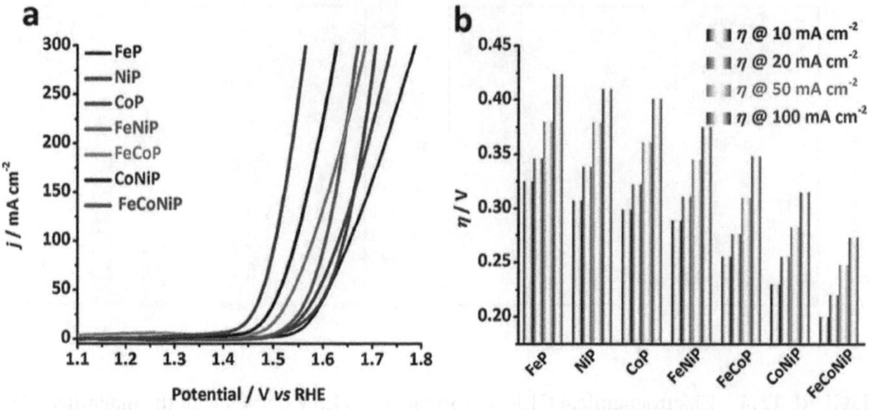

FIGURE 12.5 OER performance of different transition metal phosphide (TMP) precatalysts. (a) iR-corrected polarization curves and (b) the overpotentials required for 10, 20, 50 and 100 mA cm^{-2} current densities. (Reproduced with permission from Ref. [74].)

12.4.4 Layered Double Hydroxide (LDH)

LDHs offer a large surface area that facilitates an effectual exposure of catalytic centers and abundant active reaction sites. Due to the tunable chemical composition, LDHs provide ease of synthesis and a controllable morphology. Various LDH materials with different cationic centers were explored in the past. The NiFe-LDH was found to be the most promising candidate for OER and the performance was found in the order of NiFe > NiCo > CoCo [81]. Since LDH materials are often considered metastable and their stability is significantly influenced by synthesis parameters, the degree of crystallinity also influences the net performance. Lu et al. [82] investigated the effect of NiFe-LDHs crystallinity and found that the amorphous materials exhibit relatively higher OER performance than the crystallized ones (Figure 12.6).

While the two cationic centers are inherently essential to form LDH materials, partial substitution and cationic doping are other approaches to furnish the catalytic centers in LDH electrocatalysts. In this way, many doped LDH materials were developed that outperform the OER activity of the bare materials. For example, partial substitution of Ni for Co in the NiFe-LDH synthesized through a simple coprecipitation route advanced the OER activity. The cationic tuning of Co^{2+}/Ni^{2+} is made to obtain optimum performance of $Ni_3Co_3Fe_3$-LDH and that reached the lowest overvoltage of 250 mV [83]. Zhou et al. [84] developed Mn-doped NiFe-LDH that regulates the electronic structure of bare material favorable to OER displaying an onset potential of 1.41 V. Similarly, Wu et al. [85] formulated self-supported Ni–Fe–W LDH/NF electrocatalysts that require 247 mV overpotential to attain 100 mA cm^{-2}. Studies were also carried out to estimate the OER performances of doubly-doped LDH systems. Sun et al. [86] prepared in-situ grown Fe and B codoped NiZn LDHs over nickel foam by an ion etching strategy that presents a superior electrocatalytic performance for OER (280 mV @ 100 mA cm^{-2}).

The tuning of cationic composition seems to significantly influence the OER activities, and vacancy creation has also been a beneficial approach. Controlled creation

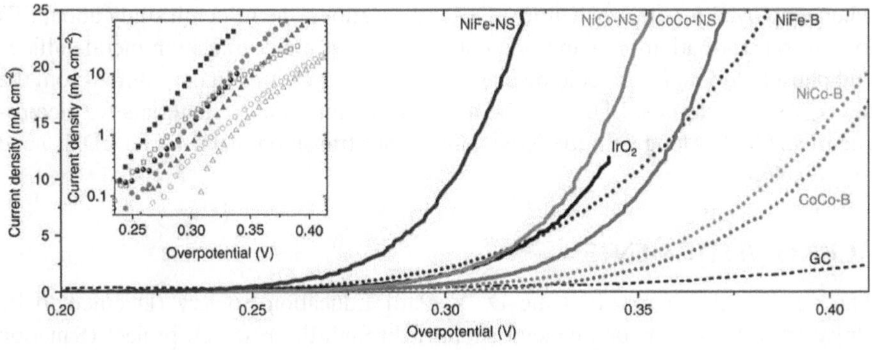

FIGURE 12.6 OER performance of exfoliated NiFe-NS, NiCo-NS and CoCo-NS nanosheets and bulk (nonexfoliated) NiFe-B, NiCo-B and CoCo-B. (Reproduced with permission from Ref. [81].)

of oxygen vacancies can help to modulate the electronic structure and OER activity of LDHs. Li et al. [87] deposited NiFe-LDH on Ni foam and demonstrated a novel route for the controlled introduction of oxygen vacancies (O-vacancies) on the surface. The NiFe-LDH/NF quantifying about 6.8% O-vacancies achieved a remarkably low overpotential of 177 mV under alkaline electrolyte.

12.5 SUMMARY AND FUTURE PERSPECTIVE

It has been well documented that to store intermittent energy from renewable sources, we need an efficient energy carrier. In this scenario, hydrogen is a promising solution. But due to sluggish kinetics, OER becomes a bottleneck for hydrogen production on an industrial scale. To increase water-splitting efficiency, an anodic half-reaction (OER) must be accelerated. In this chapter, we have summarized the progress of transition metal-based electrocatalyst for OER in recent years.

Similar to other energy-related fields, oxides of transition metals have also received much attention in the field of electrochemical water oxidation. Numerous reports are available on the single and binary metal oxides for OER application. Binary metal oxides show enhanced OER performance than corresponding single metal oxides. Also, the reports on the ternary metal oxides are limited. Even after extensive efforts, metal oxides could not compete with state-of-the-art materials available for OER. This is maybe due to their low intrinsic electric conductivity. Therefore, in the search for efficient electrocatalysts for OER, researchers also studied the sulfides of transition metals which are more conductive than the corresponding oxides. Interestingly, binary and ternary metal sulfides show outstanding electrocatalytic activity for OER with an overpotential of 207 mV (@10 mA cm^{-2}). Also, many ternary metal sulfides show both OER and HER activities, which is important for water splitting. In metal sulfides, there is a clear gap in the reports of Mn- and Fe-based sulfides. The phosphides of transition metals show OER performance similar to sulfides. The surface of both sulfides and phosphides undergoes an in-situ transformation during electrochemical OER. Hence, the metal sulfides and phosphides behave as precatalysts when employed for electrochemical OER application. The in-depth study about this in-situ transformation or changes is not reported for many transition metal sulfides and phosphides, and this could be an important research area in the future. From the new class of materials, LDHs can be more promising catalyst materials for electrochemical OER. Along with the NiFe-LDH, other transition metal-based LDHs need to be explored.

ACKNOWLEDGMENTS

The authors are thankful to the D. Y. Patil Education Society (Deemed to be University), Kolhapur for financial support through the research project (Sanction No. DYPES/DU/R&D/3101 dated 03/10/2018). Also, this work is supported by Science and Engineering Board (SERB), Department of Science and Technology, Government of India, New Delhi, through sanction number TTR/2021/000006 dated 30 March 2021.

REFERENCES

1. International Energy Outlook 2019, (2019). https://www.eia.gov/ieo.
2. P. Zhang, J. Zhang, J. Gong, Tantalum-based semiconductors for solar water splitting, *Chem. Soc. Rev.* 43 (2014) 4395–4422.
3. I. Dincer, C. Acar, Smart energy solutions with hydrogen options, *Int. J. Hydrog. Energy.* 43 (2018) 8579–8599.
4. M. Sakamoto, D. Izuhara, H. Sotouchi, F. Aoyagi, P2G System Technology development aiming at building a CO_2-free hydrogen society, *ECS Meet. Abstr*, MA2018-02, (2019) 1589.
5. R. F. Service, Hydrogen cars: Fad or the future?, *Science.* 324 (2009) 1257–1259.
6. Z. Lv, N. Mahmood, M. Tahir, L. Pan, X. Zhang, J. J. Zou, Fabrication of zero to three dimensional nanostructured molybdenum sulfides and their electrochemical and photocatalytic applications, *Nanoscale.* 8 (2016) 18250–18269.
7. P. Li, H. C. Zeng, Advanced oxygen evolution catalysis by bimetallic Ni-Fe phosphide nanoparticles encapsulated in nitrogen, phosphorus, and sulphur tri-doped porous carbon, *Chem. Commun.* 53 (2017) 6025–6028.
8. Z. L. Wang, D. Xu, J. J. Xu, X. B. Zhang, Oxygen electrocatalysts in metal-air batteries: From aqueous to nonaqueous electrolytes, *Chem. Soc. Rev.* 43 (2014) 7746–7786.
9. N. T. Suen, S. F. Hung, Q. Quan, N. Zhang, Y. J. Xu, H. M. Chen, Electrocatalysis for the oxygen evolution reaction: Recent development and future perspectives, *Chem. Soc. Rev.* 46 (2017) 337–365.
10. M. Fang, G. Dong, R. Wei, J. C. Ho, Hierarchical nanostructures: Design for sustainable water splitting, *Adv. Energy Mater.* 7 (2017) 1700559.
11. M. Gong, Y. Li, H. Wang, Y. Liang, J. Z. Wu, J. Zhou, J. Wang, T. Regier, F. Wei, H. Dai, An advanced Ni-Fe layered double hydroxide electrocatalyst for water oxidation, *J. Am. Chem. Soc.* 135 (2013) 8452–8455.
12. M. Tahir, N. Mahmood, J. Zhu, A. Mahmood, F. K. Butt, S. Rizwan, I. Aslam, M. Tanveer, F. Idrees, I. Shakir, C. Cao, Y. Hou, One dimensional graphitic carbon nitrides as effective metal-free oxygen reduction catalysts, *Sci. Rep.* 5 (2015) 12389.
13. L. Han, S. Dong, E. Wang, Transition-metal (Co, Ni, and Fe)-based electrocatalysts for the water oxidation reaction, *Adv. Mater.* 28 (2016) 9266–9291.

14. Y. Su, H. Liu, C. Li, J. Liu, Y. Song, F. Wang, Hydrothermal-assisted defect engineering in spinel Co_3O_4 nanostructures as bifunctional catalysts for oxygen electrode, *J. Alloys Compd.* 799 (2019) 160–168.
15. Q. Liu, Z. Chen, Z. Yan, Y. Wang, E. Wang, S. Wang, S. Wang, G. Sun, Crystal-plane-dependent activity of spinel Co_3O_4 towards water splitting and the oxygen reduction reaction, *ChemElectroChem.* 5 (2018) 1080–1086.
16. Y. Li, F. M. Li, X. Y. Meng, S. N. Li, J. H. Zeng, Y. Chen, Ultrathin Co_3O_4 nanomeshes for the oxygen evolution reaction, *ACS Catal.* 8 (2018) 1913–1920.
17. R. A. E. Acedera, G. Gupta, M. Mamlouk, M. D. L. Balela, Solution combustion synthesis of porous Co_3O_4 nanoparticles as oxygen evolution reaction (OER) electrocatalysts in alkaline medium, *J. Alloys Compd.* 836 (2020) 154919.
18. B. Zhang, X. Shang, Z. Jiang, C. Song, T. Maiyalagan, Z. J. Jiang, Atmospheric-pressure plasma jet-induced ultrafast construction of an ultrathin nonstoichiometric nickel oxide layer with mixed Ni^{3+}/Ni^{2+} ions and rich oxygen defects as an efficient electrocatalyst for oxygen evolution reaction, *ACS Appl. Energy Mater.* (2021) Doi: 10.1021/acsaem.1c00623.
19. Z. Yue, W. Zhu, Y. Li, Z. Wei, N. Hu, Y. Suo, J. Wang, Surface engineering of a nickel oxide-nickel hybrid nanoarray as a versatile catalyst for both superior water and urea oxidation, *Inorg. Chem.* 57 (2018) 4693–4698.
20. M. Fingerle, S. Tengeler, W. Calvet, W. Jaegermann, T. Mayer, Sputtered nickel oxide thin films on n-Si(100)/SiO_2 surfaces for photo-electrochemical oxygen evolution reaction (OER): Impact of deposition temperature on oer performance and on composition before and after OER, *J. Electrochem. Soc.* 167 (2020) 136514.
21. P. T. Babar, A. C. Lokhande, M. G. Gang, B. S. Pawar, S. M. Pawar, J. H. Kim, Thermally oxidized porous NiO as an efficient oxygen evolution reaction (OER) electrocatalyst for electrochemical water splitting application, *J. Ind. Eng. Chem.* 60 (2018) 493–497.
22. S. Kalathil, K. P. Katuri, P. E. Saikaly, Synthesis of an amorphous: Geobacter-manganese oxide biohybrid as an efficient water oxidation catalyst, *Green Chem.* 22 (2020) 5610–5618.
23. P. Plate, C. Höhn, U. Bloeck, P. Bogdanoff, S. Fiechter, F. F. Abdi, R. Van De Krol, A. C. Bronneberg, On the origin of the OER activity of ultrathin manganese oxide films, *ACS Appl. Mater. Interfaces.* 13 (2021) 2428–2436.
24. R. Phul, M. A. M. Khan, M. Sardar, J. Ahmed, T. Ahmad, Multifunctional electrochemical properties of synthesized non-precious iron oxide nanostructures, *Crystals.* 10 (2020) 751.
25. N. U. A. Babar, Y. F. Joya, H. Khalil, F. Hussain, K. S. Joya, Thin-film iron-oxide nanobeads as bifunctional electrocatalyst for high activity overall water splitting, *Int. J. Hydrogen Energy.* 46 (2021) 7885–7902.
26. Z. Zhuang, S. A. Giles, G. R. Jenness, R. Abbasi, X. Chen, B. Wang, D. G. Vlachos, Y. Yan, Oxygen evolution on iron oxide nanoparticles: The impact of crystallinity and size on the overpotential, *J. Electrochem. Soc.* 168 (2021) 034518.
27. J. Béjar, L. Álvarez-Contreras, J. Ledesma-García, N. Arjona, L. G. Arriaga, Electrocatalytic evaluation of Co_3O_4 and $NiCo_2O_4$ rosettes-like hierarchical spinel as bifunctional materials for oxygen evolution (OER) and reduction (ORR) reactions in alkaline media, *J. Electroanal. Chem.* 847 (2019) 113190.
28. N. Dalai, B. Mohanty, A. Mitra, B. Jena, Highly active ternary nickel–iron oxide as bifunctional catalyst for electrochemical water splitting, *ChemistrySelect.* 4 (2019) 7791–7796.
29. X. Gao, H. Zhang, Q. Li, X. Yu, Z. Hong, X. Zhang, C. Liang, Z. Lin, Hierarchical $NiCo_2O_4$ hollow microcuboids as bifunctional electrocatalysts for overall water-splitting, *Angew. Chemie Int. Ed.* 55 (2016) 6290–6294.

30. S. V. Devaguptapu, S. Hwang, S. Karakalos, S. Zhao, S. Gupta, D. Su, H. Xu, G. Wu, Morphology control of carbon-free spinel $NiCo_2O_4$ catalysts for enhanced bifunctional oxygen reduction and evolution in alkaline media, *ACS Appl. Mater. Interfaces.* 9 (2017) 44567–44578.
31. J. Liu, H. Yuan, Z. Wang, J. Li, M. Yang, L. Cao, G. Liu, D. Qian, Z. Lu, Self-supported nickel iron oxide nanospindles with high hydrophilicity for efficient oxygen evolution, *Chem. Commun.* 55 (2019) 10860–10863.
32. X. Chen, X. Zhang, L. Zhuang, W. Zhang, N. Zhang, H. Liu, T. Zhan, X. Zhang, X. She, D. Yang, Multiple vacancies on (111) facets of single-crystal $NiFe_2O_4$ spinel boost electrocatalytic oxygen evolution reaction, *Chem. An Asian J.* 15 (2020) 3995–3999.
33. G. Ou, F. Wu, K. Huang, N. Hussain, D. Zu, H. Wei, B. Ge, H. Yao, L. Liu, H. Li, Y. Shi, H. Wu, Boosting the electrocatalytic water oxidation performance of $CoFe_2O_4$ nanoparticles by surface defect engineering, *ACS Appl. Mater. Interf.* 11 (2019) 3978–3983.
34. Y. Huang, W. Yang, Y. Yu, S. Hao, Ordered mesoporous spinel $CoFe_2O_4$ as efficient electrocatalyst for the oxygen evolution reaction, *J. Electroanal. Chem.* 840 (2019) 409–414.
35. C. Zhang, S. Bhoyate, C. Zhao, P. Kahol, N. Kostoglou, C. Mitterer, S. Hinder, M. Baker, G. Constantinides, K. Polychronopoulou, C. Rebholz, R. Gupta, Electrodeposited nanostructured $CoFe_2O_4$ for overall water splitting and supercapacitor applications, *Catalysts.* 9 (2019) 176.
36. Z. Zhang, D. Zhou, S. Zou, X. Bao, X. He, One-pot synthesis of $MnFe_2O_4$/C by microwave sintering as an efficient bifunctional electrocatalyst for oxygen reduction and oxygen evolution reactions, *J. Alloys Compd.* 786 (2019) 565–569.
37. K. Zeng, W. Li, Y. Zhou, Z. Sun, C. Lu, J. Yan, J. H. Choi, R. Yang, Multilayer hollow $MnCo_2O_4$ microsphere with oxygen vacancies as efficient electrocatalyst for oxygen evolution reaction, *Chem. Eng. J.* (2020) 127831.
38. S. Hirai, S. Yagi, A. Seno, M. Fujioka, T. Ohno, T. Matsuda, Enhancement of the oxygen evolution reaction in Mn^{3+}-based electrocatalysts: Correlation between Jahn-Teller distortion and catalytic activity, *RSC Adv.* 6 (2016) 2019–2023.
39. L. Han, L. Guo, C. Dong, C. Zhang, H. Gao, J. Niu, Z. Peng, Z. Zhang, Ternary mesoporous cobalt-iron-nickel oxide efficiently catalyzing oxygen/hydrogen evolution reactions and overall water splitting, *Nano Res.* 12 (2019) 2281–2287.
40. S. Anantharaj, S. R. Ede, K. Sakthikumar, K. Karthick, S. Mishra, S. Kundu, Recent trends and perspectives in electrochemical water splitting with an emphasis on sulfide, selenide, and phosphide catalysts of Fe, Co, and Ni: A review, *ACS Catal.* 6 (2016) 8069–8097.
41. C. H. Lai, M. Y. Lu, L. J. Chen, Metal sulfide nanostructures: Synthesis, properties and applications in energy conversion and storage, *J. Mater. Chem.* 22 (2012) 19–30.
42. S. B. Kale, P. T. Babar, J. H. Kim, C. D. Lokhande, Synthesis of one dimensional Cu_2S nanorods using a self-grown sacrificial template for the electrocatalytic oxygen evolution reaction (OER), *New J. Chem.* 44 (2020) 8771–8777.
43. S. B. Kale, A. C. Lokhande, R. B. Pujari, C. D. Lokhande, Cobalt sulfide thin films for electrocatalytic oxygen evolution reaction and supercapacitor applications, *J. Colloid Interface Sci.* 532 (2018) 491–499.
44. H. Liu, F. X. Ma, C. Y. Xu, L. Yang, Y. Du, P. P. Wang, S. Yang, L. Zhen, Sulfurizing-induced hollowing of Co_9S_8 microplates with nanosheet units for highly efficient water oxidation, *ACS Appl. Mater. Interfaces.* 9 (2017) 11634–11641.
45. X. Zhao, J. Jiang, Z. Xue, C. Yan, T. Mu, An ambient temperature, CO_2-assisted solution processing of amorphous cobalt sulfide in a thiol/amine based quasi-ionic liquid for oxygen evolution catalysis, *Chem. Commun.* 53 (2017) 9418–9421.

46. X. Feng, Q. Jiao, T. Liu, Q. Li, M. Yin, Y. Zhao, H. Li, C. Feng, W. Zhou, Facile synthesis of Co_9S_8 hollow spheres as a high-performance electrocatalyst for the oxygen evolution reaction, *ACS Sustain. Chem. Eng.* 6 (2018) 1863–1871.
47. K. Nan, H. Du, L. Su, C.M. Li, Directly electrodeposited cobalt sulfide nanosheets as advanced catalyst for oxygen evolution reaction, *ChemistrySelect.* 3 (2018) 7081–7088.
48. S. Ju, Y. Liu, H. Chen, F. Tan, A. Yuan, X. Li, G. Zhu, In situ surface chemistry engineering of cobalt-sulfide nanosheets for improved oxygen evolution activity, *ACS Appl. Energy Mater.* 2 (2019) 4439–4449.
49. K. Wan, J. Luo, C. Zhou, T. Zhang, J. Arbiol, X. Lu, B. W. Mao, X. Zhang, J. Fransaer, Hierarchical porous Ni_3S_4 with enriched high-valence Ni sites as a robust electrocatalyst for efficient oxygen evolution reaction, *Adv. Funct. Mater.* 29 (2019) 1900315.
50. N. K. Chaudhari, A. Oh, Y. J. Sa, H. Jin, H. Baik, S. G. Kim, S. J. Lee, S. H. Joo, K. Lee, Morphology controlled synthesis of 2-D Ni–Ni_3S_2 and Ni_3S_2 nanostructures on Ni foam towards oxygen evolution reaction, *Nano Converg.* 4 (2017) 7.
51. S. B. Kale, V. C. Lokhande, S. J. Marje, U. M. Patil, J. H. Kim, C. D. Lokhande, Chemically deposited Co_3S_4 thin film: Morphology dependant electrocatalytic oxygen evolution reaction, *Appl. Phys. A Mater. Sci. Process.* 126 (2020) 206.
52. X. Shang, X. Li, W. H. Hu, B. Dong, Y. R. Liu, G. Q. Han, Y. M. Chai, Y. Q. Liu, C. G. Liu, In situ growth of Ni_xS_y controlled by surface treatment of nickel foam as efficient electrocatalyst for oxygen evolution reaction, *Appl. Surf. Sci.* 378 (2016) 15–21.
53. X. Zou, Y. Wu, Y. Liu, D. Liu, W. Li, L. Gu, H. Liu, P. Wang, L. Sun, Y. Zhang, In situ generation of bifunctional, efficient fe-based catalysts from mackinawite iron sulfide for water splitting, *Chem.* 4 (2018) 1139–1152.
54. R. B. Pujari, G. S. Gund, S. J. Patil, H. S. Park, D. W. Lee, Anion-exchange phase control of manganese sulfide for oxygen evolution reaction, *J. Mater. Chem. A.* 8 (2020) 3901–3909.
55. L. L. Feng, M. Fan, Y. Wu, Y. Liu, G. D. Li, H. Chen, W. Chen, D. Wang, X. Zou, Metallic Co_9S_8 nanosheets grown on carbon cloth as efficient binder-free electrocatalysts for the hydrogen evolution reaction in neutral media, *J. Mater. Chem. A.* 4 (2016) 6860–6867.
56. Z. F. Huang, J. Song, K. Li, M. Tahir, Y. T. Wang, L. Pan, L. Wang, X. Zhang, J. J. Zou, Hollow cobalt-based bimetallic sulfide polyhedra for efficient all-pH-value electrochemical and photocatalytic hydrogen evolution, *J. Am. Chem. Soc.* 138 (2016) 1359–1365.
57. X. Hu, R. Wang, P. Sun, Z. Xiang, X. Wang, Tip-welded ternary $FeCo_2S_4$ nanotube arrays on carbon cloth as binder-free electrocatalysts for highly efficient oxygen evolution, *ACS Sustain. Chem. Eng.* 7 (2019) 19426–19433.
58. F. Zhang, M. Cho, T. Eom, C. Kang, H. Lee, Facile synthesis of manganese cobalt sulfide nanoparticles as high-performance supercapacitor electrode, *Ceram. Int.* 45 (2019) 20972–20976.
59. J. Li, W. Xu, J. Luo, D. Zhou, D. Zhang, L. Wei, P. Xu, D. Yuan, Synthesis of 3D hexagram-like cobalt–manganese sulfides nanosheets grown on nickel foam: A bifunctional electrocatalyst for overall water splitting, *Nano-Micro Lett.* 10 (2017) 6.
60. X. Zhang, C. Si, X. Guo, R. Kong, F. Qu, A $MnCo_2S_4$ nanowire array as an earth-abundant electrocatalyst for an efficient oxygen evolution reaction under alkaline conditions, *J. Mater. Chem. A.* 5 (2017) 17211–17215.
61. S. B. Kale, A. Bhardwaj, V. C. Lokhande, D. M. Lee, S. H. Kang, J. H. Kim, C. D. Lokhande, Amorphous cobalt-manganese sulfide electrode for efficient water oxidation: Meeting the fundamental requirements of an electrocatalyst, *Chem. Eng. J.* 405 (2021) 126993.

62. X. Zhao, X. Shang, Y. Quan, B. Dong, G. Q. Han, X. Li, Y. R. Liu, Q. Chen, Y. M. Chai, C. G. Liu, Electrodeposition-solvothermal access to ternary mixed metal Ni-Co-Fe sulfides for highly efficient electrocatalytic water oxidation in alkaline media, *Electrochim. Acta.* 230 (2017) 151–159.
63. S. K. Oh, H. W. Kim, Y. K. Kwon, M. J. Kim, E. A. Cho, H. S. Kwon, Porous Co-P foam as an efficient bifunctional electrocatalyst for hydrogen and oxygen evolution reactions, *J. Mater. Chem. A.* 4 (2016) 18272–18277.
64. B. Zhang, Y. H. Lui, L. Zhou, X. Tang, S. Hu, An alkaline electro-activated Fe-Ni phosphide nanoparticle-stack array for high-performance oxygen evolution under alkaline and neutral conditions, *J. Mater. Chem. A.* 5 (2017) 13329–13335.
65. D. Li, H. Baydoun, B. Kulikowski, S. L. Brock, Boosting the catalytic performance of iron phosphide nanorods for the oxygen evolution reaction by incorporation of manganese, *Chem. Mater.* 29 (2017) 3048–3054.
66. P. Wang, F. Song, R. Amal, Y. H. Ng, X. Hu, Efficient water splitting catalyzed by cobalt phosphide-based nanoneedle arrays supported on carbon cloth, *ChemSusChem.* 9 (2016) 472–477.
67. Z. Jin, P. Li, D. Xiao, Metallic Co_2P ultrathin nanowires distinguished from CoP as robust electrocatalysts for overall water-splitting, *Green Chem.* 18 (2016) 1459–1464.
68. P. E. R. Blanchard, A. P. Grosvenor, R. G. Cavell, A. Mar, X-ray photoelectron and absorption spectroscopy of metal-rich phosphides M_2P and M_3P (M=Cr-Ni), *Chem. Mater.* 20 (2008) 7081–7088.
69. P. Wang, Z. Pu, Y. Li, L. Wu, Z. Tu, M. Jiang, Z. Kou, I. S. Amiinu, S. Mu, Iron-doped nickel phosphide nanosheet arrays: An efficient bifunctional electrocatalyst for water splitting, *ACS Appl. Mater. Interfaces.* 9 (2017) 26001–26007.
70. J. Chen, Y. Li, G. Sheng, L. Xu, H. Ye, X. Z. Fu, R. Sun, C. P. Wong, Iron-doped nickel phosphide nanosheets in situ grown on nickel submicrowires as efficient electrocatalysts for oxygen evolution reaction, *ChemCatChem.* 10 (2018) 2248–2253.
71. F. Li, Y. Bu, Z. Lv, J. Mahmood, G. F. Han, I. Ahmad, G. Kim, Q. Zhong, J. B. Baek, Porous cobalt phosphide polyhedrons with iron doping as an efficient bifunctional electrocatalyst, *Small.* 13 (2017) 1701167.
72. X. Zhang, X. Zhang, H. Xu, Z. Wu, H. Wang, Y. Liang, Iron-doped cobalt monophosphide nanosheet/carbon nanotube hybrids as active and stable electrocatalysts for water splitting, *Adv. Funct. Mater.* 27 (2017) 1606635.
73. K. Liu, C. Zhang, Y. Sun, G. Zhang, X. Shen, F. Zou, H. Zhang, Z. Wu, E. C. Wegener, C. J. Taubert, J. T. Miller, Z. Peng, Y. Zhu, High-performance transition metal phosphide alloy catalyst for oxygen evolution reaction, *ACS Nano.* 12 (2018) 158–167.
74. J. Xu, J. Li, D. Xiong, B. Zhang, Y. Liu, K. H. Wu, I. Amorim, W. Li, L. Liu, Trends in activity for the oxygen evolution reaction on transition metal (M=Fe, Co, Ni) phosphide pre-catalysts, *Chem. Sci.* 9 (2018) 3470–3476.
75. J. Ryu, N. Jung, J. H. Jang, H. J. Kim, S. J. Yoo, In situ transformation of hydrogen-evolving CoP nanoparticles: Toward efficient oxygen evolution catalysts bearing dispersed morphologies with Co-oxo/hydroxo molecular units, *ACS Catal.* 5 (2015) 4066–4074.
76. L. A. Stern, L. Feng, F. Song, X. Hu, Ni_2P as a Janus catalyst for water splitting: The oxygen evolution activity of Ni_2P nanoparticles, *Energy Environ. Sci.* 8 (2015) 2347–2351.
77. A. Dutta, A. K. Samantara, S. K. Dutta, B. K. Jena, N. Pradhan, Surface-oxidized dicobalt phosphide nanoneedles as a nonprecious, durable, and efficient oer catalyst, *ACS Energy Lett.* 1 (2016) 169–174.
78. J. Chang, Y. Xiao, M. Xiao, J. Ge, C. Liu, W. Xing, Surface oxidized cobalt-phosphide nanorods as an advanced oxygen evolution catalyst in alkaline solution, *ACS Catal.* 5 (2015) 6874–6878.

79. B. Y. Guan, L. Yu, X. W. D. Lou, General synthesis of multishell mixed-metal oxyphosphide particles with enhanced electrocatalytic activity in the oxygen evolution reaction, *Angew. Chemie - Int. Ed.* 56 (2017) 2386–2389.
80. D. D. Babu, Y. Huang, G. Anandhababu, M. A. Ghausi, Y. Wang, Mixed-metal-organic framework self-template synthesis of porous hybrid oxyphosphides for efficient oxygen evolution reaction, *ACS Appl. Mater. Interfaces.* 9 (2017) 38621–38628.
81. F. Song, X. Hu, Exfoliation of layered double hydroxides for enhanced oxygen evolution catalysis, *Nat. Commun.* 5 (2014) 1–9.
82. X. Lu, C. Zhao, Electrodeposition of hierarchically structured three-dimensional nickel-iron electrodes for efficient oxygen evolution at high current densities, *Nat. Commun.* 6 (2015) 1–7.
83. A. Guzmán-Vargas, J. Vazquez-Samperio, M. A. Oliver-Tolentino, N. Nava, N. Castillo, M. J. Macías-Hernández, E. Reguera, Influence of cobalt on electrocatalytic water splitting in NiCoFe layered double hydroxides, *J. Mater. Sci.* 53 (2018) 4515–4526.
84. D. Zhou, Z. Cai, Y. Jia, X. Xiong, Q. Xie, S. Wang, Y. Zhang, W. Liu, H. Duan, X. Sun, Activating basal plane in NiFe layered double hydroxide by Mn^{2+} doping for efficient and durable oxygen evolution reaction, *Nanoscale Horizons.* 3 (2018) 532–537.
85. L. Wu, L. Yu, F. Zhang, D. Wang, D. Luo, S. Song, C. Yuan, A. Karim, S. Chen, Z. Ren, Facile synthesis of nanoparticle-stacked tungsten-doped nickel iron layered double hydroxide nanosheets for boosting oxygen evolution reaction, *J. Mater. Chem. A.* 8 (2020) 8096–8103.
86. J. He, P. Xu, J. Sun, Fe and B codoped nickel zinc layered double hydroxide for boosting the oxygen evolution reaction, *ACS Sustain. Chem. Eng.* 8 (2020) 2931–2938.
87. X. Li, Q. Hu, Y. Xu, X. Liang, J. Feng, K. Zhao, G. Guan, Y. Jiang, X. Hao, K. Tang, Generation of oxygen vacancies in NiFe LDH electrocatalysts by ultrasound for enhancing the activity toward oxygen evolution reaction, *Carbon Resour. Convers.* 4 (2021) 76–83.

13 Innovative Approach for Real-Time *P–V* Curve Identification

Design-to-Application

Mahmadasraf A. Mulla and Ashish K. Panchal
Sardar Vallabhbhai National Institute of Technology

CONTENTS

13.1 INTRODUCTION

Increasing cost, depleting resources and threats to the environment of the fossil fuel–based electricity generations motivate the humankind to replace these technologies with renewable energy. Solar photovoltaics (PV) electricity generation is the most popular technology because of the uniform sunlight distribution on the earth's surface and the rapid decrease in the cost of the PV power generation. The Government of India launched Jawaharlal Nehru National Solar Mission (JNNSM) in 2009 to promote PV technology-related activities [1,2] The JNNSM target is to generate 100 GW by the grid-connected and the roof-top PV systems by 2022.

The PV output power varies continuously with the change in light intensity and temperature for the entire day, and on the other end, the load demands regulated power. Therefore, an electrical interface is required to perform this task, i.e. a power

DOI: 10.1201/9781003220176-13

electronics converter with its control. According to the maximum power transfer theorem, the converter performs the impedance matching between the PV and the load. The converter requires a control signal to match the impedance for different PV operating conditions. The power–voltage (P–V) curve of a PV module exhibits a concave shape with a unique peak (also known as the maximum power point (MPP)) (V_{MP}, I_{MP}) for one operating condition. The peak position moves with the change in the PV operating condition. Hence, the converter matches the impedance continuously by setting the PV voltage at V_{MP}. The controller generates a signal (a duty cycle for the converter) corresponding to every change in V_{MP} and the controller is governed by an MPP tracking (MPPT) algorithm [3,4]. This mathematical operation is generally performed in a micro-controller. The perturb-and-observe (P&O) and incremental conductance (IC) algorithms are frequently employed for the MPPT [5].

The PV behaviour is significantly influenced by the light intensity and temperature. This adds complexity to the PV power prediction, system design, system operation and control. Therefore, a basic theoretical and experimental understanding of a PV system is essential. The PV system study starts with building a basic PV circuit using the five parameters of a single-diode model. Subsequently, the basic circuit is extended to simulate a large PV array [6]. Detailed PV modelling using the single-diode-model and datasheet information is explained in [7,8]. Next, the PV array model has to be connected to the model of a DC–DC converter [9]. In [10], a theoretical assessment of three types of DC–DC converters (buck, boost and buck–boost), and in [11], a numerical analysis of a high gain DC–DC converter, are examined in connection with a PV array. Through simulation studies, the buck–boost converter is recommended for the PV to extract the maximum power in all operating conditions. In [12], an implementation of the P&O algorithm in the MATLAB/SIMULINK is demonstrated. A similar MATLAB simulation study for a stand-alone PV system and grid-integrated PV system is presented in [13–15]. These published PV works are either theoretical assessments or PV performance evaluations by the numerical simulations obtained in the MATLAB/SIMULINK platform. In the PV simulation studies, one of the drawbacks is the use of an iterative solver [16,17] (Newton-Raphson, Gauss-Seidel, Secant, False position) as the single-diode model is implicit in nature. The accuracy of the PV model is greatly influenced by the solver used. In addition, the number of steps used in an iterative solver decides the computation time, and the PV modelling becomes time-consuming. On the other hand, several assumptions in the PV model (such as series resistance/shunt resistance of the single-diode-model neglected) lead to inaccuracy in the solution. The assumptions add an error in the estimation of the maximum power of a PV module, subsequently; the PV module extension to PV array accumulates the errors in total maximum power. Thus, the PV design and analysis in the simulation studies lead to over/under-estimation. In such situations, the experimental study becomes essential to learn the PV behaviour and an estimation of the maximum power in different climatic conditions. In this direction, Fannakh et al. [18] demonstrated a case study on an experimental hardware design for the PV MPPT using fuzzy logic control algorithm. An Arduino Mega 2560 controller board in connection with MATLAB platform is used to implement the fuzzy logic. However, the complexity involved in fuzzy rules makes the MPP estimation lengthy. The use of a resistive load does not make the learner visualize maximum power transfer from the PV module. In order to avoid these disadvantages, the present work proposes a hardware implementation of a PV system that employs a simple MPPT algorithm such as P&O and IC. A 40 W module is interfaced to load via a DC–DC converter. The algorithms are implemented in ARM Cortex-M4 32-bit micro-controller. Further, a pure resistive load is replaced with a bulb bank so as to visualize the power transfer mechanism by observing the change in the illumination. The salient features of the proposed experimental setup are (i) learning a boost converter in open-loop and close-loop operation; (ii) tracing the current–voltage (*I–V*)/*P–V* curves of a PV module in real-time by continuously increasing the converter duty ratio and (iii) the realization of maximum power extraction by observing the illumination level of the bulb bank.

This chapter has the following structure: the details of the PV module characteristics and MPPT are described in Section 13.2. Section 13.3 presents the experimental prototype design along with system parameters. The details of experimentation along with results and discussion are given in Section 13.4. The conclusions are derived in Section 13.5.

13.2 PV MODULE CHARACTERISTICS AND MPPT

A PV module *I–V* relation is represented in form of the single-diode-model equation as follows [7,19–21]:

$$I = I_L - I_0\left(e^{\frac{q(V+IR_s)}{nKT}} - 1\right) - \frac{V + IR_s}{R_{sh}} \tag{13.1}$$

where q is the electronic charge = 1.602176487 E − 19, K is Boltzmann's constant = 1.3806504 E − 23, and T is the temperature in K.

The five parameters of the single-diode-model are the following: I_L is the light generated current, I_0 is the reverse saturation current, R_s is the series resistance, R_{sh} is the shunt resistance, and n is the ideality factor.

The PV output power (P_{PV}) is highly nonlinear as a function of its output voltage (V_{PV}) and exhibits a unique MPP as illustrated in the *I–V* and *P–V* curves in Figure 13.1. The output characteristics vary with the sunlight intensity and the operating temperature of the PV module as shown in Figure 13.2. Hence, the MPP also changes with a change in the operating climatic conditions [22,23].

When an electrical load is directly connected to a PV module, the PV system works with a low efficiency as the working point on the *I–V* curve is determined by the load. As illustrated in Figure 13.3, for a fixed load R_{LA}, the operating point is positioned at A. For different load $R_{LB} >> R_{LA}$, the working point moves to point B. With the load variation from R_{LA} to R_{LB}, a unique point occurs when the PV module supplies the maximum possible power. For this, the load resistance is the same as the optimal resistance ($R_L = R_{opt}$).

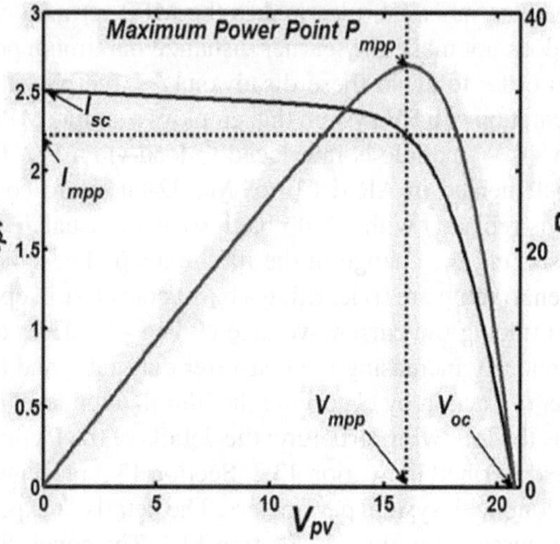

FIGURE 13.1 The *I–V* and *P–V* characteristics of a PV module.

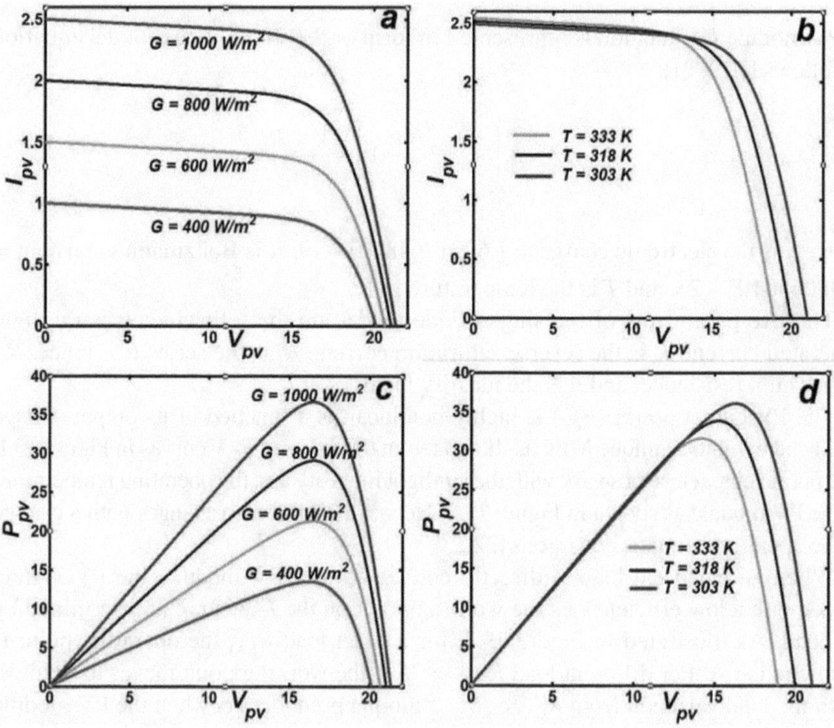

FIGURE 13.2 The *I–V* characteristics with variation in (a) sunlight intensity, (b) temperature and *P–V* characteristics with variation in (c) sunlight intensity and (d) temperature.

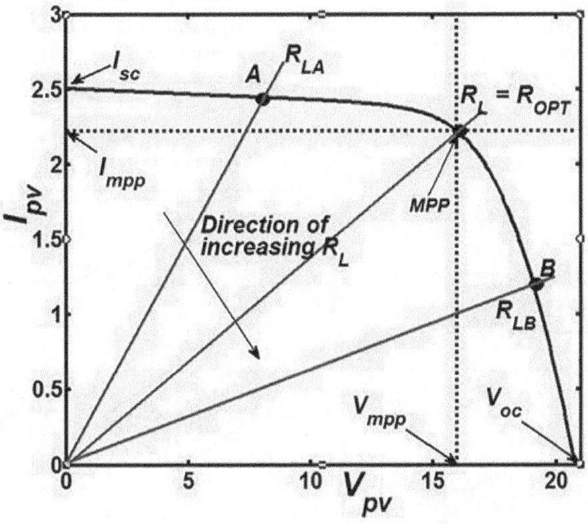

FIGURE 13.3 The PV module operations with different loads.

However, it is not always feasible to choose a constant load that is equal to the optimal load in real working conditions, and even if it is possible, the MPP moves with the changing environmental conditions. To conquer this issue, the MPPT controller has to be employed to compel the PV module's operation to the MPP. The MPPT controller makes the PV module provide more than 97% of the power with proper optimization. The MPPT matches the load resistance with the optimal resistance of the PV [24] by adjusting the PV voltage at V_{mpp}. Generally, the MPPT operation is performed with help of a converter by adjusting the input voltage to the MPP. Thus, it provides impedance matching for the MPP [24]. When the PV module and the load are interfaced by a boost converter, the operating point relies on the effective impedance R_{in}, R_{in} is a function of the load impedance R_L and converter duty ratio (α). Therefore, α is tuned to change R_{in} to match R_{opt} for all climatic conditions.

In order to decide α corresponding to V_{mpp}, a numerical process is required to be performed by a micro-controller. On the *P–V* characteristic (Figure 13.1) for the MPP (V_{mpp}, P_{mpp}), the rate of change of P_{PV} as a function of V_{PV} is zero, i.e. $dP_{\text{PV}}/dV_{\text{PV}}=0$. This condition is used to determine correct α. Normally, the P&O and IC algorithms are used to determine duty ratio α [5].

As illustrated in Figure 13.4a, the P&O algorithm compares a new power $P_{\text{PV}}(n)$ with an old power $P_{\text{PV}}(n-1)$ along with a new voltage $V_{\text{PV}}(n)$ with an old $V_{\text{PV}}(n-1)$. Then, the voltage V_{PV} is increased/decreased as shown in the flowchart to reach the MPP. For the IC algorithm, a new current $I_{\text{PV}}(n)$ is compared with an old current $I_{\text{PV}}(n-1)$ along with a new voltage $V_{\text{PV}}(n)$ with an old $V_{\text{PV}}(n-1)$. Then, the IC ($\Delta I_{\text{PV}}(n)/\Delta V_{\text{PV}}(n)$) is compared with the instantaneous conductance ($-I_{\text{PV}}(n)/V_{\text{PV}}(n)$) as depicted in the flowchart in Figure 13.4b. Finally, the voltage V_{PV} is increased/decreased to reach the MPP. An experimental prototype developed for studying real-time curve tracing and MPPT is discussed in following Section 13.3.

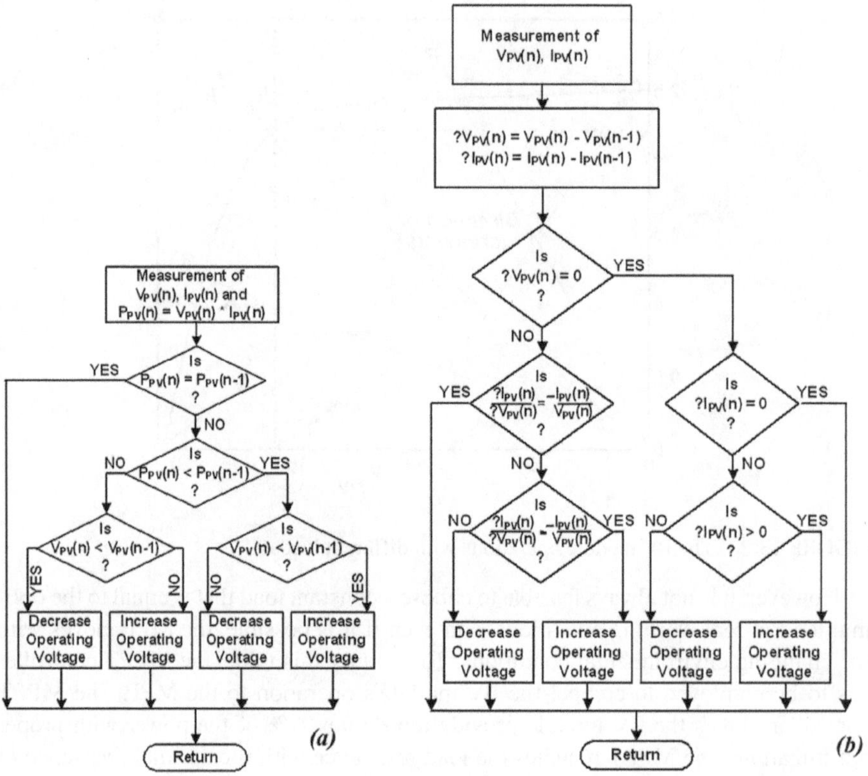

FIGURE 13.4 The flow chart for MPPT methods: (a) the perturb-and-observe and (b) the incremental conductance.

13.3 EXPERIMENTAL PROTOTYPE AND SYSTEM PARAMETERS

An experimental prototype of a boost converter-based solar MPPT is developed. A 40 W polycrystalline silicon PV module (WS40 manufactured by Waaree Energies Pvt. Limited, India) is connected to a set of 40 W bulb bank through a DC–DC boost converter as illustrated in Figure 13.5a. The control circuit is realized using STM32F407VGT6, which is ARM Cortex-M4 32-bit micro-controller. In the prototype hardware, the bulb bank is specifically used to visualize the variation in the bulb intensity with a variation in the PV output power.

Table 13.1 lists the components used in the experimental setup, and Figure 13.5b displays the photograph of the experimental setup mentioning different components. The PV current/voltage and the boost converter output current/voltage signals are obtained and interfaced at the analogue-to-digital converter (ADC) of the micro-controller. A control signal for the relay and the semiconductor switch of the boost converter are generated from the micro-controller programme. The micro-controller also sends all signals serially that are collected in MATLAB-based utility for the visual observations and analysis [25].

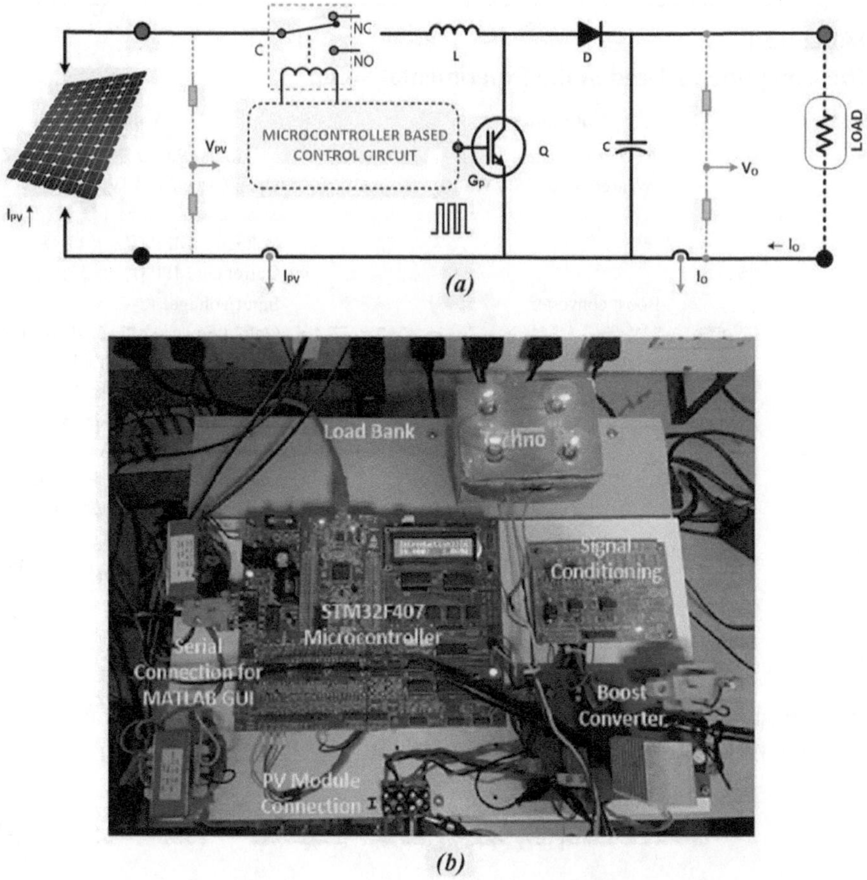

FIGURE 13.5 Experimental setup: (a) PV module connected to a load through a DC–DC boost converter and (b) photograph of the setup.

The STM32F407VGT6 is a high-end ARM core, Cortex-M4 32-bit micro-controller working with a clock frequency of 168 MHz [26]. The Cortex-M4 core has a floating-point unit that handles single-precision data-processing instructions and supports a full set of DSP instructions. Four numbers of 12-bit, ADC channels are configured to obtain two voltages and two currents. The boost converter is implemented using IGBT STGW30NC120HD and diode RHRP30120. The gate pulses provided from the micro-controller are optically isolated and processed by the driver circuit before it is connected to IGBT. The details of boost converter design and control circuit implementation are discussed in the following sub-sections.

13.3.1 Boost Converter for MPPT

The basic structure of a DC–DC boost converter used in the PV system for MPPT is shown in Figure 13.5a. A relay is connected between PV system and boost converter

TABLE 13.1
The Components Used in the Experimental Setup

Sr. No.	Component	Value
1	40 W solar panel (Waaree Model No. WS40)	Maximum Power [P_{mpp}]: 40.00 W Open-Circuit Volt [V_{oc}]: 21.37 V Short-Circuit Current [I_{sc}]: 2.50 A Voltage at MPP [V_{mpp}]: 17.18 V Current at MPP [I_{mpp}]: 2.30 A
2	Boost converter	Input Voltage: 12–22 V Output Power/Voltage: 40 W/36 V Switching Frequency: 40 kHz Output Voltage Ripple: 2% Output Current Ripple: 2% Inductor Value: 5 mH Output Capacitor Value: 47 uF IGBT: STGW30NC120HD Diode: RHRP30120
	Controller	ARM Cortex-M4 micro-controller STM32F407VGT MCU @168 MHz
4	Load	40 W bulb bank

to establish an open-circuit point during curve tracing. The output voltage of the boost converter can be presented as follows:

$$V_o = \frac{1}{1-\alpha} V_{\text{pv}} \quad \text{and} \quad I_o = (1-\alpha) I_{\text{pv}} \tag{13.2}$$

with V_o and I_o are the boost converter output voltage and current. The V_{pv} and I_{pv} are PV module voltage and current, respectively.

Dividing both the sides in Eq. (13.2), the impedance characteristics are described as

$$R_{\text{in}} = (1-\alpha)^2 R_L \tag{13.3}$$

where R_{in} and R_L are the input and the load resistances of the boost converter, respectively.

The boost converter works with two modes, the continuous conduction mode (CCM) and the discontinuous conduction mode (DCM). In the CCM, the current through the inductor is always positive. Moreover, the CCM is the most appropriate mode for higher efficiency and better operations of the switches and the passive elements.

The inductance L acts as a filter to smooth out the ripple contents in the currents of the converter. The L_{bo} value specifies the boundary condition for L for a specific α and a selected switching frequency f so as to ensure the converter operation in CCM. The boundary value for the inductance is expressed as

$$L_{bo} = \frac{V_{i\max}\alpha_{\min}}{\Delta I_L\ f} \tag{13.4}$$

The capacitance C reduces the ripple voltage contents in the converter output. The C_{bo} value specifies the boundary condition for C for a given α and selected switching frequency f to decrease ripple in the converter output voltage to the desired value [27]. The boundary value for the capacitance is expressed as

$$C_{bo} = \frac{I_o\ \alpha_{\max}}{\Delta V_o f} \tag{13.5}$$

where V_{imax} is the maximum converter input voltage; α_{min} and α_{max} are the minimum and maximum operating duty ratios, respectively.

13.3.2 Design of 40 W Boost Converter for MPPT

The 40 W boost converter required for MPPT of 40 W panel is designed as per the following steps:

1. Design Specifications: input voltage range 12–22 V, output voltage = 36 V, switching frequency f = 40 kHz, the ripple in the output voltage (ΔV_O) = 2%, and the ripple in inductor current (ΔI_L) = 2%.
2. For a 40 W panel having V_{mpp} and I_{mpp} as listed in Table 13.1, 18 V and 40 W bulb bank is selected.
3. Using Eq. (13.4), the calculated minimum value of the inductor is 3.21 mH. The selected value of the inductor used in the prototype is 5 mH, 3 Amp.
4. Considering the selected voltage ripple in Eq. (13.5), the minimum value of the capacitor is calculated as 25.72 μF. The nearby selected value of the capacitor is 47 μF, 50 V.

These values of energy storage elements will meet the designed specification of current and voltage ripple and also assure the operation of the converter in CCM mode.

13.3.3 Control Circuit Implementation

The control circuit is an important building block to operate the boost converter in different modes. The control circuit is designed to operate the boost converter in the following four modes:

1. Open-Loop Mode: In this mode, a user sets a duty ratio and the boost converter is operated with the set duty ratio.
2. Closed-Loop Mode: In this mode, a user sets the desired output voltage and the controller decides the required duty ratio.
3. *I–V*/*P–V* Characteristics Mode: In this mode, the PV module characteristics are traced in the present climatic conditions.

4. MPPT Mode: In this mode, the PV module operates at the MPP using P&O and IC algorithms.

The block diagram of the control scheme is shown in Figure 13.6. Totally, four analogue quantities – the PV current, the PV voltage, the boost converter output current and voltage – are sensed and are interfaced with the controller. The aforementioned four modes of operation are given for selection. A reference input for open-loop mode is a duty ratio and the desired converter output is the reference for closed-loop mode. There is no reference input for *I–V*/*P–V* characteristic and MPPT mode.

The details of control circuit development are shown in Figure 13.6. In the open-loop mode, the duty ratio is compared with the high-frequency carrier wave to generate the Pulse Width Modulation (PWM) pulses. In the closed-loop mode, the reference value of boost converter output voltage is compared with the actual voltage, and the error is processed through proportional–integral (PI) controller to decide the duty ratio. In curve tracing mode, the duty ratio is varied from 0% to 100%, and PV voltage and currents are captured. Furthermore, for the open-circuit point, the relay is operated to disconnect the boost converter from panel terminal. In the last mode,

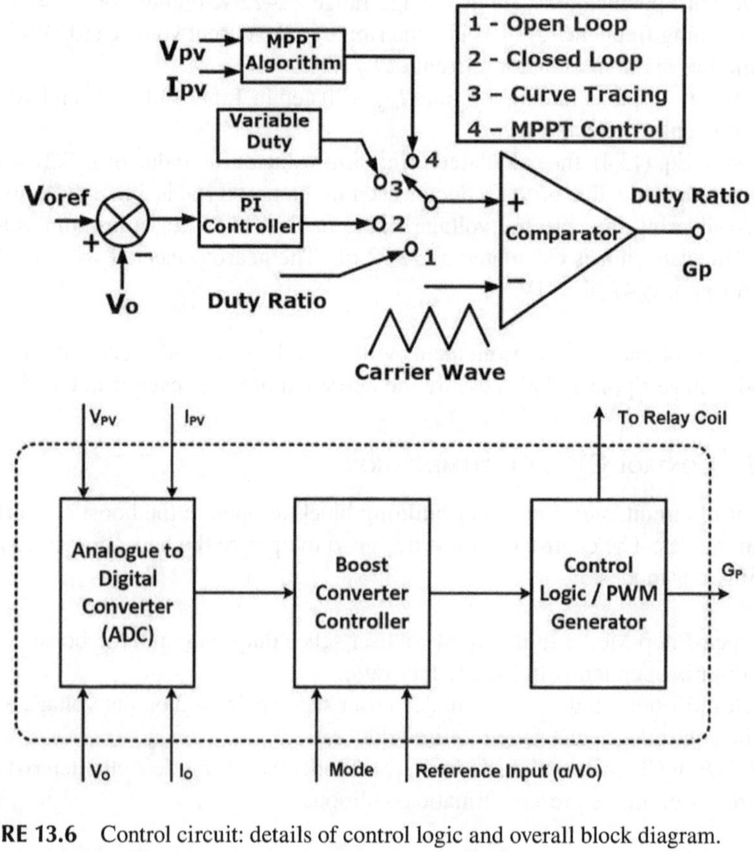

FIGURE 13.6 Control circuit: details of control logic and overall block diagram.

i.e. MPPT mode, the P&O or IC algorithms are implemented to decide the duty ratio, which is compared with the carrier wave to generate the PWM pulses.

13.4 RESULTS AND DISCUSSION

With a view to demonstrate the performance of the boost converter, the performance of the PV module and the boost converter-based solar MPPT, the aforementioned four modes of operations are formed.

13.4.1 Boost Converter in an Open Loop

In this mode of operation, the duty ratio of the boost converter is varied manually in small steps and its effect on the output of PV module and the output of the boost converter is observed. A MATLAB user interface is developed to operate the boost converter in an open loop. Figure 13.7 shows the waveforms of the boost converter for two different duty ratios 30% and 70% when the boost converter is operated in the open-loop mode. It is observed that the converter output voltage is 19.5 and 44.0 V for 30% and 70% duty ratios, respectively. Therefore, by adjusting the duty ratio of the converter, the output voltage is controlled.

By operating the boost converter with different duty ratios, the effective impedance at the panel terminal is varying. The intensity of the bulb bank connected across the output reflects the power variations. For a particular value of duty ratio, the intensity is maximum, which needs to be adjusted automatically throughout the day for MPPT.

The variation of PV output voltage, current and power with duty ratio can also be used to plot the *I–V*/*P–V* characteristic at working ambient conditions, i.e. radiation and temperature. One additional reading of open-circuit voltage of the panel is required for plotting the complete *I–V*/*P–V* characteristic.

13.4.2 Boost Converter in a Closed Loop

Normally, the boost converter is used for providing a regulated output voltage from the variable input voltage. In this mode of operation, the boost converter duty ratio is automatically adjusted by the controller to produce the required output voltage. An error in the reference output voltage and the sensed output voltage of the boost converter is processed by PI controller to produce the required duty ratio for a gate pulse of the boost converter. For each value of the desired output voltage, the PV voltage, PV current, output voltage, and output current are observed.

The MATLAB user interfaces were developed to operate the boost converter in a closed loop. The waveforms of the boost converter are shown in Figure 13.8. For the desired output voltage of 20 V, the boost converter duty ratio is automatically set to 31% (Figure 13.8a); and for a desired output voltage of 40 V, the boost converter duty ratio is automatically set to 66% (Figure 13.8b). The automatically adjusted value of the duty ratio is calculated using the positive width measured on the oscilloscope.

The variation of the output power of PV module can be analysed further to understand that such control of the boost converter does not result in extracting the

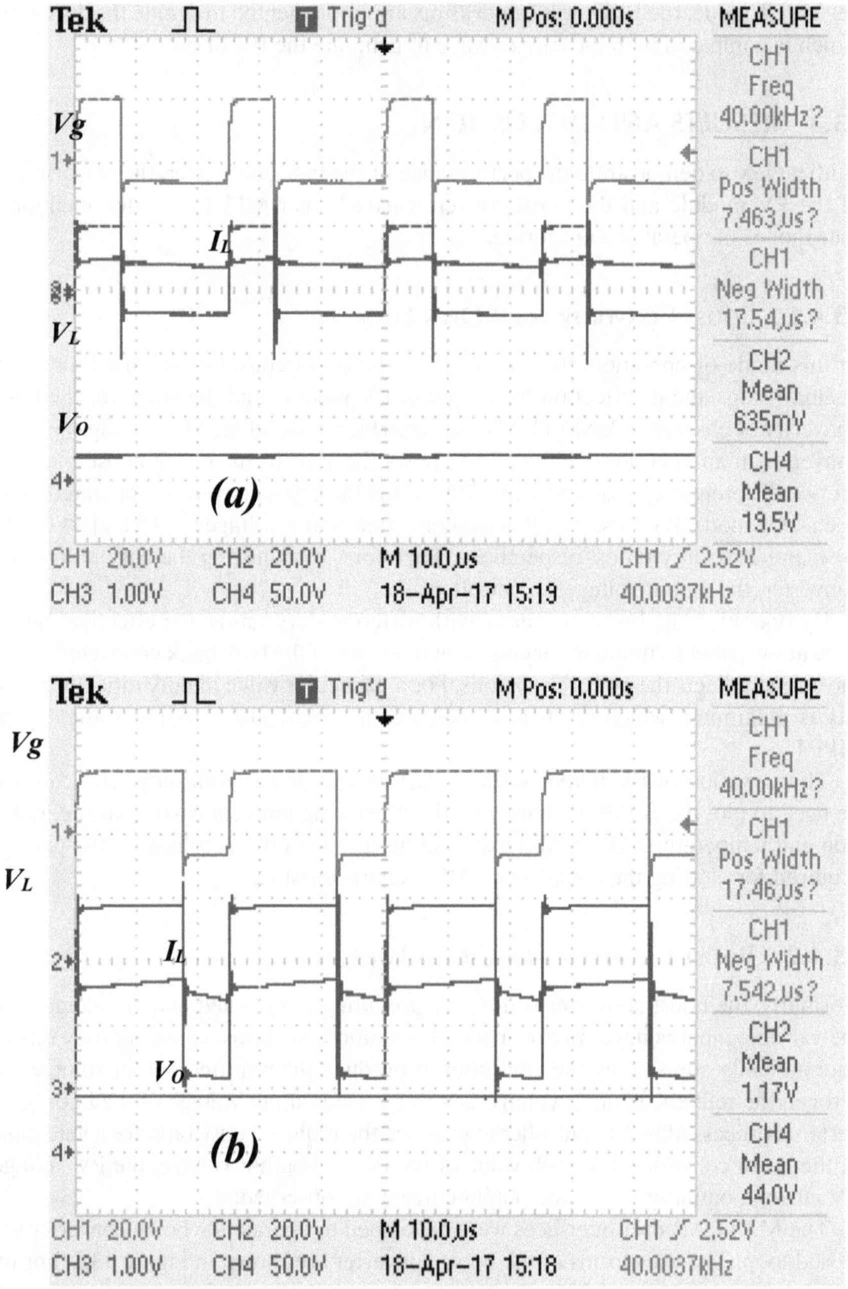

FIGURE 13.7 The DSO snapshots of gate pulse (V_g), inductor voltage (V_L), inductor current (I_L) and output voltage (V_o) for boost converter operating in an open-loop with duty ratio (a) 30% and (b) 70%.

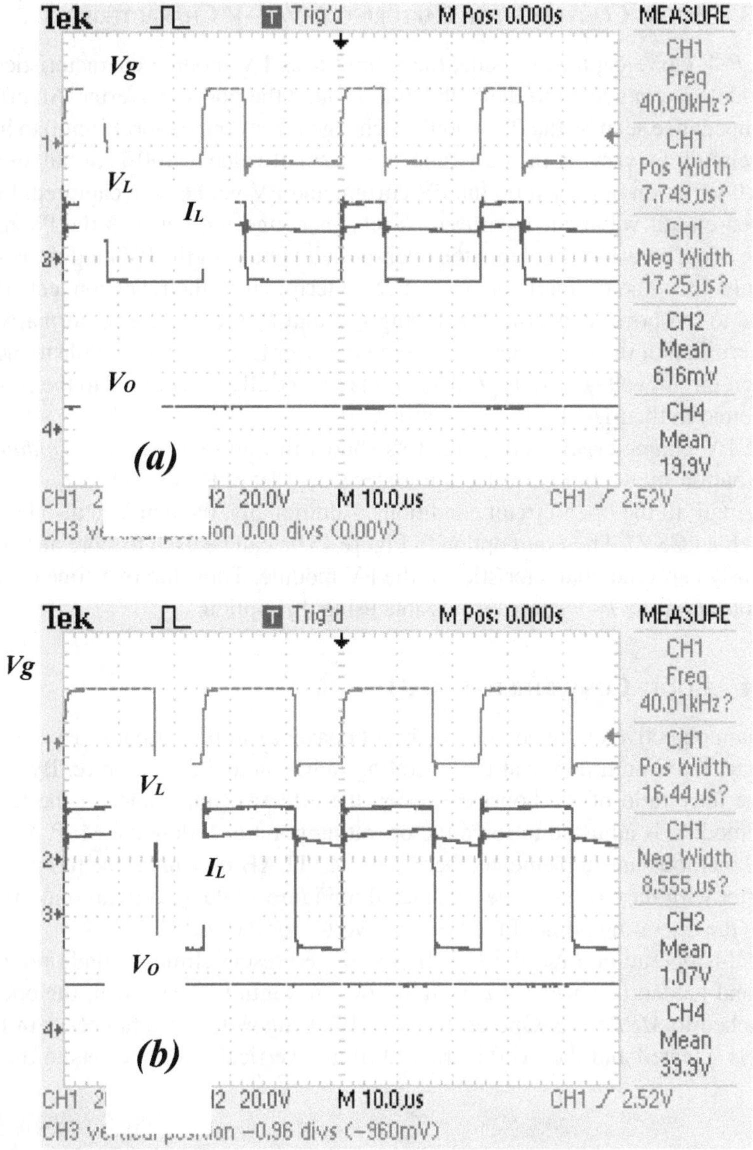

FIGURE 13.8 The DSO snapshots of gate pulse (V_g), inductor voltage (V_L), inductor current (I_L) and output voltage (V_o) for boost converter operating in a closed loop for output voltage (a) 20 V and (b) 40 V.

maximum power from the PV module. A small independent experiment to study the effect of PI controller parameters on the closed-loop performance of boost converter can also be studied using this mode of operation.

13.4.3 Boost Converter for Capturing *I–V/P–V* Characteristics

In the *P–V* curve capturing mode, the setup traces PV module characteristics. The PV module is connected to the bulb bank through the boost converter. An effective load impedance seen by the PV module is changed from zero (short-circuit) to infinite (open-circuit) by varying the duty ratio from a small value to 100%. Simultaneously, for each variation of duty ratio the PV current and PV voltage are captured. Finally, the open-circuit voltage is measured. A relay is connected between the PV module and the boost converter to obtain the open-circuit voltage of the PV module as shown in Figure 13.5 The normally connected (NC) terminal of the relay connects the PV module to the boost converter for tracing *I–V* and *P–V* curves. The normally open (NO) terminal of the relay opens the boost converter from the PV module to measure the open-circuit voltage. All (*V*, *I*) data points are serially transmitted to the computer and plotted in the MATLAB-based utility.

The PV characteristics are plotted as shown in Figure 13.9a. The *Update P–V Curve* button shown in Figure 13.9a initiates a new set of PV data (V_{PV}, I_{PV}) from the short-circuit to the open-circuit conditions. Additionally, the data can also be stored in data file (.CSV). The *Load* option in Figure 13.9a can be used to store and plot the previously captured characteristics of the PV module. Thus, the real-time capturing and plotting of the *P–V* curves is possible using this option.

13.4.4 Boost Converter for MPPT

The main objective of connecting the boost converter to the output of PV module is to extract maximum power at the operating radiation and temperature. By controlling the duty ratio of the boost converter, the effective impedance at the terminal of PV module is adjusted to force the operating point to follow the MPP. The P&O and IC methods are implemented here. The MATLAB user interface to operate the converter with the MPPT mode is depicted in Figure 13.9b. A vertical pink line represents the operating point of the boost converter for the MPPT.

For this operation, first, the *P–V* curve, in the present climatic condition, is captured and loaded using *P–V Curve* tab shown in Figure 13.9a. Then, the operation is switched to *MPPT* tab. One of the two MPPT algorithms (right corner in Figure 13.9b) is selected and the MPP is tracked till the vertical pink line sets to the MPP.

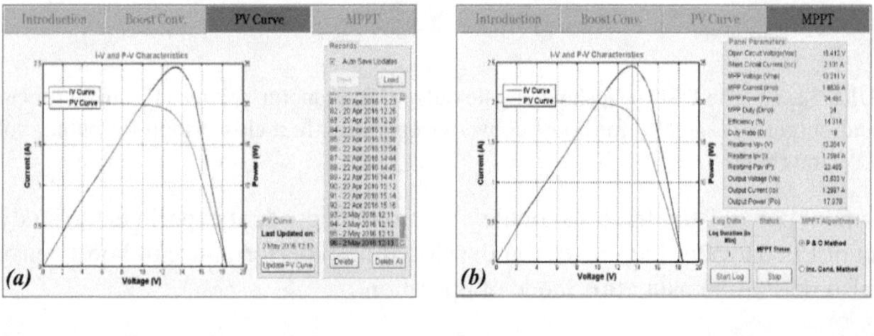

FIGURE 13.9 The MATLAB user interface for (a) plotting the *I–V/P–V* characteristics and (b) MPPT mode.

The bulb bank helps to visually observe the MPPT process. The varying bulb intensity with the varying boost converter voltage makes the students understand the MPPT process.

As the PV module voltage varies from zero to V_{MP} to V_{OC}, the bulb intensity increases, reaches maximum and then further reduces. To understand the MPPT process, three operating points on the *P–V* curve are selected: (i) left to the MPP (point A), (ii) on the MPP (point B) and (iii) right to the MPP (point C) as shown in Figure 13.10a. Three duty ratios are set in the open-loop mode: (i) 75% for the operation to the left of MPP (point A), (ii) 65% for the MPP operation (point B) and (iii) 2% for the PV operation to the right side of the MPP (point C). The corresponding three photographs of the bulbs bank load are displayed in Figure 13.10b–d. It is clearly observed that when the PV voltage is set to the MPP, the bulb intensity is with full brightness (Figure 13.10c) than in the other two cases (Figure 13.10b and d).

A new method of MPP can be implemented on the same laboratory prototype, and comparative performance with existing MPPT methods can also be studied. These laboratory exercises enable students to validate the concept of PV module characteristics, effect of controlling different loads at the output terminal of PV module and the process to extract MPP using a boost converter experimentally.

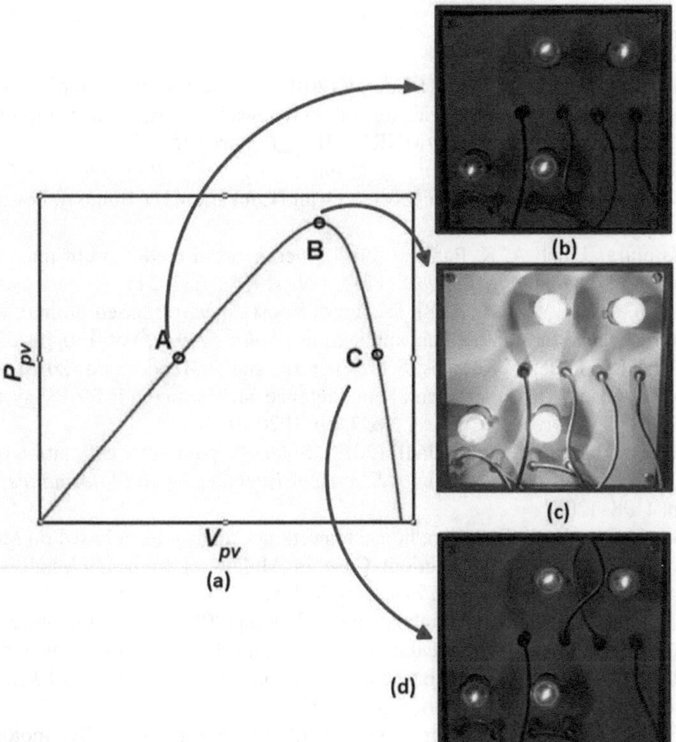

FIGURE 13.10 The MPPT operation: (a) The *P–V* curve mentioning three operating points and photographs of the bulbs with duty ratios (b) 75%, (c) 65% and (d) 2%.

13.5 CONCLUSIONS

The idea presented in this chapter focuses on an experimental study of a PV system rather than a numerical simulation study. In order to fulfil this objective, (i) the PV module characteristics, its behaviour in different operating conditions and MPPT algorithms are explained. (ii) A complete components design of a DC–DC boost converter is detailed. (iii) In order to implement the different modes of operation, a control scheme is developed to drive the boost converter. The specific feature of the work is the implementation of the P&O and IC MPPT algorithms using ARM Cortex-M4 micro-controller. During open- and closed-loop operations of the boost converter, the DSO snapshots of the gate pulses, inductor voltage and current, and output voltage of the converter are also demonstrated. The MATLAB interfacing plots give the evidences for successful *P–V* curve tracing and MPPT in real-time. During the MPPT mode, the highest intensity shown by the bulb bank proves the successful implementation of the P&O and IC algorithms to extract maximum power from the PV module. The hardware prototype developed here can be hands-on laboratory experimentation for teaching/learning PV module MPPT along with boost converter operation. In fact, a new MPPT algorithm can be implemented/tested/compared using this prototype.

REFERENCES

1. Ministry of New & Renewable Energy (2010, October). Govt. of India, India, Human resources development strategies for Indian renewable energy sector. http://www.mnre.gov.in/file-manager/UserFiles/MNRE_HRD_Report.pdf, 2010.
2. C. S. Solanki (2010). *Lab Training Manual on Teach a 1000 Teachers Program PV.* National Centre for Photovoltaic Research and Education, IIT Bombay, Powai, Mumbai, India, pp. 35–36.
3. G. Kumarand, and A. K. Panchal (2014). Geometrical prediction of maximum power point for photovoltaics. *Applied Energy*, Vol. 119, pp. 237–245.
4. R. Ayop, and C. W. Tan (2018). Design of boost converter based on maximum power point resistance for photovoltaic applications. *Solar Energy*, Vol. 160, pp. 322–335.
5. D. Sera, L. Mathe, T. Kerekes, S. V. Spataru, and R. Teodorescu (2013). On the perturb-and-observe and incremental conductance MPPT methods for PV systems. *IEEE Journal of Photovoltaics*, Vol. 3, No. 3, pp. 1070–1078.
6. N. Rawat, P. Thakur, and U. Jadli (2019). Solar PV parameter estimation using multi-objective optimisation. *Bulletin of Electrical Engineering and Informatics*, Vol. 8, No. 4, pp. 1198–1205.
7. N. Saad et al. (2019). Solar irradiance uncertainty management based on Monte Carlo-beta probability density function: Case in Malaysian tropical climate. *Bulletin of Electrical Engineering and Informatics*, Vol. 8, No. 3, pp. 1135–1143.
8. S. J. Patel, G. Kumar, A. K. Panchal, and V. Kheraj (2015). Maximum power point computation using current–voltage data from open and short circuit regions of photovoltaic module: A teaching learning based optimization approach. *Journal of Renewable and Sustainable Energy*, Vol. 7, No. 4, p. 043112.
9. D. S. Nayak, and R. Shivarudraswamy (2020). Solar fed BLDC motor drive for mixer grinder using a buck-boost converter. *Bulletin of Electrical Engineering and Informatics*, Vol. 9, No. 1, pp. 48–56.

10. J. M. Enrique, J. M. Andújar, and M. A. Bohorquez (2010). A reliable, fast and low cost maximum power point tracker for photovoltaic applications. *Solar Energy*, Vol. 84, No. 1, pp. 79–89.
11. A. Amir, A. Amir, H. S. Che, A. El. Khateb, and N. A. Rahim (2019). Comparative analysis of high voltage gain DC-DC converter topologies for photovoltaic systems. *Renewable Energy*, Vol. 136, pp. 1147–1163.
12. J. J. Nedumgatt, K. B. Jayakrishnan, S. Umashankar, D. Vijayakumar, and D. P. Kothari (2011, December). Perturb and observe MPPT algorithm for solar PV systems-modelling and simulation. In *2011 Annual IEEE India Conference*, Hyderabad, India (pp. 1–6), IEEE.
13. A. Paul, B. K. Dey, N. Mandal, and A. Bhattacharjee (2016). MATLAB/Simulink model of stand-alone Solar PV system with MPPT enabled optimized Power conditioning unit. In *2016 IEEE 7th Annual Information Technology, Electronics and Mobile Communication Conference (IEMCON)*, Vancouver, BC, Canada (pp. 1–6), IEEE.
14. B. Fekkak, M. Menaa, and B. Boussahoua (2018). Control of transformer less grid-connected PV system using average models of power electronics converters with MATLAB/Simulink. *Solar Energy*, Vol. 173, pp. 804–813.
15. A. Colmenar-Santos, T. Guinduláin-Argandoña, E. Rosales-Asensio, E. L. Molina-Ibáñez, and J. J. Blanes-Peiró (2018). Simulation of modeling of multi-megawatt photovoltaic plants with high voltage direct current grid integration. *Solar Energy*, Vol. 166, pp. 28–41.
16. R. Abbassi, A. Abbassi, M. Jemliand, and S. Chebbi (2018). Identification of unknown parameters of solar cell models: A comprehensive overview of available approaches. *Renewable and Sustainable Energy Reviews*, Vol. 90, pp. 453–474.
17. S. Raj, A. Kumar Sinha, and A. K. Panchal (2013). Solar cell parameters estimation from illuminated I-V characteristic using linear slope equations and Newton-Raphson technique. *Journal of Renewable and Sustainable Energy*, Vol. 5, No. 3, pp.033105.
18. M. Fannakh, M. L. Elhafyaniand, and S. Zouggar (2018). Hardware implementation of the fuzzy logic MPPT in an Arduino card using a Simulink support package for PV application. *IET Renewable Power Generation*, Vol. 13, No. 3, pp. 510–518.
19. P. I. Muoka, M. E. Haque, A. Gargoom, and M. Negnevitsky (2014). DSP-based hands-on laboratory experiments for photovoltaic power systems. *IEEE Transactions on Education*, Vol. 58, No. 1, pp. 39–47.
20. G. Verbič, C. Keerthisingheand, and A. C. Chapman (2017). A project-based cooperative approach to teaching sustainable energy systems. *IEEE Transactions on Education*, Vol. 60, No. 3, pp. 221–228.
21. S. J. Patel, A. K. Panchal, and V. Kheraj (2014). Extraction of solar cell parameters from a single current–voltage characteristic using teaching learning based optimization algorithm. *Applied Energy*, Vol. 119, pp. 384–393.
22. C. S. Solanki. *Solar Photovoltaics: Fundamentals, Technologies and Applications*, PHI Learning Pvt. Ltd., New Delhi, 2015.
23. G. Kumar, M. B. Trivedi, and A. K. Panchal (2015). Innovative and precise MPP estimation using P–V curve geometry for photovoltaics. *Applied Energy*, Vol. 138, pp. 640–647.
24. S. Kolsi, H. Sametand, and M. B. Amar (2014). Design analysis of DC-DC converters connected to a photovoltaic generator and controlled by MPPT for optimal energy transfer throughout a clear day. *Journal of Power and Energy Engineering*, Vol. 2 No. 1, p. 27.
25. S. Choi, and M. Saeedifar (2011). An educational laboratory for digital control and rapid prototyping of power electronic circuits. *IEEE Transactions on Education*, Vol. 55, No. 2, pp. 263–270.

26. STMicroelectronics "RM0090 Reference manual," September 2016. www.st.com/resource/en/reference_manual/DM00031020.pdf Assessed July 2018.
27. S. Panigrahi, and A. Thakur (2019). Modelling and simulation of three phases cascaded H-bridge grid-tied PV inverter. *Bulletin of Electrical Engineering and Informatics*, Vol. 8, No. 1, pp. 1–9.

14 Superhydrophobic Coatings of Silica NPs on Cover Glass of Solar Cells for Self-Cleaning Applications

Rajaram S. Sutar
Raje Ramrao Mahavidyalaya

Sampat G. Deshmukh
SKN Sinhgad College of Engineering

A. M. More
Commerce and Vinayakrao Patil Science College

A. K. Bhosale and Sanjay S. Latthe
Raje Ramrao Mahavidyalaya

CONTENTS

DOI: 10.1201/9781003220176-14

14.1 INTRODUCTION

Solar cell is a promising and widely used energy device due to its environment-friendly and renewability characteristics. Such energy devices absorb incident sunlight and convert it into electric energy. However, the solar cell is contaminated due to the accumulation of organic and inorganic dust particles over a long period. These dust particles prevent incident light to reach the solar cell panels, consequently, the output efficiency gets decreased [1,2]. Paudyal and Shakya [3] have studied the effect of dust deposition density ranging from 0.1047 to 9.6711 g m^{-2} on output efficiency of solar modules during a period of 148 days in outdoor exposure. The 9.6711 g m^{-2} dust deposition has reduced the efficiency of the solar panel up to 29.76% with respect to the module that is cleaned on daily basis. Adinoyi et al. [4] have experienced 50% decrease in output power of photovoltaic (PV) module, when it was kept without cleaning for a period of 6 months in outdoor conditions. Results of this literature confirm that dust accumulation could significantly affect the output power of solar modules. Therefore, it is a challenging task to keep the output efficiency of the solar panels at maximum for a long period without frequent cleaning of the surface.

A lotus leaf grabbed the attention of scientists and inspired them to develop a self-cleaning surface, which exhibits a highly nonwetting property with a water contact angle of nearly 160° and a sliding angle of less than 5° [5,6]. Currently, many efforts have been taken to develop lotus leaf–like self-cleaning superhydrophobic coating to protect the surface of windshields of vehicles, the entire body of vehicles, window and door glasses, skyscrapers, solar cell panels, fabrics, sports shoes, and so on [7–10]. When the suspension of SiO_2 nanomaterial in alcohol was sprayed on paper at room temperature and atmospheric pressure, the hydrophobicity of the coatings was found dependent on the size of nanoparticles (NPs) as well as the aggregation of NPs, which again depends on the type of alcohol used in the coating solution [11]. Datta et al. [12] have fabricated self-cleaning superhydrophobic coating on solar cell cover glass by depositing sol–gel processed silica NPs coating and grafting a monolayer of fluoroalkylsilane on it. Almari et al. [13] have investigated the efficiency of solar PV panels after coating with hydrophobic SiO_2 nanomaterial. This investigation concludes that output power was increased by 15% more than the dusty panels and 5% more than the uncoated panels, which were cleaned manually every day. Zhi et al. [14] have prepared antireflective, transparent, and self-cleaning superhydrophobic surface through dipping in silica NP solution. Such coating showed static water contact angle of 157.9° and contact angle hysteresis of 1.2°. Wang et al. [15] have sprayed the paint-like suspension of silica NPs and 1H, 1H, 2H, 2H-perfluorooctyltriethoxysilane (FAS) onto the polydimethylsiloxane (PDMS) coated glass substrate to obtain robust self-cleaning superhydrophobic coating. Zhang et al. [16] have prepared superhydrophobic surface on glass substrate by spraying the suspension of hydrophobic silica NPs and curing at room temperature without any modification or surface treatment. The damaged superhydrophobicity of coating was recovered by simply respraying the suspension. Lazauskas et al. [17] have fabricated a transparent, self-cleaning superhydrophobic surface by dropping the dispersion of hexamethyldisilazane (HMDS) and SiO_2 NPs in ethanol on microscopic glass slides and dried at room temperature to evaporate ethanol.

Herein, we have prepared a superhydrophobic coating on the glass slide and covered the glass of solar cell panel by spraying the suspension of hydrophobic SiO_2 NPs. The aggregated SiO_2 NPs on the glass surface formed a hierarchical dual scale rough structure, which is responsible for the high water contact angle $160 \pm 2°$ and low sliding angle 6°. The prepared superhydrophobic coating was sustained for water jet impact test and mechanical durability test. The rolling water droplets cleaned the contaminated solar cell panel, which confirmed that the prepared superhydrophobic coating exhibits self-cleaning ability.

14.2 EXPERIMENTAL SECTION

14.2.1 Materials

Hydrophobic silica NPs (surface area $210\,m^2g^{-1}$) were purchased from AEROSIL Company, RX 300-5, Japan. Hexane (puriss for synthesis) was bought from Spectrochem, Mumbai, India. Micro glass slides (Blue Star, India) and solar cell panels (local market) were procured.

14.2.2 Preparation of Superhydrophobic

The glass slide and cover glass of the solar cell panel were cleaned with the laboratory detergent (Molyclean 02 Neutral, from Molychem, India) to eliminate surface contaminants. The suspension of hydrophobic silica NPs was prepared by dispersing it in hexane ($5\,mg\,mL^{-1}$) using ultrasonication for 20 minutes. The prepared suspension was applied on a glass slide and cover glass of solar cell panel using a spray coating method. The successive ten layers of SiO_2 NPs were applied from 10 cm with a time interval of 1 minute between each layer. Then, the coatings were kept in air at room temperature to evaporate hexane. The process of deposition of suspension of SiO_2 NPs on the substrate is shown in Figure 14.1.

14.2.3 Characterization

The surface microstructure of the coated substrate was observed by scanning electron microscope (SEM, JEOL, JSM-7600F). The water contact angles and sliding angles were measured three times at various locations, and the mean value was noted as final value using a contact angle meter (HO-IAD-CAM-01, Holmarch Opto-Mechatronics Pvt. Ltd., India). The self-cleaning property of the coating was investigated by spreading the fine chalk particles as dust on the coating. The mechanical stability of coating was evaluated by adhesive tape peeling, sandpaper abrasion, and water jet impact test.

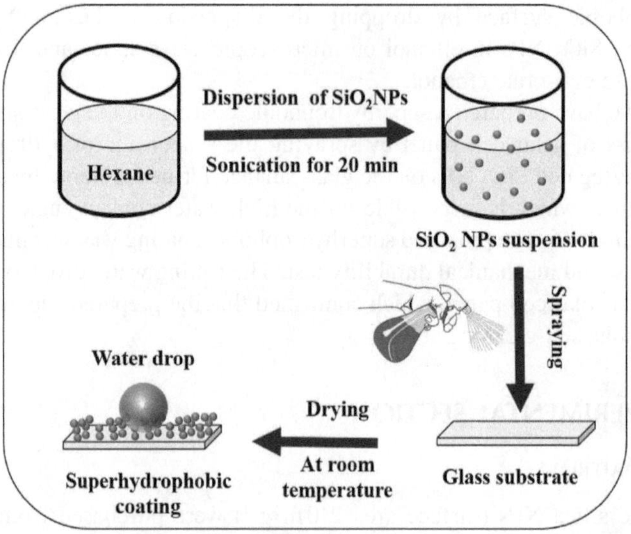

FIGURE 14.1 Schematic of experimental procedure for the preparation of superhydrophobic coating on the substrate.

14.3 RESULT AND DISCUSSION

14.3.1 Surface Structure and Wettability

In the beginning, the primary coatings were done at a laboratory scale on micro glass slide. Figure 14.2a reveals color-dyed water drops on a coated glass slide. The water drop attains a spherical shape with a water contact angle of $160 \pm 2°$ on prepared superhydrophobic coating and a sliding angle of less than 6°. The optical image of water drop on the superhydrophobic coating received from the contact angle meter is shown in Figure 14.2b.

The SEM image confirms the suspension of SiO_2 NPs was uniformly covered on the surface of the glass slide. The rough porous microstructure was attained from aggregated SiO_2 NPs (Figure 14.2c). Numerous nanovoids were observed on the coating, which is evidence of nanoscale roughness. However, in high-magnification SEM image (Figure 14.2d), it is clearly observed that the rough and porous microstructure of the superhydrophobic coating is a result of aggregated SiO_2 NPs. Low surface energy of hydrophobic NPs and micro/nanostructure of coating are two important key factors of the superhydrophobic coating. Zhang et al. [16] have observed a similar surface structure in the spray deposited coating of hydrophobic silica NPs. Such micro-/nanoscale hierarchical rough surface allows more air to be trapped underneath the water drop, and hence the solid–liquid contact area will be effectively minimized.

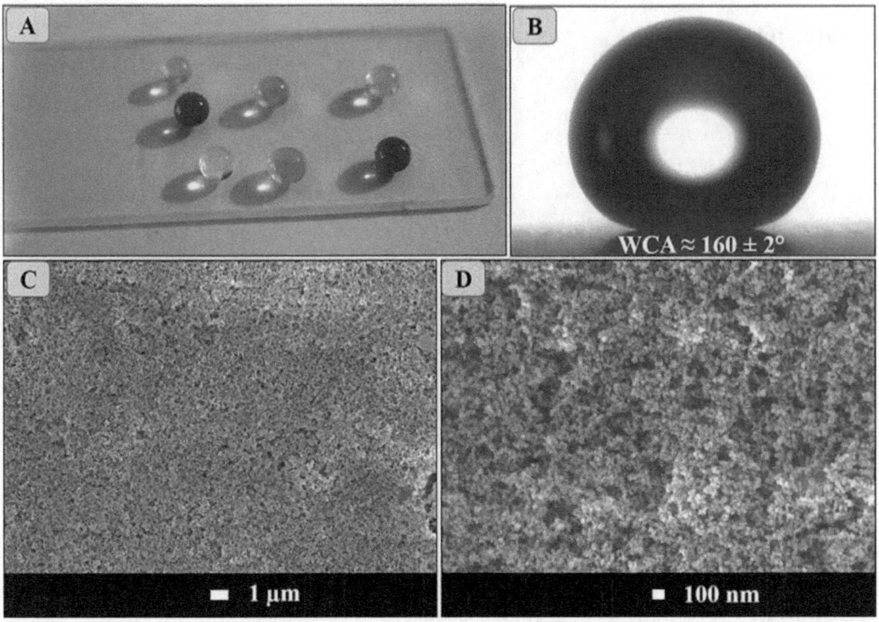

FIGURE 14.2 (a) The photograph of color-dyed water drops on coated micro glass slide, (b) the optical image of a water drop on the superhydrophobic coating, and (c) low- and (d) high-magnification SEM images of superhydrophobic coating.

14.3.2 Durability of Superhydrophobic Coating

The mechanical durability of the superhydrophobic coatings is essential key factor for its commercial applications. However, the fragile hierarchical structure of superhydrophobic coatings can be easily destroyed by mechanical rubbing and hence gives poor durability. In the literature, the adhesive tape peeling and sandpaper abrasion tests are mostly used to determine the durability of the prepared coatings. The adhesive tape with adhesion strength of 4 N m^{-1} was gently applied on the prepared superhydrophobic coating. A 50 g metal disk was rolled on it to ensure good contact between tape and coating. The adhesive tape was slowly peeled off and quickly water contact angle and sliding angle were measured on the coating. The coating exhibited superhydrophobicity even after four times of the tape peeling test was carried out. This adhesive tape peeling test confirmed that the silica NPs are firmly adhered to the glass substrate. In recent work, Dessouky et al. [18] have observed that the hydrophobic silica NPs–coated metals lost their superhydrophobicity after single adhesive tape test and the water contact angle reduced from 155° to 118°. For further evaluation of the mechanical stability of the coating, a 50 g weight-loaded superhydrophobic glass slide was placed on sandpaper (600 grit) and dragged linearly with a speed of 5 mm s^{-1} for 30 cm [15]. We observed that the superhydrophobic coating was completely scratched and lost superhydrophobicity. The continuous water jet

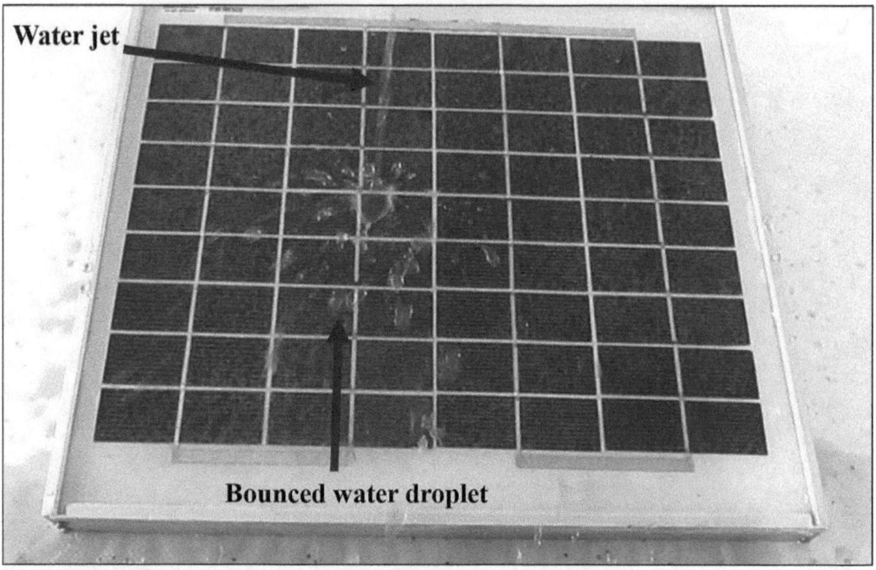

FIGURE 14.3 The water jet bouncing on coated solar cell panel.

was applied on coated solar cell panel from the height of 10 cm to demonstrate the remarkable water resistance [19]. The water jet bounces up when it falls on a coated solar cell panel (Figure 14.3). The air trapped in a rough structure of the coating could not allow the water to enter inside it. However, after 5 minutes of water jet impacting, water drops stuck on the coating, which confirms that the coating material got removed from the cover glass due to water jet impact.

14.3.3 Self-Cleaning Property

Self-cleaning is one of the most important properties of superhydrophobic surface in practical applications, which can automatically clean the surface through rolling water drops. Mostly solar cell panels are installed in large areas for the generation of electric power. They can be covered by various types of contaminants, subsequently reducing the output power of solar cell panels. The frequent cleaning of solar panels serves challenging problem in the whole world. The mechanical/chemical methods of cleaning are time-consuming and cost-effective. The superhydrophobic coating on the cover glass of the solar cell panel can be a solution and water drops on the coated cover glass can eventually roll off by collecting the water drops; hence, the solar cell panels are self-cleaned. Fine particles of chalk dust were spread randomly on coated solar cell panel to investigate self-cleaning behavior (Figure 14.4a). Dust contaminated solar cell panel kept at an inclination of 30° from the horizontal plane and water droplets poured on it [20]. The dust particles cannot be removed from the uncoated solar panel with a water droplet. However, on superhydrophobic coated solar panel, the rolling water drops take off the dust particles on their way and eventually cleaned the surface of the solar panel. Figure 14.4b reveals a dust contaminated solar panel

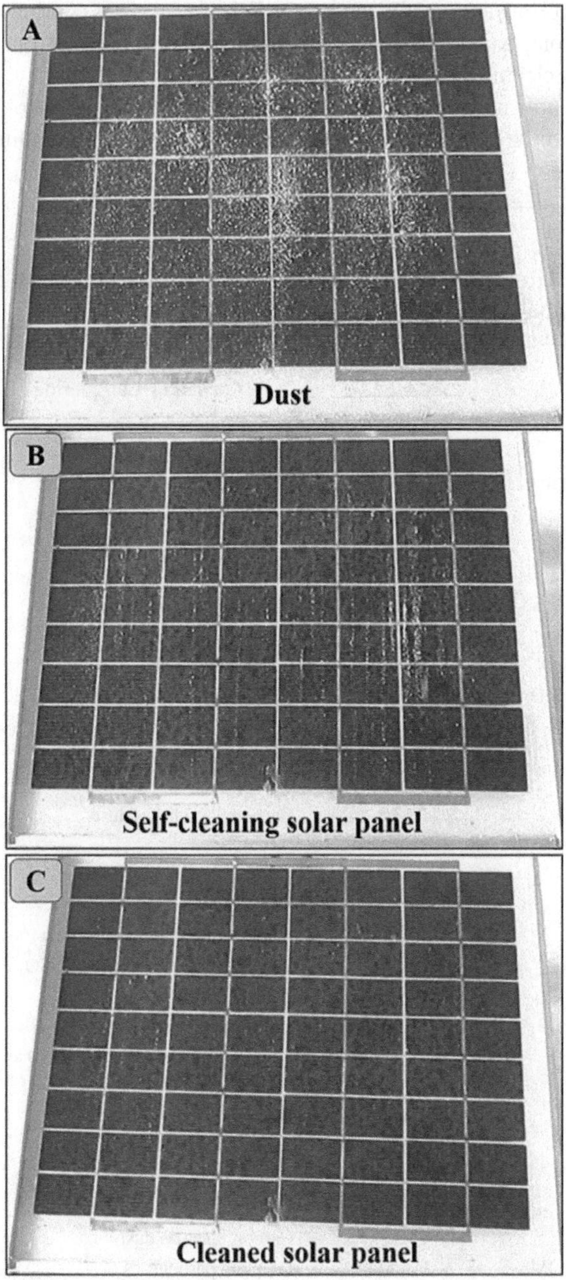

FIGURE 14.4 (a) Randomly spread dust particles on superhydrophobic cover glass of solar panel, (b) self-cleaning behavior, and (c) cleaned solar panel.

cleaned by the rolling action of water drops. When water was continuously poured on contaminated solar panel, almost all dust particles are removed from the solar panel without contact cleaning (Figure 14.4c).

Literature confirms that the SiO_2 particle–based superhydrophobic coatings are more efficient in the self-cleaning phenomenon. Due to high bonding energy, SiO_2

TABLE 14.1
Summary of a Literature Review on the Self-Cleaning Property of SiO_2 NPs–Based Superhydrophobic Coatings

Materials	WCA/SA (°)	Self-Cleaning Performance Carried by Considering	References
PDMS/SiO_2	156.4/5	Sand particles as dust	[21]
Linear low density polyethylene (LLDPE)/ SiO_2	170/3.8	Charcoal powder as dust	[22]
Polystyrene/SiO_2	158/9	Charcoal powder as dust	[23]
Hydroxy acrylic resin/ Hexamethyl disilazane (HMDS) SiO_2	170/2	Hydrophilic fly-ash and hydrophobic carbon nanotube particles as dust	[24]
Polybenzoxazine/SiO_2	167/5	Graphite powders as dust	[25]
Polyvinyl chloride/SiO_2	169/6	Soil-based muddy water as dust	[26]
Poly(methyl methacrylate) (PMMA)/SiO_2	165/4	Soil, muddy water and chalk particles as dust	[27]
SiO_2/ methyltrichlorosilane (TMCS)	153/8	Carbon particles as dust	[28]
SiO_2	162/5	Chalk dust as dust	[29]
SiO_2/epoxy resin	155/5	Sand and muddy water as dust	[30]
SiO_2 and Fluorinated Epoxy	158.6/10	Carbon black powder and $CuSO_4$ powder as dust	[31]
Polyethylene wax/SiO_2	163/9	Hydrophobic solvent (MnO powder partially wetted by oil) as dust	[32]
PDMS/SiO_2	152/10	Carbon-powder as dust	[33]
Poly(methyl methacrylate)/SiO_2	159/1<0	Oil Red O powder as dust	[34]
PDMS/SiO_2	165/<10	Carbon black as dust	[35]
SiO_2	152/10	Carbon black as dust	[36]
Fluorinated-SiO_2/PDMS	156.5/2	Graphite particles as dust	[37]
Methyl phenyl silicone resin/SiO_2	162/2	Silicon carbide particles as dust	[38]
Polytetrafluoroethylene/ SiO_2	153/5	Methyl blue powder as dust	[39]
PMMA/SiO_2	165/7	Mud particles as dust	[40]
SiO_2	160/6	Fine particles of chalk dust	Present study

NPs show regular and ordered structure and good mechanical strength with high thermal and chemical resistance. Moreover, silica-based coatings are highly transparent [41]. Table 14.1 clarifies that the hydrophobic SiO_2 NPs–based superhydrophobic coatings show high water repellency and self-cleaning property. Various types of dust contaminants are picked away by rolling water droplets from the coating surface and left clean surface.

14.4 CONCLUSION

The suspension of SiO_2 NPs was successively deposited on the cover glass of solar cell panel by spray technique. The prepared superhydrophobic SiO_2 NPs coating has revealed a water contact angle of $160° \pm 2°$ along with a sliding angle of 6°. The coating was stable up to four cycles of adhesive tape peeling test and exhibited poor stability for sandpaper abrasion test. The dust particles were effectively removed by the rolling action of water drops, which concludes that the prepared superhydrophobic SiO_2 NPs coating exhibits excellent self-cleaning property. Thus, such coating may be useful for solar cell manufacturing industries due to its simple, low-cost technique and excellent self-cleaning property.

HIGHLIGHTS

- The silica NPs were used for the preparation of superhydrophobic coatings.
- The modest and inexpensive spray deposition technique is utilized.
- The coating layer was applied on cover glass of the solar cell panel.
- The coating exhibits good mechanical stability.
- The coating on the cover glass of solar cell panel revealed excellent self-cleaning capability.

ACKNOWLEDGMENTS

This work is financially supported by DST – INSPIRE Faculty Scheme, Department of Science and Technology (DST), Govt. of India [DST/INSPIRE/04/2015/000281]. SSL acknowledges financial assistance from the Henan University, Kaifeng, P. R. China. We greatly appreciate the support of the National Natural Science Foundation of China (21950410531).

REFERENCES

1. Mani, M. and Pillai, R. (2010). Impact of dust on solar photovoltaic (PV) performance: Research status, challenges and recommendations. *Renewable and Sustainable Energy Reviews*, 14(9): pp. 3124–3131.
2. Sarver, T., Al-Qaraghuli, A. and Kazmerski L.L. (2013). A comprehensive review of the impact of dust on the use of solar energy: History, investigations, results, literature, and mitigation approaches. *Renewable and Sustainable Energy Reviews*, 22: pp. 698–733.
3. Paudyal, B.R. and Shakya, S.R (2016). Dust accumulation effects on efficiency of solar PV modules for off grid purpose: A case study of Kathmandu. *Solar Energy*, 135: pp. 103–110.

4. Adinoyi, M.J. and Said, S.A. (2013). Effect of dust accumulation on the power outputs of solar photovoltaic modules. *Renewable Energy*, 60: pp. 633–636.
5. Barthlott, W. and Neinhuis, C. (1997). Purity of the sacred lotus, or escape from contamination in biological surfaces. *Planta*, 202(1): pp. 1–8.
6. Neinhuis, C. and Barthlott, W. (1997). Characterization and distribution of water-repellent, self-cleaning plant surfaces. *Annals of Botany*, 79(6): pp. 667–677.
7. Dalawai, S.P., Aly, M.A.S., Latthe, S.S., Xing, R., Sutar, R.S., Nagappan, S., Ha, C.S., Sadasivuni, K.K. and Liu, S. (2020). Recent advances in durability of superhydrophobic self-cleaning technology: A critical review. *Progress in Organic Coatings*, 138: p. 105381.
8. Davis, A., Yeong, Y.H., Steele, A., Bayer, I.S. and Loth, E. (2014). Superhydrophobic nanocomposite surface topography and ice adhesion. *ACS Applied Materials & Interfaces*, 6(12): pp. 9272–9279.
9. Latthe, S.S., Sutar, R.S., Kodag, V.S., Bhosale, A.K., Kumar, A.M., Sadasivuni, K.K., Xing, R. and Liu, S. (2019). Self–cleaning superhydrophobic coatings: Potential industrial applications. *Progress in Organic Coatings*, 128: pp. 52–58.
10. Milionis, A., Sharma, C.S., Hopf, R., Uggowitzer, M., Bayer, I.S. and Poulikakos, D. (2019). Engineering fully organic and biodegradable superhydrophobic materials. *Advanced Materials Interfaces*, 6(1): p. 1801202.
11. Ogihara, H., Xie, J., Okagaki, J. and Saji, T. (2012). Simple method for preparing superhydrophobic paper: Spray-deposited hydrophobic silica nanoparticle coatings exhibit high water-repellency and transparency. *Langmuir*, 28(10): pp. 4605–4608.
12. Datta, A., Singh, V.K., Das, C., Halder, A., Ghoshal, D. and Ganguly, R. (2020). Fabrication and characterization of transparent, self-cleaning glass covers for solar photovoltaic cells. *Materials Letters*, 277: p. 128350.
13. Alamri, H.R., Rezk, H., Abd-Elbary, H., Ziedan, H.A. and Elnozahy, A. (2020) Experimental Investigation to improve the energy efficiency of solar PV panels using hydrophobic SiO_2 nanomaterial. *Coatings*, 10(5): p. 503.
14. Zhi, J. and Zhang, L.-Z. (2018) Durable superhydrophobic surface with highly antireflective and self-cleaning properties for the glass covers of solar cells. *Applied Surface Science*, 454: pp. 239–248.
15. Wang, P., Liu, J., Chang, W., Fan, X., Li, C. and Shi, Y. (2016). A facile cost-effective method for preparing robust self-cleaning transparent superhydrophobic coating. *Applied Physics A*, 122(10): pp. 1–10.
16. Zhang, C., Kalulu, M., Sun, S., Jiang, P., Zhou, X., Wei, Y. and Jiang, Y. (2019). Environmentally safe, durable and transparent superhydrophobic coating prepared by one-step spraying. *Colloids and Surfaces A: Physicochemical and Engineering Aspects*, 570: pp. 147–155.
17. Lazauskas, A., Jucius, D., Puodžiukynas, L., Guobienė, A. and Grigaliūnas, V. (2020). SiO_2-based nanostructured superhydrophobic film with high optical transmittance. *Coatings*, 10(10): p. 934.
18. El Dessouky, W.I., Abbas, R., Sadik, W.A., El Demerdash, A.G.M. and Hefnawy, A. (2017). Improved adhesion of superhydrophobic layer on metal surfaces via one step spraying method. *Arabian Journal of Chemistry*, 10(3): pp. 368–377.
19. Torun, I., Celik, N., Hancer, M., Es, F., Emir, C., Turan, R. and Onses, M.S. (2018). Water impact resistant and antireflective superhydrophobic surfaces fabricated by spray coating of nanoparticles: Interface engineering via end-grafted polymers. *Macromolecules*, 51(23): pp. 10011–10020.
20. Chi, F., Liu, D., Wu, H. and Lei, J. (2019). Mechanically robust and self-cleaning antireflection coatings from nanoscale binding of hydrophobic silica nanoparticles. *Solar Energy Materials and Solar Cells*, 200: p. 109939.

21. Gong, X. and He, S. (2020). Highly durable superhydrophobic polydimethylsiloxane/silica nanocomposite surfaces with good self-cleaning ability. *ACS Omega*, 5(8): pp. 4100–4108.
22. Satapathy, M., Varshney, P., Nanda, D., Mohapatra, S.S., Behera, A. and Kumar, A. (2018). Fabrication of durable porous and non-porous superhydrophobic LLDPE/SiO2 nanoparticles coatings with excellent self-cleaning property. *Surface and Coatings Technology*, 341: pp. 31–39.
23. Pawar, P.G., Xing, R., Kambale, R.C., Kumar, A.M., Liu, S. and Latthe, S.S. (2017). Polystyrene assisted superhydrophobic silica coatings with surface protection and self-cleaning approach. *Progress in Organic Coatings*, 105: pp. 235–244.
24. Hu, C., Chen, W., Li, T., Ding, Y., Yang, H., Zhao, S., Tsiwah, E.A., Zhao, X. and Xie, Y. (2018) Constructing non-fluorinated porous superhydrophobic SiO_2-based films with robust mechanical properties. *Colloids and Surfaces A: Physicochemical and Engineering Aspects*, 551: pp. 65–73.
25. Zhang, H., Lu, X., Xin, Z., Zhang, W. and Zhou, C. (2018). Preparation of superhydrophobic polybenzoxazine/SiO_2 films with self-cleaning and ice delay properties. *Progress in Organic Coatings*, 123: pp. 254–260.
26. Sutar, R.S., Kalel, P.J., Latthe, S.S., Kumbhar, D.A., Mahajan, S.S., Chikode, P.P., Patil, S.S., Kadam, S.S., Gaikwad, V.H., Bhosale, A.K. and Sadasivuni, K.K. (2020). ICAMS-2020, Jath, India, Superhydrophobic PVC/SiO_2 coating for self-cleaning application. *Macromolecular Symposia*. Wiley Online Library.
27. Sutar, R.S., Gaikwad, S.S., Latthe, S.S., Kodag, V.S., Deshmukh, S.B., Saptal, L.P., Kulal, S.R. and Bhosale, A.K. (2020). ICAMS-2020, Jath, India, Superhydrophobic nanocomposite coatings of hydrophobic silica NPs and poly (methyl methacrylate) with notable self-cleaning ability. *Macromolecular Symposia*. Wiley Online Library.
28. Gurav, A.B., Xu, Q., Latthe, S.S., Vhatkar, R.S., Liu, S., Yoon, H. and Yoon, S.S. (2015). Superhydrophobic coatings prepared from methyl-modified silica particles using simple dip-coating method. *Ceramics International*, 41(2): pp. 3017–3023.
29. Zou, X., Tao, C., Yang, K., Yang, F., Lv, H., Yan, L., Yan, H., Li, Y., Xie, Y., Yuan, X. and Zhang, L. (2018). Rational design and fabrication of highly transparent, flexible, and thermally stable superhydrophobic coatings from raspberry-like hollow silica nanoparticles. *Applied Surface Science*, 440: pp. 700–711.
30. Peng, W., Gou, X., Qin, H., Zhao, M., Zhao, X. and Guo, Z. (2018). Creation of a multifunctional superhydrophobic coating for composite insulators. *Chemical Engineering Journal*, 352: pp. 774–781.
31. Huang, X. and Yu, R. (2021). Robust superhydrophobic and repellent coatings based on micro/nano SiO_2 and fluorinated epoxy. *Coatings*, 11(6): p. 663.
32. Guan, Y., Yu, C., Zhu, J., Yang, R., Li, X., Wei, D. and Xu, X. (2018). Design and fabrication of vapor-induced superhydrophobic surfaces obtained from polyethylene wax and silica nanoparticles in hierarchical structures. *RSC Advances*, 8(44): pp. 25150–25158.
33. Chang, H., Tu, K., Wang, X. and Liu, J. (2015). Fabrication of mechanically durable superhydrophobic wood surfaces using polydimethylsiloxane and silica nanoparticles. *RSC Advances*, 5(39): pp. 30647–30653.
34. Latthe, S.S., Terashima, C., Nakata, K., Sakai, M. and Fujishima, A. (2014). Development of sol–gel processed semi-transparent and self-cleaning superhydrophobic coatings. *Journal of Materials Chemistry A*, 2(15): pp. 5548–5553.
35. Liu, P., Yu, H., Hui, F., Villena, M.A., Li, X., Lanza, M. and Zhang, Z. (2020). Fabrication of 3D silica with outstanding organic molecule separation and self-cleaning performance. *Applied Surface Science*, 511: pp. 145537.
36. Liu, S., Latthe, S.S., Yang, H., Liu, B. and Xing, R. (2015). Raspberry-like superhydrophobic silica coatings with self-cleaning properties. *Ceramics International*, 41(9): pp. 11719–11725.

37. Wu, Y., Shen, Y., Tao, J., He, Z., Xie, Y., Chen, H., Jin, M. and Hou, W. (2018). Facile spraying fabrication of highly flexible and mechanically robust superhydrophobic F-SiO_2@ PDMS coatings for self-cleaning and drag-reduction applications. *New Journal of Chemistry*, 42(22): pp. 18208–18216.
38. Bhushan, B. and Multanen, V. (2019). Designing liquid repellent, icephobic and self-cleaning surfaces with high mechanical and chemical durability. *Philosophical Transactions of the Royal Society A*, 377(2138): p. 20180270.
39. He, J., Zhao, Y., Yuan, M., Hou, L., Abbas, A., Xue, M., Ma, X., He, J. and Qu, M. (2020). Fabrication of durable polytetrafluoroethylene superhydrophobic materials with recyclable and self-cleaning properties on various substrates. *Journal of Coatings Technology and Research*,17 pp. 755–763.
40. Meena, M.K., Sinhamahapatra, A. and Kumar, A. (2019). Superhydrophobic polymer composite coating on glass via spin coating technique. *Colloid and Polymer Science*, 297(11): pp. 1499–1505.
41. Tian, P. and Guo, Z. (2017). Bioinspired silica-based superhydrophobic materials. *Applied Surface Science*, 426: pp. 1–18.

15 Carbonaceous Composites of Rare Earth Metal Chalcogenides

Synthesis, Properties and Supercapacitive Applications

Dhanaji B. Malavekar, Shital B. Kale and Chandrakant D. Lokhande
D.Y. Patil Education Society

CONTENTS

DOI: 10.1201/9781003220176-15

15.1 INTRODUCTION

Rapid economic and social development in the last few decades has increased per capita energy consumption. Therefore, the need for sustainable and environmentally friendly energy sources is emphasized in the current days [1]. Electrical energy is one of the forms that can be easily generated from renewable energy sources [2]. However, the irregularities in the production of electricity further pressed the demand for electricity storage systems. Electricity can be stored in various types of batteries and supercapacitors (SCs) [3]. The SCs possess high energy density than dielectric capacitors and higher power density than batteries. Prolonged cycle life, large efficiency, and rapid charge–discharge ability are the advantages of SCs. However, the lower energy density of the SCs blocks off its practical application [4]. With the slight modification in the electrode material, the energy density of the SCs possibly elevated up to a sufficient level that can be used in portable electronic devices, electrical vehicles, military equipment, aerospace, and aviation [5,6].

The supercapacitive electrode materials are sorted as an electric double-layer capacitor (EDLC) and pseudocapacitor. Because of the generation of Helmholtz double layer at an electrode–electrolyte interface, the specific power and cycle life of the SCs based on EDLC materials are higher than pseudocapacitors. All of the carbon allotropes such as aerogels, carbon nanotubes (CNTs), activated charcoal (AC), graphene oxide (GO), and graphene show EDLC type behaviour [7,8]. The pseudocapacitive electrode stores energy via reversible redox reactions. Transition metal compounds and conducting polymers also show pseudocapacitive behaviour [9,10].

The pseudocapacitors show much greater specific capacitances than EDLCs. However, the lower electrical conductivity of these compounds becomes a barricade in the commercialization of redox SCs [11,12]. The expansion and contraction occurring in the charging–discharging process are very complicated due to electron transfer. Therefore, damage caused to crystallinity and the surface textures of the electrode are the major challenges to the pseudocapacitors [13,14]. Many researchers have developed various methods such as composition with the carbon-based material, preparation of bimetallic compounds, and doping of another metal in the host compound. In this chapter, SC electrode materials constituting at least one carbon component with the rare earth metal compound are discussed [15,16].

To improve electrochemical performance, physico-chemical features of the composite electrode are responsible. The major impact of carbon-based materials on physical properties is the main cause of the change in electrochemical charge storage behaviour. Most carbonaceous materials are good conductors of electricity, can be easily configured, and are eco-friendly [17]. Various carbon allotropes are used in several electrochemical devices because of their stability at various temperatures and in various electrolytes irrespective of the pH. The cost-effectiveness of the carbon materials makes it favourable for electrode material in applied electrochemistry. The wide variation in the dimensions of carbon materials from 0D (AC, carbon dots) to 3D foams makes it suitable for different applications [18]. To improve the energy density of SCs, composite electrode materials incorporating the favourable merits of EDLCs and pseudocapacitors have been developed, which sustain the high power feature of carbon material and meanwhile possess high energy density, which comes from reversible redox reactions of pseudocapacitive material.

The rare earth elements cover 17 elements from the modern periodic table comprising scandium and yttrium, as well as 15 elements from the lanthanide series. These elements play important roles in various crucial applications. The physico-chemical characteristics of the rare earth element compounds resemble the trivalent metal oxides [19]. The unique 4-f electron structure, strong spin–orbit coupling and large atomic and magnetic moments, and ability to change oxidation states under different chemical conditions are the most highlighted properties of rare earth metals. Rare earth metals are well known since the 1960s, and rare earth metal compounds are used for high-tech military applications, powerful magnets, refining crude oil, and important components of electric vehicles [20]. Furthermore, composite electrode materials could be used to fabricate symmetric or asymmetric supercapacitors (SSCs or ASCs) [21]. The ASCs can attain higher specific energy and power [22,23].

Few review articles are available summarising electrochemical studies of composites of the rare earth metal compounds. These articles address several aspects of rare earth metal compound–based SCs. However, the chapter comprising electrochemical properties of various carbonaceous materials composited with rare earth metal–based compounds is not available to date. This chapter intent to broadly describe the major role of carbonaceous materials in the rare earth metal–based electrodes reported so far in the literature, their synthesis and characterization, and impact of the composition. The schematic of some of the carbon allotropes used for the composition with rare earth metal chalcogenides is shown in Figure 15.1.

15.2 PRINCIPLE AND MECHANISM OF SUPERCAPACITOR

Growing electricity generation from renewable energy sources has created a demand for electricity storage technologies. Plenty of investigation has been conducted to increase the specific energy of SCs in the last few decades. These efforts are based on the fundamental understanding of the working principle of SCs. Two distinct working principles were put forward to describe the charge storage process in SCs.

15.2.1 Electric Double-Layer Capacitance (EDLC)

As an electrode immersed in the electrolyte, the layer of charges formed spontaneously due to a simple arrangement of charges is called the electric double layer. In this process, the charge is primarily accumulated at the interface by the non-faradaic process. The charge stored in EDLC type SCs is 100,000 times more than the dielectric capacitors. The conventional illustration of the charge storage principle is shown in Figure 15.2. The electrochemical characteristics of EDLC SCs are controlled by the physico-chemical interactions between electrodes, outer surface, and electrolyte [24]. The effective width of the electric double layer depends on the concentration of electrolyte, the radius of electrolyte ions, the charge on electrolyte ions,

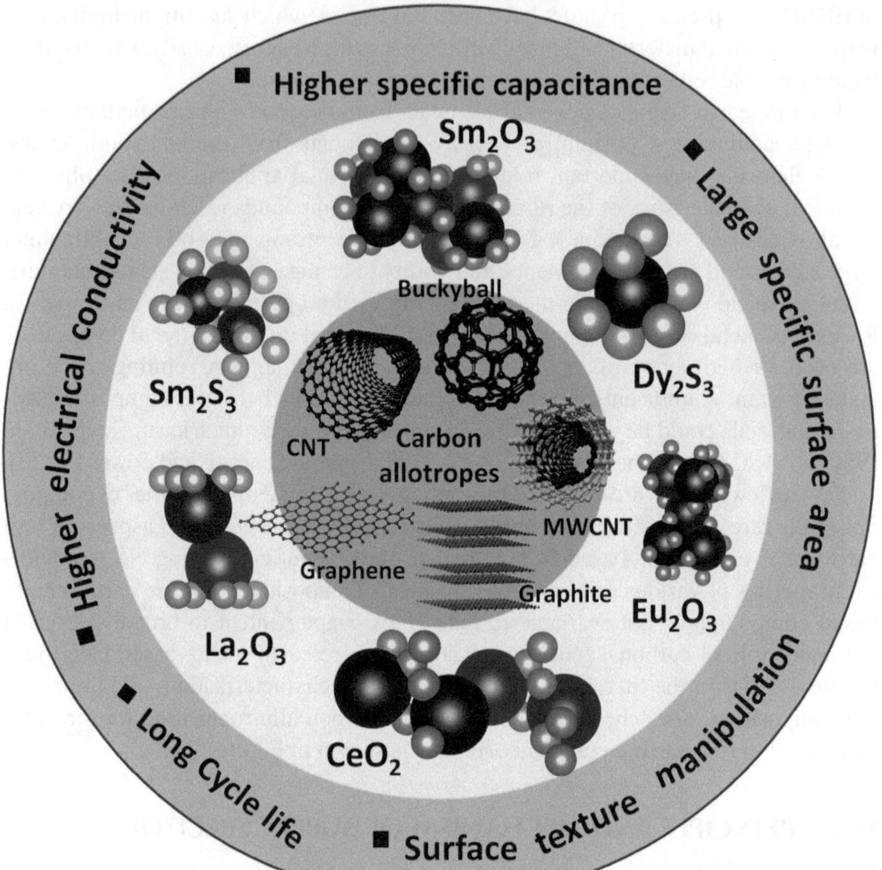

FIGURE 15.1 Schematic representation of carbon allotropes supported rare earth metal chalcogenide materials for SCs.

and hydration sphere (5–10 Å). The materials possessing high specific surface area (SSA) and higher conductivity such as AC, graphene, CNTs, and aerogels are some of the most extensively studied active materials for EDLC. The SSA of materials in between 800 and 3,000 $m^2 g^{-1}$ can give capacitance up to 550 F g^{-1}. On the contrary, the finite electrical conductivity and inaccessibility of all the surface sites are the obstacles to attain such high specific capacitance and have been restrained to 50–350 F g^{-1}.

15.2.2 Pseudocapacitor

Compared to EDLCs, the electroactive materials store charges that involve reversible faradaic reactions and are called pseudocapacitors. In this process, the valance state of electroactive material changes from one state to another as a consequence of

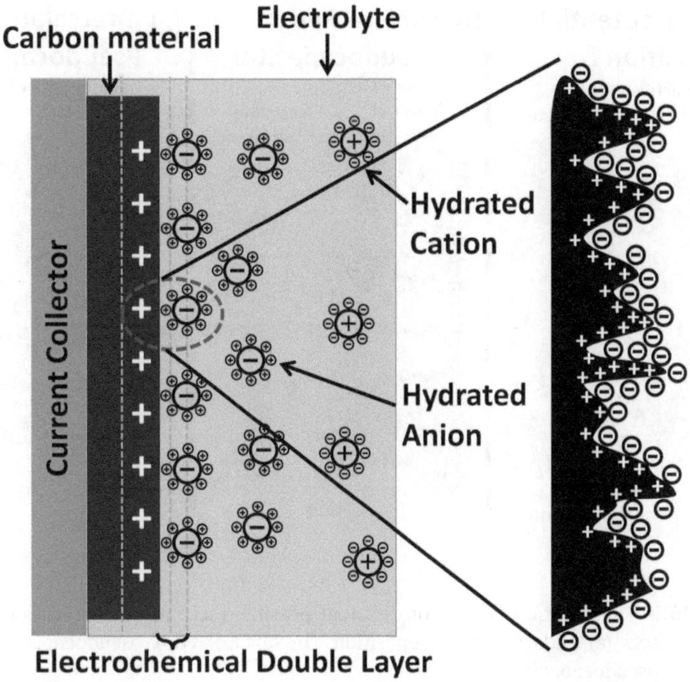

FIGURE 15.2 Schematic illustration of working principle for an electrochemical double-layer capacitance.

electron transfer. The first reported pseudocapacitive material was ruthenium oxide (RuO_2) [25,26]. The rectangular shape of the voltammogram of RuO_2 thin film is the characteristics of capacitive nature but this process is 'Faradaic'. This mechanism is the intermediate between the EDLC and the diffusion-controlled redox reactions involved in battery-type electroactive materials. The extents of the charge stored by the electrode depend on the coefficient of diffusion and the diffusion time. This charge storage process involves different faradaic mechanisms such as underpotential deposition, redox pseudocapacitance, and intercalation pseudocapacitance. The former one occurs upon the external face of the electrode causing deposition of electrolyte ions at the potentials positive to their redox reversible redox potential. In redox pseudocapacitance, some extents of electrode material undergo reversible redox reaction via electron transfer, and oxidized species are electrochemically adsorbed onto the surface (e.g. RuO_2 [27,28] and MnO_2 [29,30]). Intercalation of ions into redox-active material without modification in crystal structure and in the time scale of EDLC is called intercalation pseudocapacitance (e.g. Nb_2O_5) [31]. Three different mechanisms for charge storage in pseudocapacitors are presented in Figure 15.3a–c.

The operational potential window of the pseudocapacitive materials between 0.3 and 0.6 V affects specific capacitance and specific energy. The maximum attainable capacitance depends on the number of electrons involved in the reversible redox reactions and electrolyte ion density [32].

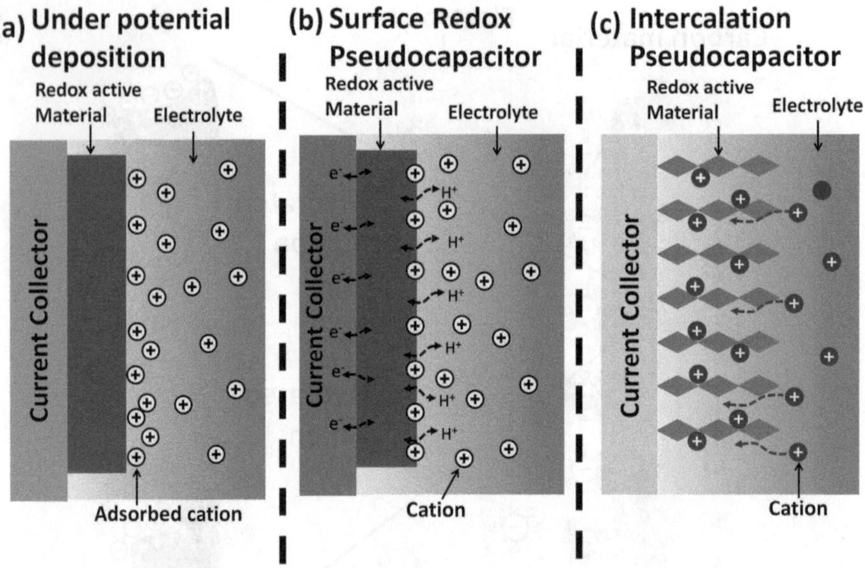

FIGURE 15.3 Schematic diagram of different possible mechanisms for various types of pseudocapacitors (a) underpotential deposition, (b) surface redox pseudocapacitor, and (c) intercalation pseudocapacitor.

15.3 FACTORS AFFECTING SUPERCAPACITOR PERFORMANCE

15.3.1 Chemical Composition of Material

The SC performance is predominantly determined by the chemical composition of the active material. The nature of the charge transfer mechanism, efficiency, stability, and range of specific energy and power are decided by the chemical composition of the electrode material. The adsorbed water can assist the ion transporting process. The supercapacitive properties can be enhanced by optimum heating of electrode material, as the heat treatment removes physically bound water content [33,34].

15.3.2 Electrolyte

The electrolyte is an important component of SC device and its chemical and physical characteristics affect the device's performance, including specific capacitance, rate performance, power density, cycling stability, and safety [35]. The interaction between electrode and electrolyte at the interface in all electrochemical processes influences surface morphology and state and internal structure of electrode material. To date, no electrolyte can accomplish the ideal requirements of the SC device. The ionic conductivity of water-based electrolytes is high but has low energy density and power density, less stability, and leakage problems. On the other hand, organic electrolytes have high working potential but lack higher ionic conductivity compared to aqueous electrolytes, and the problem of leakage with safety concerns still remains.

The solid-state electrolytes can overcome the concerns of liquid electrolytes. The chemical structure, reaction mechanism, and progress in electrolyte development occupy an important place for the fabrication of efficient and safety of SC devices [36].

15.3.3 Temperature

As the viscosity of the electrolyte is related to the temperature, the SC temperature plays a crucial role in the charge–discharge operations of the SCs. The EDLC electrode materials are thermally stable, but the crystal structure and reactivity of pseudocapacitive materials depend on the temperature of the system [37]. The capacitance, efficiency, and stability of SC depend on the working temperature. Generally, at higher temperatures, specific capacitance increases, but stability decreases significantly [38].

15.3.4 Crystal Structure and Crystallinity

Various types of crystal structures were prepared and tested for supercapacitive applications. Some efforts were made to increase the inter planer distance of the crystal so it could facilitate fast intercalation of ions into the layered crystal structure. Nanocrystalline or amorphous materials possess a higher electroactive surface area. For higher crystalline material, a lower electroactive surface area and higher electrical conductivity are expected. Hence, intercalation–deintercalation, migration, and diffusion of ions are restricted [39,40].

15.3.5 Morphology

Normally, the morphology of material strongly affects the performance of the SC electrode. To improve ionic transport inside the electrode, open architecture and morphology of the electrode play a central role. Therefore, the electrodes based on different dimensions have different electrochemical performances [41,42].

15.3.6 Specific Surface Area and Pore Structure

Usually, higher SSA efficiently offers higher electrochemical active sites. The appropriate pore size distribution on the surface of the electrode has a crucial role in the process of charge storage for both EDLCs and pseudocapacitors [43]. Chmiola et al. [44] found that pore diameter should be greater than the diameter of solvated electrolyte ions. With the rise in pore size, the volumetric capacitance decreases. The microporous structure with pores <1 nm helps to improve capacitance.

15.3.7 Thickness of the Electrode

The mechanical integrity, specific capacitance, and internal resistance (IR) depend on the thickness of the electrode material. Tsay et al. [45] mentioned that with increasing thickness of electrode, the specific capacitance of carbon electrode decreases as intercalation of electrolyte ions into electrode becomes more difficult. Liu et al. [46] found a nonlinear correlation between IR and the thickness of the electrode.

15.4 RARE EARTH METAL CHALCOGENIDES–BASED CARBONACEOUS COMPOSITES

15.4.1 Cerium Chalcogenides Composites

The ability of cerium to undergo reversible redox reactions in electrochemical reactions made cerium chalcogenides (specially, cerium oxide) and its composites the most studied rare earth metal chalcogenide for catalysis and energy storage applications. The composite materials are prepared in order to increase SSA, conductivity, and to manipulate surface morphology. Deng et al. [47] prepared cerium oxide/Multi Wall Carbon Nanotubes (MWCNTs) composites at different weight ratios by hydrothermal method. It was observed that CeO_2/MWCNTs nanocomposite with a 1:1 weight ratio shows higher specific capacitance compared to MWCNTs, CeO_2, and other composite electrodes. The Coulombic efficiency for the composite electrode increases significantly and achieved a capacitance of 455.6 F g^{-1} at charge–discharge current density of 1 A g^{-1}. The composite of CeO_2 with graphene was prepared by physical mixing method by Sarpoushi et al. [48]. The electrochemical characteristics of electrodes were evaluated in NaCl, NaOH, and KOH electrolytes. The electrode of composite material exhibited capacitance of 11.09 F g^{-1} in NaCl electrolyte. This study confirmed that manipulation of composition of the electrode with the addition of graphene enhances electrochemical performance and the capacity depends on the extent of charge separation and Faradic reactions.

In the efforts to add one more pseudocapacitive component to the rare earth elements chalcogenide composite, in situ polymerization and electrochemical reduction methods were used to prepare a ternary composite material graphene–CeO_2/polyaniline. It was observed that the variation in the composition of CeO_2 and PANI effectively alters the supercapacitive characteristics of the composite material in acidic electrolyte. The best electrochemical performance at the charging current density 1 A g^{-1} was 454.8 F g^{-1} for a mass ratio of 1:4 [49].

CeO_2/graphene composite was synthesised by Wang et al. [50] in which CeO_2 was anchored on graphene. The SSA of composite material decreased from 487 to 194 m^2g^{-1} as compared with graphene, but higher specific capacitance (208 F g^{-1}) was achieved for a composite material with excellent cycle life (~100%) after 1,000 cycles.

Ji et al. [51] synthesised reduced graphene oxide (rGO)/CeO_2 composites using a two-step method, a thermal treatment preceded by a self-assembly method. The introduction of rGO prevented aggregation as well as controlled morphology of CeO_2. At a sweep rate of 0.005 V s^{-1}, the composite electrode exhibited a capacitance of 265 F g^{-1}. The composite electrode maintained 96.2% of its initial capacity over 1,000 cyclic voltammetry (CV) cycles at a voltage sweep rate of 0.8 V s^{-1}. Such type of nanocomposites offers great promise for SC applications. A composite of CeO_2 and AC was prepared by mechanical mixing. The composite material containing

10 wt% CeO_2 exhibited a capacitance of 162 F g^{-1}. The SSC device fabricated using a composite electrode retained 99% capacitance after 1,000 galvanostatic charge–discharge (GCD) cycles at 0.02 A cm^{-2} current density and exhibited a specific power of 3,500 W kg^{-1} at a discharging current of 18 mA cm^{-2} [52].

Deng et al. [53] synthesized CeO_2 and CeO_2 decorated GO using the facile solution-based co-precipitation method. At the mole ratio of 1:4, CeO_2 decorated GO showed specific capacitance of 382.94 F g^{-1} at 3 A g^{-1}. The increase in capacitance was due to the contributions of the good electrical conductivity of GO and the pseudocapacitance of CeO_2. Moreover, about 86% of initial capacitance was stored after 500 electrochemical charge–discharge cycles. Padmanathan et al. [54] reported CeO_2 nanostructures synthesized via hydrothermal method. The CeO_2 nanorods supported by carbon microstructure achieved capacitance of 644 F g^{-1} at the current of 0.5 A g^{-1}. Dezfuli et al. [55] reported a composite of CeO_2–rGO prepared by sonochemical route to shoot up the specific capacitances of SSCs. The specific capacitance of 211 F g^{-1} was obtained for CeO_2–rGO composite electrodes at a sweep rate of 0.002 V s^{-1}. After 4,000 repeated cycles, the capacity increased to a value of 105.6%.

15.4.2 Lanthanum Chalcogenides Composites

Lanthanum-based chalcogenides were most studied for the electrochemical SC among the other REEs. Rajagopal et al. [56] prepared lanthanum oxide (La_2O_3) and rGO composite by hydrothermal method and achieved areal capacitance of 889.29 F cm^{-2} much greater than 67.47 F cm^{-2} of La_2O_3. The electrochemical stability of La_2O_3 nanoparticles was 61.39% after 1,000 cycles at the sweep rate of 0.1 V s^{-1}, and the introduction of rGO increased the capacitance retention ability of the electrode up to 84%. The nanocomposite of rGO/La_2O_3 prepared by a simple reflux process followed by reduction using N_2H_4 displayed a high specific capacitance of 156.25 F g^{-1} at 0.1 A g^{-1} and outstanding electrochemical cyclability. The material retained 78% of its initial Coulombic efficiency over 500 cycles. The outstanding electrochemical performance of composite electrode material was credited to the increased interactive area between rGO–La_2O_3 and electrolyte [57].

15.4.3 Samarium Chalcogenide Composites

Aside from cerium and lanthanum chalcogenides, other rare earth metal chalcogenides have been evaluated for their electrochemical properties concerning supercapacitive application, having economic disadvantages, but the electrochemical activities are good that make them fascinating for the purpose. These electrodes showed specific capacitance in between 81 and 403 F g^{-1} along with good cyclability. Dezfuli et al. [58] synthesized Sm_2O_3 nanoparticles using sonochemical method and composited them onto the surface of GO through a self-assembly procedure in different compositions of GO. The reduction of GO to rGO was carried out using hydrazine (N_2H_4) as a reducing agent. Observation of broad X-ray diffraction (XRD) peaks confirmed the formation of low crystalline samarium oxide nanoparticles. The unchanged morphology of Sm_2O_3 nanoparticles suggested that composition and reduction processes do not affect morphology. Also, the advantage of such type of synthesis process was the prevention of restacking of rGO sheets. The composite electrode exhibited a much higher specific capacitance of 321 F g^{-1} at the sweep rate of 0.002 V s^{-1} compared to pristine Sm_2O_3 nanoparticles (151 F g^{-1}) in 0.5 M Na_2SO_4 electrolyte. Additionally, the composite material revealed 99% cyclability of its specific capacitance over 4,000 cycles. The stability study suggests that the quantity of rGO in the electrode has a great influence on the electrochemical cyclability of the material, and hence the composition of the optimal quantity of rGO in Sm_2O_3 is decisive for improvement in cyclic stability. The resistive parameters of the composite material were also superior to that of pristine Sm_2O_3 nanoparticles. The charge transfer resistance was decreased to 1.83 Ω compared to other composites with a lower and higher concentration of rGO.

Among the composite materials, most of the times, the pseudocapacitive materials were anchored on carbon backbone. Despite that, they often undergo severe performance fading after repeated cycling over 500 times due to shrinkage and swelling of pseudocapacitive material exposed to electrolyte. To overcome this problem and achieve higher intercalation of electrolyte ions, the GO, Sm_2S_3, and GO/Sm_2S_3 composite thin films were successfully deposited by Ghogare et al. [59] using successive ionic layer adsorption and reaction (SILAR) method. The schematic of the method is shown in Figure 15.4a. Introduction of GO in the preparation method made nanocrystalline Sm_2S_3.

The SEM images (Figure 15.4 b and c) displayed the generation of number of mesopores on the film surface with an average pore radius of 21.79 nm. The modification of electrodes through alternate layers of Sm_2O_3 and GO sheets acted as the best strategy. The cracked-mud-like surface morphology of Sm_2S_3 changed to porous nanostrips like-morphology after the addition of GO. The composite electrode exhibited a higher specific capacitance (360 F g^{-1}) which is ~six times that of Sm_2S_3 electrode. The capacitive retention of GO/Sm_2S_3 electrode was 88.14% over 2,000 repeated

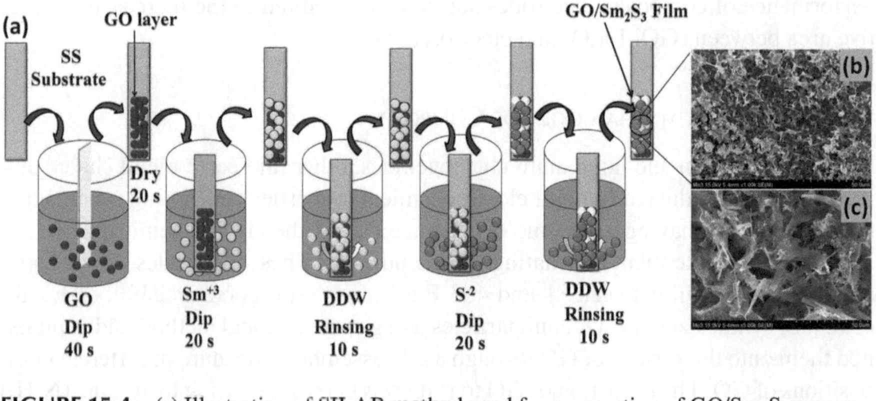

FIGURE 15.4 (a) Illustration of SILAR method used for preparation of GO/Sm_2S_3 composite thin films on stainless steel substrate, (b and c) SEM images of GO/Sm_2S_3. (Reproduced from ref. 59 © 2020, Elsevier.)

cycles at a sweep rate of 0.1 V s^{-1}. The structures have shown a high value of specific capacitance and rate capability and good electrochemical cycling stability due to the synergistic effect of both Sm_2O_3 and GO. The series resistance of Sm_2S_3 electrode decreased from 3.57 to 1.14 Ω cm^{-2} with the composition.

15.4.4 Europium Chalcogenides Composites

The synthesis of Eu_2O_3 nanoparticles and composition of them with rGO sheets through self-assembly is reported by Naderi et al. [60]. Further, the electrochemical studies proved that Eu_2O_3 decorated on rGO sheets has a capacitance of 313 F g^{-1} at a sweep rate of 0.002 V s^{-1}. After 4,000 cycles, the Eu_2O_3/rGO showed electrochemical cycling stability of 96.5%.

Nanorods of Eu_2O_3 were synthesised using the hydrothermal method and blended with the sheet-like rGO structure by employing a sonochemical procedure through a self-assembly methodology. Due to the effect of composition, the capacitance of Eu_2O_3 nanorods improved to 403 F g^{-1} from 288 F g^{-1}. The CV and GCD behaviours of Eu_2O_3, rGO, and Eu_2O_3/rGO electrodes (Figure 15.5a and b) indicated that

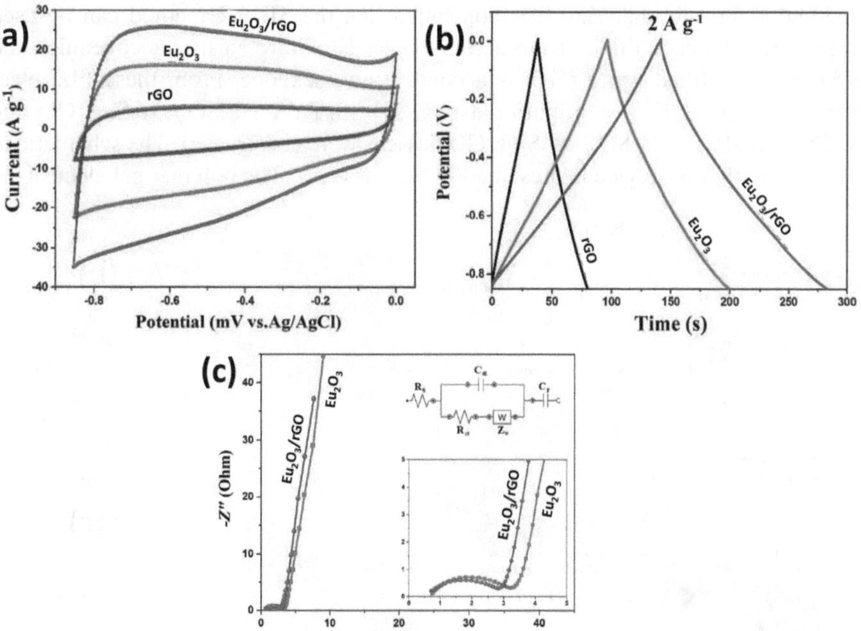

FIGURE 15.5 (a) The CV curves at a potential sweep rate of 0.05 V s^{-1}, (b) the GCD plots at current density of 2 A g^{-1} of Eu_2O_3, rGO, and Eu_2O_3/rGO, and (c) Nyquist plots of Eu_2O_3 and Eu_2O_3/rGO. (Reproduced with permission from ref. 61 © 2020, The Royal Society of Chemistry.)

Eu_2O_3/rGO showed a good electrochemical response. This proved that in rare earth metal compounds the electrochemical performance can be enhanced by the composition with rGO. In most composites, the rare earth chalcogenide serves as the pseudocapacitive material and the associated carbon allotrope as a charge transfer mediator. The Nyquist plots of Eu_2O_3 and Eu_2O_3/rGO (Figure 15.5c) confirm that R_s and R_{ct} of composite material decreased significantly. The values of R_s were brought down from 0.75 to 0.71 Ω, and the values of R_{ct} were reduced from 2.61 to 2.19 Ω with the composition of Eu_2O_3 and rGO. The stability of Eu_2O_3 nanorods enhanced from 95.5% to 96.8% after 5,000 cycles [61].

15.4.5 Dysprosium Chalcogenides Composites

A highly controlled SILAR method was employed by Lokhande's group to prepare Dy_2S_3@rGO composite. The composition of Dy_2S_3 with rGO in thin-film form enhanced SSA up to 78 $m^2 g^{-1}$. Bagwade et al. [62] studied the electrochemical behaviour of Dy_2S_3 and Dy_2S_3@rGO composite in 1 M aqueous Na_2SO_4 and observed that the R_s of Dy_2S_3 electrode decreased from 1.95 to 1.1 Ω cm^{-2} for Dy_2S_3@rGO. For Dy_2S_3@rGO thin film, the specific capacitance improved to 517 F g^{-1} than that of Dy_2S_3 thin film (256 F g^{-1}). By comparing the electrochemical results of the pristine and composite material, it is concluded that the SILAR method can be used to prepare nanocrystalline electroactive materials of rare earth chalcogenides and effectively modified using rGO as a conducting backbone. From these film electrodes, flexible SSC of configuration rGO@Dy_2S_3/PVA–Na_2SO_4/Dy_2S_3@rGO and ASC; MnO_2/PVA–Na_2SO_4/Dy_2S_3@rGO devices were constructed. The schematic of fabrication of ASC device is presented in Figure 15.6a. The polymer gel electrolyte,

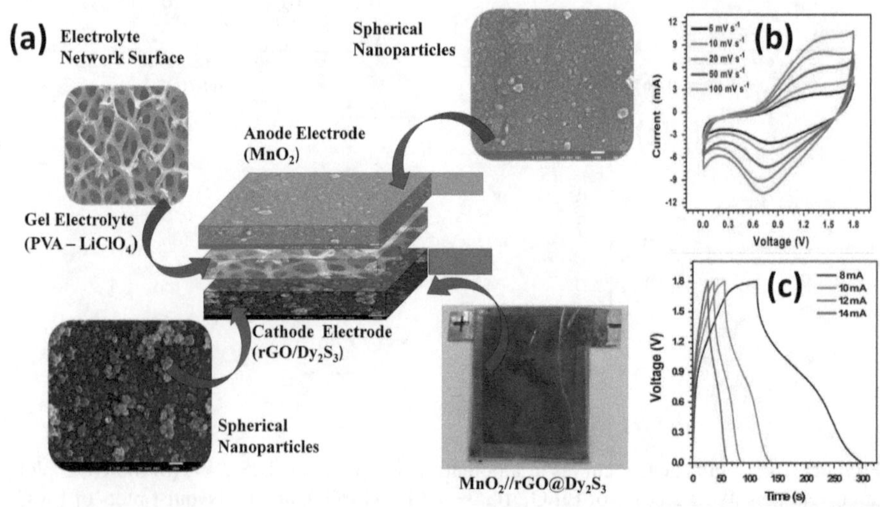

FIGURE 15.6 (a) Schematic representation of fabrication of asymmetric solid-state MnO_2// rGO@Dy_2S_3 device, (b) the CV curves, and (c) the GCD curves. (Reproduced from ref. 62 © 2020, Elsevier.)

PVA–Na_2SO_4, acts as a separator as well. The voltammograms and GCD curves of ASC device are presented in Figure 15.6b and c, respectively. The specific capacitance for SSC device was 81 F g^{-1} and for ASC, 118 F g^{-1} at 0.005 V s^{-1} sweep rate, which was sustained when twisted by 165°. The ASC device provided specific energy of 41 Wh kg^{-1} at 1,330 W kg^{-1}.

15.5 SUMMARY AND CONCLUSIONS

The chapter discusses research development in the carbon composites of rare earth metal compounds for SC electrodes. The pristine rare earth metal chalcogenides suffer from the lower electroactive area, limited electrical conductivity, and poor cyclability, which restrict their use as an electroactive material in SCs. The composite electrodes provide high charge storage per unit mass as well as high specific energy. Individual carbon allotrope has various advantages such as 0D carbons, providing the flexibility for tuning porosity. Compared to 0D carbon 1D carbon nanostructures have relatively smaller SSA. However, they assist in faster charge transport in SC electrodes due to the more electrical conductivity. The high electronic conductivity, surface area, and mechanical strength can be achieved through the composition of 2D carbon nanostructures. To increase electrode–electrolyte interactions, the surface texture of rare earth metal chalcogenides–based composite materials plays an important role. The synthesis method and surface texture along with electrochemical parameters of carbon composites of rare earth metal chalcogenides are summarised in Table 15.1. The highest specific capacitance was obtained for La_2O_3/rGO (889.29 F cm^{-2}) followed by CeO_2/CNTs (818 F g^{-1}).

ACKNOWLEDGEMENT

This work is financially supported by D. Y. Patil Education Society, Kolhapur (Sanction no. DYPES/DU/R&D/3101) and Science and Engineering Board (SERB), New Delhi (Sanction no. TTR/2021/000006).

TABLE 15.1
Electrochemical Performance of Carbon Composites of Rare Earth Metal Chalcogenides

Sr. No.	Electrode Material	Synthesis Method	Morphology	Electrolyte	Specific Capacitance ($F\ g^{-1}$)	Stability (%) (Cycles)	R_s	R_{ct}	References
1	CeO_2/MWCNTs	Hydrothermal	Aggregated nanoparticles	6 M KOH	455.6 @ 1 A g^{-1}	81.1 (2,000)	–	–	46
2	CeO_2/graphene.	Ultrasonication	Nano-flakes	3 M NaCl	11.09 @ 0.005 V s^{-1}	137 (500)	8.351 Ω	732 Ω	47
3	rGO-CeO_2/PANI	Sacrificial template	Nanosheets with porous structure	1 M H_2SO_4	454.8 @ 1.0 A g^{-1}	70.23 (10,000)	–	2.12 Ω	48
4	CeO_2/graphene	Co-precipitation	Nanosheets	3 M KOH	208 @ 0.0031 mA cm^{-2}	100 (1,000)	1.13 Ω	–	49
5	rGO/CeO_2	self-assembly followed by thermal treatment	Nanosheets	3 M KOH	265 @ 0.005 V s^{-1}	96.2 (1,000)	–	–	50
6	CeO_2/AC	Co-precipitation	Nanoparticle	1 M H_2SO_4	162 @ 0.01 V s^{-1}	99 (1000)	0.55 Ω	0.13 Ω	51
7	CeO_2/GO	Co-precipitation	Nanoparticle	6 M KOH	382.94 @ 3 A g^{-1}.	86.05 (500)	–	–	52
8	CeO_2/C	Hydrothermal	Nanorods coated nano-plates	3 M KOH	644 @ 0.5 A g^{-1}.	91.4 (2000)	–	0.5 Ω	53
9	CeO_2/CNTs	Ultrasound and hydrothermal	Nanoparticles	2 M KOH	818 @ 0.001 V s^{-1}.	95.3 (2,000)	–	–	54
10	CeO_2/rGO	Sonochemical	Nanorods coated nanosheets	0.5 M Na_2SO_4	211 @ 0.002 V s^{-1}	105.6 (4,000)	0.36 Ω	7.5 Ω	55

(Continued)

TABLE 15.1 (*Continued*)
Electrochemical Performance of Carbon Composites of Rare Earth Metal Chalcogenides

Sr. No.	Electrode Material	Synthesis Method	Morphology	Electrolyte	Specific Capacitance ($F g^{-1}$)	Stability (%) (Cycles)	R_s	R_{ct}	References
11	La_2O_3/rGO	Hydrothermal	Nanosheets	1 M Na_2SO_4	889.29 F cm^{-2} @ 0.01 V s^{-1}	84 (1,000)	23.32 Ω	–	56
12	La_2O_3/rGO	Reflux	Nanorods coated nanosheets	3 M KOH	156.25 @ 0.1 A g^{-1}	78 (500)	0.45 Ω	0.5 Ω	57
13	Sm_2O_3/rGO	Sonochemical	Nanoparticles coated nanosheets	0.5 M Na_2SO_4	321 @ 0.002 V s^{-1}	99 (4,000)	1.09 Ω	1.83 Ω	58
14	Sm_2O_3/GO	SILAR	Nano-strips like	1 M Na_2SO_4	360 @ 0.005 V s^{-1}	88.14 (2,000)	3.86 Ω cm^{-2}	4.59 Ω cm^{-2}	59
15	Eu_2O_3/rGO	Sonochemical	Layered and wrinkled	3 M KCl	313 @ 0.002 V s^{-1}	96.5 (4,000)	1.53 Ω	4.03 Ω	60
16	Eu_2O_3/rGO	Hydrothermal	Nanosheets	3 M KOH	403 @ 0.002 V s^{-1}	96.8 (5,000)	0.71 Ω	2.19 Ω	61
17	Dy_2S_3/rGO	SILAR	Nano-strips like	1 M Na_2SO_4	60.5 @ 0.005 V s^{-1}	88 (5,000)	1.1 Ω cm^{-2}	2.4 Ω cm^{-2}	62

REFERENCES

1. Down, M. P., Rowley-Neale, S. J., Smith, G. C., & Banks, C. E. (2018). Fabrication of graphene oxide supercapacitor devices. *J. Phys. Chem. Solids*, 1, 707–714.
2. Hasan, M. M., Islam, M. D., & Rashid, T. U. (2020). Biopolymer-based electrolytes for dye-sensitized solar cells: A critical review. *Energy & Fuels*, 34, 15634–15671.
3. Zhu, Q., Zhao, D., Cheng, M., Zhou, J., Owusu, K. A., Mai, L., & Yu, Y. (2019). A new view of supercapacitors: Integrated supercapacitors. *Adv. Energy Mater.*, 9, 1901081.
4. Dou, Q., & Park, H. S. (2020). Perspective on high-energy carbon-based supercapacitors. *Energy Environ. Mater.*, 3, 286–305.
5. Ci, J., Cao, C., Kuga, S., Shen, J., Wu, M., & Huang, Y. (2017). Improved performance of microbial fuel cell using esterified corncob cellulose nanofibers to fabricate air-cathode gas diffusion layer. *ACS Sustain. Chem. Eng.*, 5, 9614–9618.
6. Malavekar, D. B., Lokhande, V. C., Mane, V. J., Ubale, S. B., Patil, U. M., & Lokhande, C. D. (2020). Enhanced energy density of flexible asymmetric solid state supercapacitor device fabricated with amorphous thin film electrode materials. *J. Phys. Chem. Solids*, 141, 109425.
7. Dousti, B., Choi, Y. I., Cogan, S. F., & Lee, G. S. (2020). A high energy density 2D microsupercapacitor based on an interconnected network of a horizontally aligned carbon nanotube sheet. *ACS Appl. Mater. Interf.*, 12, 50011–50023.
8. Avasthi, P., Kumar, A., & Balakrishnan, V. (2019). Aligned CNT forests on stainless steel mesh for flexible supercapacitor electrode with high capacitance and power density. *ACS Appl. Nano Mater.*, 2, 1484–1495.
9. Zhao, Y., Liu, J., Hu, Y., Cheng, H., Hu, C., Jiang, C., Jiang, C., Cao, A., & Qu, L. (2013). Highly compression-tolerant supercapacitor based on polypyrrole-mediated graphene foam electrodes. *Adv. Mater.*, 25, 591–595.
10. Li, M., Tang, Z., Leng, M., & Xue, J. (2014). Flexible solid-state supercapacitor based on graphene-based hybrid films. *Adv. Funct. Mater.*, 24, 7495–7502.
11. Asen, P., Shahrokhian, S., & Zad, A. I. (2018). Transition metal ions-doped polyaniline/graphene oxide nanostructure as high performance electrode for supercapacitor applications. *J. Solid State Electrochem.*, 22, 983–996.
12. Meng, Q., Cai, K., Chen, Y., & Chen, L. (2017). Research progress on conducting polymer based supercapacitor electrode materials. *Nano Energy*, 36, 268–285.
13. Dubal, D. P., Ayyad, O., Ruiz, V., & Gomez-Romero, P. (2015). Hybrid energy storage: The merging of battery and supercapacitor chemistries. *Chem. Soc. Rev.*, 44, 1777–1790.
14. Liu, R., Zhou, A., Zhang, X., Mu, J., Che, H., Wang, Y., Wang, T., Zhang, Z., Kou, Z., & Kou, Z. (2021). Fundamentals, advances and challenges of transition metal compounds-based supercapacitors. *Chem. Eng. J.*, 412, 128611.
15. Sk, M. M., Yue, C. Y., Ghosh, K., & Jena, R. K. (2016). Review on advances in porous nanostructured nickel oxides and their composite electrodes for high-performance supercapacitors. *J. Power Sources*, 308, 121–140.
16. Yang, Z., Ren, J., Zhang, Z., Chen, X., Guan, G., Qiu, L., Zhang, Y., Peng, H., & Peng, H. (2015). Recent advancement of nanostructured carbon for energy applications. *Chem. Rev.*, 115, 5159–5223.
17. Wang, J., Zhang, X., Li, Z., Ma, Y., & Ma, L. (2020). Recent progress of biomass-derived carbon materials for supercapacitors. *J. Power Sources*, 451, 227794.
18. Borenstein, A., Hanna, O., Attias, R., Luski, S., Brousse, T., & Aurbach, D. (2017). Carbon-based composite materials for supercapacitor electrodes: A review. *J. Mater. Chem. A*, 5, 12653–12672.
19. Ghosh, S., Anbalagan, K., Kumar, U. N., Thomas, T., & Rao, G. R. (2020). Ceria for supercapacitors: Dopant prediction, and validation in a device. *Appl. Mater. Today*, 21, 100872.

20. Jadhav, S. B., Malavekar, D. B., Bulakhe, R. N., Patil, U. M., In, I., Lokhande, C. D., & Pawaskar, P. N. (2021). Dual-functional electrodeposited vertically grown Ag-La_2O_3 nanoflakes for non-enzymatic glucose sensing and energy storage application. *Surf. Interf.*, 23, 101018.
21. Kim, P., Anderko, A., Navrotsky, A., & Riman, R. E. (2018). Trends in structure and thermodynamic properties of normal rare earth carbonates and rare earth hydroxycarbonates. *Minerals*, 8, 106.
22. Zaka, A., Hayat, K., & Mittal, V. (2021). Recent trends in the use of three-dimensional graphene structures for supercapacitors. *ACS Appl. Electron. Mater.*, 3, 574–596.
23. Zhang, M., He, L., Shi, T., & Zha, R. (2018). Nanocasting and direct synthesis strategies for mesoporous carbons as supercapacitor electrodes. *Chem. Mater.*, 30, 7391–7412.
24. Simon, P., & Gogotsi, Y. (2013). Capacitive energy storage in nanostructured carbon–electrolyte systems. *Acc. Chem. Res.*, 46, 1094–1103.
25. Trasatti, S., & Buzzanca, G. (1971). Ruthenium dioxide: A new interesting electrode material. Solid state structure and electrochemical behaviour. *J. Electroanal. Chem. Interf. Electrochem.*, 29, A1–A5.
26. Ardizzone, S., Fregonara, G., & Trasatti, S. (1990). "Inner" and "outer" active surface of RuO_2 electrodes. *Electrochim. Acta*, 35, 263–267.
27. Patil, U. M., Kulkarni, S. B., Jamadade, V. S., & Lokhande, C. D. (2011). Chemically synthesized hydrous RuO_2 thin films for supercapacitor application. *J. Alloys Compd.*, 509, 1677–1682.
28. Makino, S., Ban, T., & Sugimoto, W. (2015). Towards implantable bio-supercapacitors: Pseudocapacitance of ruthenium oxide nanoparticles and nanosheets in acids, buffered solutions, and bioelectrolytes. *J. Electrochem. Soc.*, 162, A5001.
29. Mane, V. J., Malavekar, D. B., Ubale, S. B., Lokhande, V. C., & Lokhande, C. D. (2020). Manganese dioxide thin films deposited by chemical bath and successive ionic layer adsorption and reaction deposition methods and their supercapacitive performance. *Inorg. Chem. Commun.*, 115, 107853.
30. Chodankar, N. R., Dubal, D. P., Lokhande, A. C., Patil, A. M., Kim, J. H., & Lokhande, C. D. (2016). An innovative concept of use of redox-active electrolyte in asymmetric capacitor based on MWCNTs/MnO_2 and Fe_2O_3 thin films. *Sci. Rep.*, 6, 1–14.
31. Kong, L., Zhang, C., Wang, J., Qiao, W., Ling, L., & Long, D. (2015). Free-standing T-Nb_2O_5/graphene composite papers with ultrahigh gravimetric/volumetric capacitance for Li-ion intercalation pseudocapacitor. *Acs Nano*, 9, 11200–11208.
32. Augustyn, V., Come, J., Lowe, M. A., Kim, J. W., Taberna, P. L., Tolbert, S. H., Abruña, H. D., Simon, P., & Dunn, B. (2013). High-rate electrochemical energy storage through Li^+ intercalation pseudocapacitance. *Nat. Mater.*, 12, 518–522.
33. Prasad, K. R., & Miura, N. (2004). Electrochemically synthesized MnO_2-based mixed oxides for high performance redox supercapacitors. *Electrochem. Commun.*, 6, 1004–1008.
34. Kim, B. K., Sy, S., Yu, A., & Zhang, J. (2015). Electrochemical supercapacitors for energy storage and conversion. In Jinyue Yan (Ed), *Handbook of Clean Energy Systems*, 1–25.
35. Pal, B., Yang, S., Ramesh, S., Thangadurai, V., & Jose, R. (2019). Electrolyte selection for supercapacitive devices: A critical review. *Nanoscale Adv.*, 1, 3807–3835.
36. Zhong, C., Deng, Y., Hu, W., Qiao, J., Zhang, L., & Zhang, J. (2015). A review of electrolyte materials and compositions for electrochemical supercapacitors. *Chem. Soc. Rev.*, 44, 7484–7539.
37. Wang, F., Wu, X., Yuan, X., Liu, Z., Zhang, Y., Fu, L., Zhu, Y., Zhou, Q., Wu, Y. & Huang, W. (2017). Latest advances in supercapacitors: From new electrode materials to novel device designs. *Chem. Soc. Rev.*, 46, 6816–6854.

38. Sivaraman, P., Mishra, S. P., Potphode, D. D., Thakur, A. P., Shashidhara, K., Samui, A. B., & Bhattacharyya, A. R. (2015). A supercapacitor based on longitudinal unzipping of multi-walled carbon nanotubes for high temperature application. *RSC Adv.*, 5, 83546–83557.
39. Cheng, J. P., Gao, S. Q., Zhang, P. P., Wang, B. Q., Wang, X. C., & Liu, F. (2020). Influence of crystallinity of $CuCo_2S_4$ on its supercapacitive behavior. *J. Alloys Compd.*, 825, 153984.
40. Malavekar, D. B., Kale, S. B., Lokhande, V. C., Patil, U. M., Kim, J. H., & Lokhande, C. D. (2020). Chemically Synthesized Cu_3Se_2 film based flexible solid-state symmetric supercapacitor: Effect of reaction bath temperature. *J. Phys. Chem. C.*, 124, 28395–28406.
41. Subramanian, V., Zhu, H., Vajtai, R., Ajayan, P. M., & Wei, B. (2005). Hydrothermal synthesis and pseudocapacitance properties of MnO_2 nanostructures. *J. Phys. Chem. B*, 109(43), 20207–20214.
42. Verma, M. L., Minakshi, M., & Singh, N. K. (2014). Structural and electrochemical properties of nanocomposite polymer electrolyte for electrochemical devices. *Ind. Eng. Chem. Res.*, 53, 14993–15001.
43. Kondrat, S., Perez, C. R., Presser, V., Gogotsi, Y., & Kornyshev, A. A. (2012). Effect of pore size and its dispersity on the energy storage in nanoporous supercapacitors. *Energy Environ. Sci.*, 5, 6474–6479.
44. Chmiola, J., Yushin, G., Gogotsi, Y., Portet, C., Simon, P., & Taberna, P. L. (2006). Anomalous increase in carbon capacitance at pore sizes less than 1 nanometer. *Science*, 313, 1760–1763.
45. Tsay, K. C., Zhang, L., & Zhang, J. (2012). Effects of electrode layer composition/thickness and electrolyte concentration on both specific capacitance and energy density of supercapacitor. *Electrochim. Acta*, 60, 428–436.
46. Liu, X., Dai, X., Wei, G., Xi, Y., Pang, M., Izotov, V., Klyui, N., Havrykov, D., Ji, Y., Guo, Q., & Han, W. (2017). Experimental and theoretical studies of nonlinear dependence of the internal resistance and electrode thickness for high performance supercapacitor. *Sci. Rep.*, 7(1), 1–8.
47. Deng, D., Chen, N., Li, Y., Xing, X., Liu, X., Xiao, X., & Wang, Y. (2017). Cerium oxide nanoparticles/multi-wall carbon nanotubes composites: Facile synthesis and electrochemical performances as supercapacitor electrode materials. *Phys. E*, 86, 284–291.
48. Sarpoushi, M. R., Nasibi, M., Golozar, M. A., Shishesaz, M. R., Borhani, M. R., & Noroozi, S. (2014). Electrochemical investigation of graphene/cerium oxide nanoparticles as an electrode material for supercapacitors. *Mater. Sci. Semicond. Process.*, 26, 374–378.
49. Xie, A., Tao, F., Li, T., Wang, L., Chen, S., Luo, S., & Yao, C. (2018). Graphene-cerium oxide/porous polyaniline composite as a novel electrode material for supercapacitor. *Electrochim. Acta*, 261, 314–322.
50. Wang, Y., Guo, C. X., Liu, J., Chen, T., Yang, H., & Li, C. M. (2011). CeO_2 nanoparticles/graphene nanocomposite-based high performance supercapacitor. *Dalton Trans.*, 40, 6388–6391.
51. Ji, Z., Shen, X., Zhou, H., & Chen, K. (2015). Facile synthesis of reduced graphene oxide/CeO_2 nanocomposites and their application in supercapacitors. *Ceram. Int.*, 41, 8710–8716.
52. Aravinda, L. S., Bhat, K. U., & Bhat, B. R. (2013). Nano CeO_2/activated carbon based composite electrodes for high performance supercapacitor. *Mater. Lett.*, 112, 158–161.
53. Deng, D., Chen, N., Xiao, X., Du, S., & Wang, Y. (2017). Electrochemical performance of CeO_2 nanoparticle-decorated graphene oxide as an electrode material for supercapacitor. *Ionics*, 23, 121–129.

54. Padmanathan, N., & Selladurai, S. (2014). Shape controlled synthesis of CeO_2 nanostructures for high performance supercapacitor electrodes. *RSC Adv.*, 4, 6527–6534.
55. Dezfuli, A. S., Ganjali, M. R., Naderi, H. R., & Norouzi, P. (2015). A high performance supercapacitor based on a ceria/graphene nanocomposite synthesized by a facile sonochemical method. *RSC Adv.*, 5, 46050–46058.
56. Rajagopal, R., & Ryu, K. S. (2018). Facile hydrothermal synthesis of lanthanum oxide/hydroxide nanoparticles anchored reduced graphene oxide for supercapacitor applications. *J. Ind. Eng. Chem.*, 60, 441–450.
57. Zhang, J., Zhang, Z., Jiao, Y., Yang, H., Li, Y., Zhang, J., & Gao, P. (2019). The graphene/lanthanum oxide nanocomposites as electrode materials of supercapacitors. *J. Power Sources*, 419, 99–105.
58. Dezfuli, A. S., Ganjali, M. R., & Naderi, H. R. (2017). Anchoring samarium oxide nanoparticles on reduced graphene oxide for high-performance supercapacitor. *Appl. Surf. Sci.,* 402, 245–253.
59. Ghogare, T. T., Lokhande, V. C., Ji, T., Patil, U. M., & Lokhande, C. D. (2020). A graphene oxide/samarium sulfide (GO/Sm_2S_3) composite thin film: Preparation and electrochemical study. *Surf. Interfaces*, 19, 100507.
60. Naderi, H. R., Ganjali, M. R., & Dezfuli, A. S. (2018). High-performance supercapacitor based on reduced graphene oxide decorated with europium oxide nanoparticles. *J. Mater. Sci.: Mater. Electron.*, 29, 3035–3044.
61. Aryanrad, P., Naderi, H. R., Kohan, E., Ganjali, M. R., Baghernejad, M., & Dezfuli, A. S. (2020). Europium oxide nanorod-reduced graphene oxide nanocomposites towards supercapacitors. *RSC Adv.*, 10, 17543–17551.
62. Bagwade, P. P., Malavekar, D. B., Ghogare, T. T., Ubale, S. B., Mane, V. J., Bulakhe, R. N., In, I. & Lokhande, C. D. (2021). A high performance flexible solid-state asymmetric supercapacitor based on composite of reduced graphene oxide@ dysprosium sulfide nanosheets and manganese oxide nanospheres. *J. Alloys Compd.*, 859, 157829.

16 Low-Stress Abrasion Response of Heat-Treated LM25–SiCp Composite

Raj Kumar Singh
Rewa Engineering College

Sanjay Soni and Pushyamitra Mishra
Maulana Azad National Institute of Technology

CONTENTS

16.1 INTRODUCTION

Despite a lot of literature surveys available on aluminium matrix composites, many researchers are eager to study them more. This is mainly due to its practical use and commercialisation in several applications like marine, defence, automotive, aerospace and thermal organisations [1]. The machinery and equipment that are performing in a sandy environment suffer from severe abrasive wear. In order to prevent such types of machines and equipment from failure, understanding the low-stress abrasive wear mechanism is essential. The sand particles are significantly harder than the metals. Due to the small size of the sand particle, it is challenging to prevent them from going inside the machine, for instance, filtering. In all respects, the wear of the materials by sand particles is a severe problem. Therefore, the investigation of the low-stress (three-body) abrasion behaviour of the materials gets extremely important.

DOI: 10.1201/9781003220176-16

Al-Si alloys have appeared as energy and cost-effective alternative for various structural and tribological applications due to their several exciting characteristics. However, the main limitation of the alloys is that they suffer from their low wear properties exceptionally against higher applied loads and abrasive sizes during high-stress (two-body) abrasion [2]. Various research works have been conducted to determine the material removal mechanism of silicon-based aluminium alloys and their composite under sliding wear and two-body abrasion condition [3–19]. For sliding wear condition, Al-Si-based composite has a more significant seizure and wear resistance at higher sliding speeds and applied loads [3–14], but for high-stress abrasion condition, composite provides better wear properties than the matrix alloy up to specific abrasive size and applied load [15–19].

Few researchers have examined the three-body (low-stress) abrasion response of the different composites. Prasad et al. [20] examined the low-stress abrasion properties of LM6-flyash composite compared to the base alloy. The stir-squeeze casting technique was used for preparing the composite, and the variation of the flyash was 5%–12.5% by weight. The wear tests were performed at three varying applied loads (2, 3 and 4 kg), wheel speeds (250, 300, and 350 rpm), test speeds (200–400 rpm). AFS 50–70 seconds and of 200 μm was used as abrasives. The wear results have shown that the LM6-flyash composites have excellent wear properties than the matrix alloy because of the existence of flyash particles. Chandra et al. [21] studied the dry sand abrasion behaviour of 7,075-albite particle composite and compared it with the matrix alloy. The stir casting techniques were used to synthesise the composites, and the size and quantity of the reinforcement were 90–150 μm and 0–10 wt%, respectively. The wear parameter for the experiments was applied loads (2–10 N), fixed 200 rpm wheel speed and fixed 30 minute time. 200 μm of silica sand was used as abrasives. The results exhibited the uniform dispersion of the reinforcements in the matrix. The wear resistance of the composite was superior to the matrix alloy. Zheng et al. [22] investigated the three-body abrasive wear response of Cr26 iron-ZTA particle composite. For the synthesis of the composites, 5–15 mass% of contents of binder (Ti alloy) were chosen. The three-body abrasion test was performed on block-on-ring apparatus at 30 rpm speed and 1–3 kg applied loads. The results exhibited that the wear resistance of composite was increased by 5% because of the presence of the ZTA particles in the Cr26 iron. Ceramic particle-reinforced composite showed superior wear resistance by increasing the binder contents for 1 kg of the applied load; however, wear resistance was decreased for 2 and 3 kg of the applied load because the interfacial bonding strength between reinforcement and matrix decreased. Radhika et al. [23] investigated the three-body abrasion behaviour of Al–TiC composite and compared it to base materials. The centrifugal casting route reinforced the 10% TiC particles (size: 50 μm) with functionally graded LM25 alloy for preparing the functionally graded aluminium composite. The wear test is performed at the sliding distance of 1–13 mm, sliding speed of 100–200 rpm and an applied load of 28–99 N. The results exhibited that the abrasive wear rate was increased with increasing the loads; however, in case of sliding speed, the trend was reversed. The functionally graded composite has superior wear resistance than the LM25 alloy. If the applied load increased, the abrasion wear transition from mild to severe. Hence, the composite may be used to make cylinder liner in the cylinder blocks to prevent the three-body abrasion. Canakci [24] studied the three-body abrasion response of AA2014–B_4C composites compared to the base materials. The stirrer used casting route was used to synthesise the composite involving the AA2014 matrix and B_4C particles with a mean size of 85 μm in the varying proportions of 0, 2, 3, 6, 9 and 12 vol.%. The wear tests were performed on a modified block-on-disk tribometer at varying sliding times 200–2,080 seconds, constant applied load 92 N and fixed rotational speed 100 rpm. The results exhibited that the AA2014–B_4C composite has excellent wear properties than the AA2014 alloy because of the presence of boron carbide. Micro-ploughing and micro-cutting are leading the material removal mechanism. Due to the presence of boron carbide in the composite, micro-cutting and micro-ploughing were reduced. Modi et al. [25] examined the three-body abrasion response of Zn-Al alloy–Al_2O_3 composite compared with the matrix material. The liquid metallurgical route was used to prepare the composite involving the Zn-Al matrix and Al_2O_3 particles with a size of 75–150 μm in the proportion of 10 wt.%. The wear experiments were performed at 36 and 108 N applied load and fixed rotational speed of 100 rpm. The 50–100 μm of silica sand, SiC and zircon particle were separately used as the abrasives. The results exhibited that the composite has superior wear resistance than the base materials, irrespective of test conditions because alumina protected the composite's softer surface. Up to steady-state condition, the wear rate of the materials was decreased with increasing the sliding distances. The materials experienced a minimum material loss in zircon abrasives and maximum materials loss in silicon carbide abrasives. Shankar et al. [26] examined the abrasion response of Al8100–Gr composites and compared the matrix materials. The stirrer used casting techniques was used to synthesise the composites. The mean size and varying graphite particle weight percentages were 50 μm and 2–8, respectively. The 400 grit size of silica sand was used as the abrasives. The wear tests were conducted at varying applied loads of 10, 20 and 30 kg, different time periods 10, 20 and 30 minutes and a fixed rotational speed of 200 rpm. The results showed that the Al8100–Gr composite has superior wear properties than the base material. The abrasion rate of the materials was enhanced by increasing the applied loads. Sridhar et al. [27] examined the three-body abrasion response of A356-bottom ash composite and matrix alloy after heat treatment and compared it to the as-cast state. The liquid metallurgical route was used to synthesise the composite. The abrasive wear experiments were conducted at a different weight percentage of reinforcements (2%–10%) and applied loads (2–10 N), constant 200 rpm wheel speed and fixed 30 minute time. 200 μm of silica sand was used as abrasives. The abrasion results showed that increasing the reinforcement wt% and decreasing the applied loads increase the composite's wear resistance and indicate that the heat-treated materials have better wear resistance than the as-cast condition. However, the low-stress (three-body) abrasion response of silicon-based aluminium alloy reinforced with SiCp composite has not been investigated thoroughly. Therefore, very little is known about the low-stress abrasion response of the Al-Si-SiCp composite.

The silicon carbide particle (SiCp) has higher hardness and is easily available at a lower price. Therefore, there is a chance to make SiCps reinforced composite for wear-resistive applications at a lower cost. There is a necessity to investigate the abrasion response to assess the efficiency of these composites. The high-stress abrasion behaviour of Al-Si-SiCp composite was studied by Singh et al. [14]. The current investigation is focused on the low-stress abrasion behaviour of the Al-Si-SiCp composite.

The current work investigates the effect of abrading distance, wheel speed and applied load on the abrasion response of Al-Si-SiCp composite in as-received and after heat treatment under the low-stress abrasive wear. Furthermore, the worn-out surface of specimens was analysed by a field-emission scanning electron microscopy (FESEM) attached with energy-dispersive X-ray spectroscopy (EDS) to understand the wear mechanism during the low-stress abrasion process.

16.2 EXPERIMENTS

16.2.1 Synthesis of the Materials

The LM25 (hypoeutectic Al-7.2Si) alloy and Al-7.2Si-SiC composite were chosen in the present investigation. The constituent elements (in wt.%) involved in hypoeutectic alloy and their composite were examined by an optical emission analyser and presented in Table 16.1. In brief, in the stir casting process of composite, the preheated SiCps with 20–40 μm size (Grindwell Norton Ltd., Bangalore, India) and 10% wt were dispersed in the melt of base alloy by making a vortex through a mechanical stirrer. This melt was hardened (by air-cooled) into a hardened steel die. The die was in the form of a disk with dimensions Ø120 mm × 12 mm. A detailed explanation of the synthesis process of materials and heat treatment process was mentioned by Singh et al. [14].

16.2.2 Microstructure Analysis

The structural analysis of the abrasive medium and microstructural study of hypoeutectic alloy and their composite in the as-received and after heat treatment was done using FESEM attached with EDS. The wear (worn) surfaces of the specimens were also analysed by using FESEM equipped with EDS.

16.2.3 Evaluation of Densities and Hardnesses

The densities of samples were evaluated by utilising the Archimedes principle. The hardness of the samples was measured by using a Vickers hardness tester. The samples were fabricated and tested as per the ASTM E-92 standard [28]. The samples were metallographically polished to achieve a perfectly parallel surface for precise

TABLE 16.1
The Chemical Composition of Specimen [14]

	Elements (in wt.%)							
	Si	Mg	Fe	Mn	Cu	Ni	Al	SiCp
Hypoeutectic alloy	7.2	0.5	0.4	0.3	0.2	0.1	Rest	–
Composite	7.2	0.5	0.4	0.3	0.2	0.1	Rest	10

hardness measurements. A diamond pyramid indenter was impressed on the materials at 5 kgf of applied load for 5 seconds dwell time. Five indents were made on the test specimen surface to avoid the segregation effect of the particle, and the mean hardness value was determined.

16.2.4 Low-Stress Abrasion

The abrasion experiments were conducted using a dry abrasion tester (TR-50, Ducom Instruments, Bangalore, India) as per the ASTM G65 standard [29]. The wear test was performed using the rectangular specimen in the dimension of 76 mm × 25.4 mm × 12.7 mm. Silica sand particles with a size of 0.15–0.6 mm were used as the abrasives. Before the abrasion test, abrasives were moved through a sieve tray to verify the size and distribution of the particles. Most of the sand particles had a size of 0.3 mm. The abrasive medium was permitted to move downward from the funnel into the rotating wheel/specimen contact region. The rotating wheel with a diameter of 228.6 mm was rimmed from a 12.7 mm thick chlorobutyl rubber. The wear experiments were performed at 100 and 200 rpm rotational wheel speed (equivalent to 1.2 and 2.4 m s^{-1} linear wheel speed, respectively) and applied load of 5 and 20 N. The feed rate of the abrasive medium was 5.9 g s^{-1}. The weight loss measurement was done every 60 seconds interval (equivalent to a linear abrading distance of 72 m). The wear tests were performed up to 240 seconds (equal to a linear abrading distance of 288 m) [30]. The samples were cleaned with acetone and weighed using a microbalance of precision 0.00001 g (AUW220D, Shimadzu, Japan) before and later the test. The volume loss was computed by weight loss (g) per unit density (g mm^{-3}). The wear (abrasion) rate is defined by the ratio of loss of volume (mm^3) and abrading distance (m).

16.3 RESULT AND DISCUSSION

16.3.1 Microstructure Characterisation

Figure 16.1a reveals the morphology of silica sand particles which have an angular shape with sharpening edges. Figure 16.1b shows EDS for spectrum 1, which confirms the elements of the sand particles. The microstructure of the base alloy exhibits

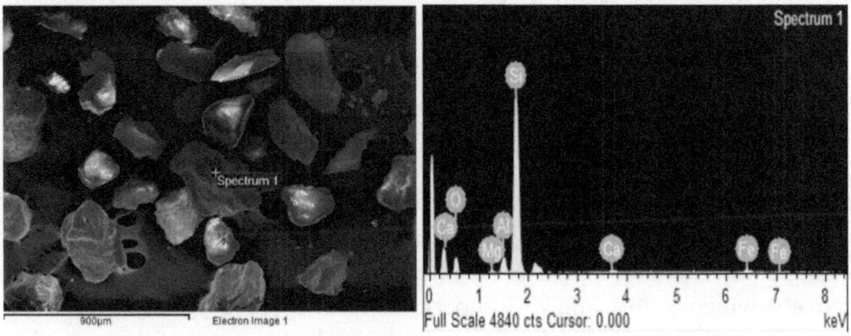

(a) (b)

FIGURE 16.1 The morphology of silica sand (a) and EDS analysis result of the spectrum (b).

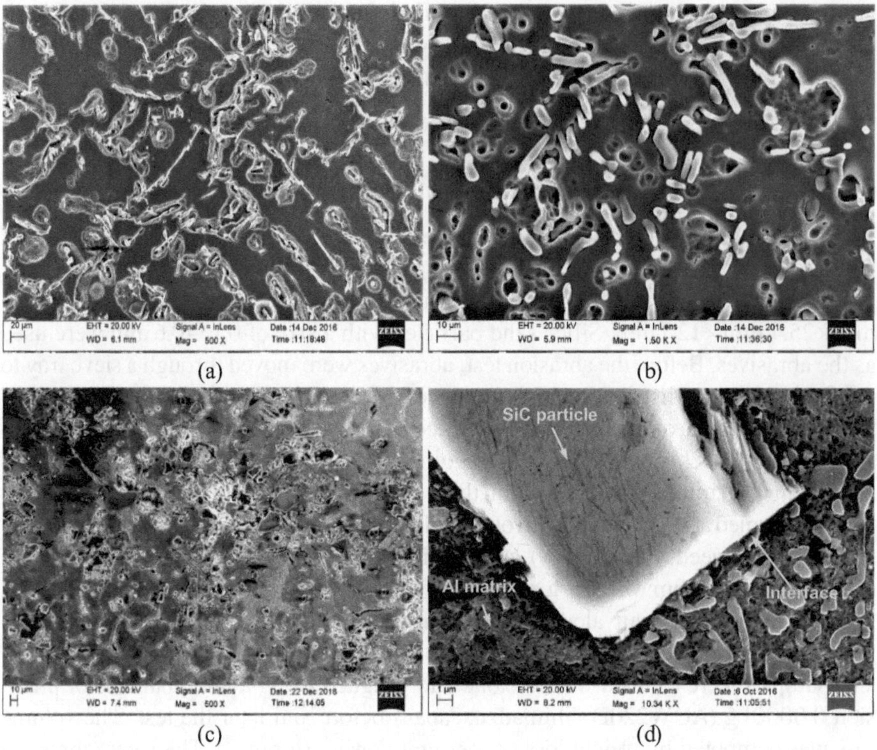

FIGURE 16.2 The microstructure of base alloy and composite in as-cast (a, c) and after heat treatment (b, d) state, respectively. Here, (d) reveals the interfacial bonding of as-cast LM25–SiCp.

the uniform distribution of needle-shaped eutectic silicon throughout the base alloy (see Figure 16.2a). The LM25–SiCp composite microstructure shows a homogeneous distribution of dispersoid particles in the matrix, no pits, and superior bonding of the LM25–SiC interface (see Figure 16.2c). The needle-shaped silicon changes into the smooth-edged silicon after the heat treatment and also improved the uniformity of dispersion of dispersoids in the matrix (see Figure 16.2b). The interfacial bonding of as-cast LM25–SiCp is exhibited in Figure 16.2d.

16.3.2 Density and Hardness Analysis

The specimen's density was registered as 2.67 and 2.74 g cm^{-3}, respectively. The LM25–SiCp composite density was more significant than the base alloy because SiCps have a higher density than the aluminium alloy. The hardness of the LM25–SiCp composite in as-cast and after heat treatment was registered as 149 and 163 HV, respectively, which is 99% and 77% superior to its alloy, respectively. This is because of the presence of SiCp in the aluminium matrix. The hardness of the base alloy and composite was registered 75 and 149 HV, which is 23% and 9% improved by heat treatment; this results from the formation of the intermetallics precipitate [14].

16.3.3 Low-Stress Abrasion

The abrasion rate of materials is shown in Table 16.2 and Figure 16.3 regarding abrading distances at different applied loads and wheel speeds. The abrasion rate of the composite has been compared to the base value of their alloy with the effects of abrading distance.

For 72 m abrading distance, 1.2 m s^{-1} wheel speed and 5 N applied load, abrasion rate of the as-received LM25 alloy and composite was 0.227 and 0.059 $mm^3 m^{-1}$, respectively, and for 288 m abrading distance, the abrasion rate was 0.211 and 0.049 $mm^3 m^{-1}$, respectively, as exhibited in Figure 16.3a. The abrasion rate is decreased by 74% and 76.78% at 72 m and 288 m abrading distance, respectively, for as-cast composite. For 72 m abrading distance, 1.2 m s^{-1} wheel speed and 5 N applied load, the abrasion rate of heat-treated base material and composite was 0.185 and 0.038 $mm^3 m^{-1}$, respectively, and for 288 m abrading distance, the abrasion rate was 0.158 and 0.027 $mm^3 m^{-1}$, respectively, as exhibited in Figure 16.3a. The abrasion rate is decreased by 79.46% and 82.91% at 72 and 288 m abrading distance, respectively, for heat-treated composite. For 72 m abrading distance, 1.2 m s^{-1} wheel speed and 20 N applied load, the abrasion rate of the as-received LM25 alloy and LM25–SiCp composite was 0.592 and 0.227 $mm^3 m^{-1}$, respectively, and for 288 m abrading distance, the abrasion rate was 0.500 and 0.126 $mm^3 m^{-1}$, respectively, as exhibited in Figure 16.3b. The abrasion rate of the as-cast composite is decreased by 61.66% and 74.8% at 72 and 288 m sliding distance, respectively. For 72 m abrading distance, 1.2 m s^{-1} wheel speed and 20 N applied load, the abrasion rate of heat-treated base material and composite was 0.452

TABLE 16.2
Abrasive Wear Rate ($mm^3 m^{-1}$) Experimental Data

S. No.	Wheel Speed (rpm)	Applied Load (N)	Abrading Distance (m)	LM25 Alloy (AC)	LM25 Alloy (HT)	LM25 Composite (AC)	LM25 Composite (HT)
1	100	5	72	0.22737	0.18519	0.05886	0.03767
2	100	5	144	0.22498	0.17010	0.05268	0.03284
3	100	5	216	0.21295	0.16455	0.05049	0.02807
4	100	5	288	0.21119	0.15801	0.04875	0.02747
5	100	20	72	0.59160	0.45204	0.22680	0.20106
6	100	20	144	0.56310	0.40608	0.16527	0.14639
7	100	20	216	0.51939	0.38528	0.13765	0.11400
8	100	20	288	0.50026	0.36634	0.12570	0.09648
9	200	5	72	0.42998	0.37484	0.20019	0.17074
10	200	5	144	0.39958	0.33263	0.13716	0.11165
11	200	5	216	0.38261	0.32407	0.12893	0.08327
12	200	5	288	0.36824	0.31644	0.10586	0.07661
13	200	20	72	1.00083	0.80914	0.28938	0.24746
14	200	20	144	0.73900	0.63592	0.26514	0.20744
15	200	20	216	0.70443	0.60653	0.25532	0.19157
16	200	20	288	0.69139	0.55872	0.24336	0.18619

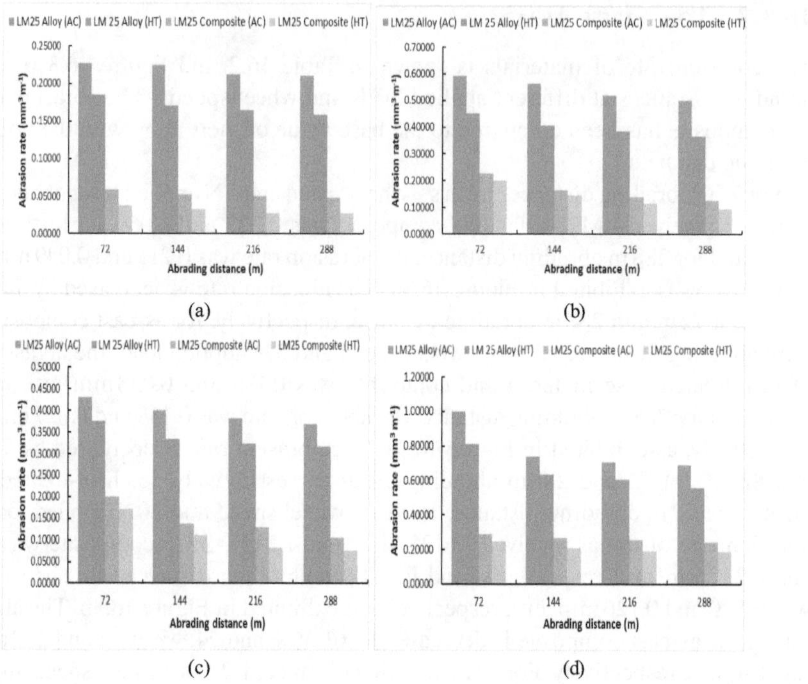

FIGURE 16.3 The abrasion rate of samples concerning the abrading distance, (a, b) applied load of 5 and 20 N and wheel speed of 1.2 m s^{-1} respectively, and (c, d) applied load of 5 and 20 N and wheel speed of 2.4 m s^{-1}, respectively.

and 0.201 $mm^3 m^{-1}$, respectively, and for 288 m abrading distance, the abrasion rate was 0.366 and 0.096 $mm^3 m^{-1}$, respectively, as exhibited in Figure 16.3b. The abrasion rate is decreased by 55.53% and 73.77% at 72 m and 288 m abrading distance, respectively, for heat-treated composite.

For 72 m abrading distance, 2.4 m s^{-1} wheel speed and applied load of 5 N, the abrasion rate of the as-received base material and composite was 0.43 and 0.2 $mm^3 m^{-1}$, respectively, and for 288 m abrading distance, the abrasion rate was 0.368 and 0.106 $mm^3 m^{-1}$, respectively, as exhibited in Figure 16.2c. The abrasion rate is decreased by 53.49% and 71.2% at 72 and 288 m abrading distance, respectively, for the as-cast composite. For 72 m abrading distance, 2.4 m s^{-1} wheel speed and 5 N applied load, the abrasion rate of heat-treated base material and composite was 0.375 and 0.170 $mm^3 m^{-1}$, respectively, and for 288 m abrading distance, the abrasion rate was 0.316 and 0.077 $mm^3 m^{-1}$, respectively, as exhibited in Figure 16.3c. The abrasion rate is decreased by 54.67% and 75.63% at 72 m and 288 m abrading distance, respectively, for heat-treated composite. For 72 m abrading distance, 2.4 m s^{-1} wheel speed and 20 N applied load, the abrasion rate of the as-received LM25 alloy and LM25–SiCp composite was 1.000 and 0.289 $mm^3 m^{-1}$, respectively, and for 288 m abrading distance, the abrasion rate was 0.691 and 0.243 $mm^3 m^{-1}$, respectively, as exhibited in Figure 16.3d. The abrasion rate is decreased by 71.1% and 64.83% at 72 and 288 m abrading distance, respectively, for the as-cast composite. For 72 m abrading distance, 2.4 m s^{-1} wheel speed and 20 N applied load, the abrasion rate of heat-treated base alloy and composite was 0.809 and 0.247 $mm^3 m^{-1}$, respectively, and for 288 m abrading distance, the abrasion rate was 0.559 and 0.186 $mm^3 m^{-1}$, respectively, as exhibited in Figure 16.3d. The abrasion rate is decreased by 69.47% and 66.72% at 72 and 288 m abrading distance, respectively, for heat-treated composite.

The heat-treated base alloy and composite at 1.2 m s^{-1} wheel speed and 5 N applied load, the percentage improvement of abrasion rate was 18.5 and 35.59, respectively, at 72 m abrading distance of 288 m was 25.11 and 44.9, respectively. However, at 1.2 m s^{-1} wheel speed and 20 N applied load, the percentage improvement of the abrasion rate was 23.52 and 11.45, respectively, at 72 m abrading distance, and for 288 m abrading distance, it was 26.8 and 23.81, respectively. After heat treatment condition, the base material and composite at 5 N applied load and 2.4 m s^{-1} wheel speed, the percentage improvement of abrasion rate was 12.79 and 15, respectively, at 72 m abrading distance, and 14.13 and 26.67 in case of 288 m abrading distance, respectively. However, at 20 N applied load and 2.4 m s^{-1} wheel speed, the percentage improvement of abrasion rate was 19.1 and 14.53, respectively, at 72 m abrading distance, and 19.1 and 23.46, respectively, for 288 m abrading distance.

It is found that the abrasion rate of LM25 alloy and LM25–SiCp composite decreased with enhancing the abrading distances, as shown in Figure 16.3. This is attributed to the increasing abrasion-induced work hardening of the matrix in the sub-surface region, consequently more plastic deformation, and large extent of dispersion of harder reinforcement phase over the specimen surface and also deterioration of cutting ability of sand particles [31]. The LM25–SiCp composite demonstrated excellent wear properties compared to the base alloy regardless of the wheel speed, abrading distance and applied load [20–26]. This is because of the distribution of SiC particles in the matrix, which protects the matrix against the sand particles. The wear resistance of the hypoeutectic alloy and composite increased by the heat treatment process [27,14]; this results from the formation of the intermetallic precipitate.

The abrasion rate of both the as-received and after heat treatment of the materials was enhanced by increasing wheel speeds and applied loads, as exhibited in Figure 16.3. This attributes to increase in the penetration depth (as a result, either pits or grooves) with increase in applied loads and wheel speeds. In the abrasion process, the pits and grooves are developed on the wear surfaces by removing materials due to rolling motion and scratching/cutting action of the abrasive medium, respectively, as shown in Figure 16.5. Scratches on the worn (wear) surface are caused by the sand particles, which increased with increasing the applied load and wheel speed. During the abrasion process, the rolling motion of sand particles is higher influencing, which leads to more quantity of pits formation irrespective of the materials. The appearance of the grooves increases with the wheel speed and applied load. The SiCp are exposed due to the removal of the composite matrix, and these particles prevent the rolling and abrasion actions of the sand particles. Due to the rolling motion, sometimes the particles of sand get embedded between the SiCps, which take part neither in rolling action nor in abrasion action, consequently in a lower abrasion rate. These facets are attributed to the low abrasion rate in the LM25–SiCp composite than the hypoeutectic alloy, regardless of the wheel speed and applied load.

16.3.4 Abrasive Worn Surface

The abrasive worn surface and EDS results of as-cast LM25 alloy at applied load, abrading distance and liner speed for lower and higher conditions are shown in Figure 16.4. In the low-stress abrasion process, the sand particles either fall on the sample surface or pierce the sample surfaces due to wear scars for a short distance, and it is relying on the size of abrasive (sand particle) and applied load [32]. The loose sand particles are free to move between the contact region of the sample surface and

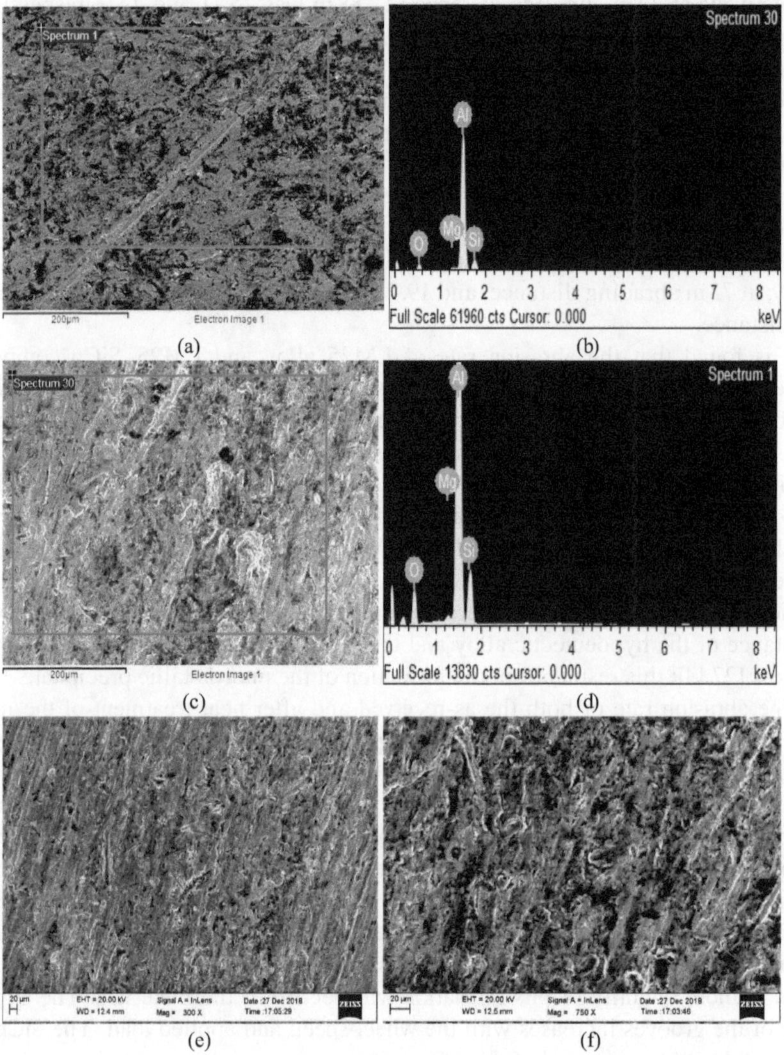

FIGURE 16.4 Wear surfaces of the as-cast base alloy (a, e) for an applied load of 5 N, 1.2 m s^{-1} wheel speed and abrading distance of 72 m and (b) EDS analysis result of spectrum 30 respectively, and (c, f) for an applied load of 20 N, 2.4 m s^{-1} wheel speed and 288 m abrading distance and (d) EDS analysis result of spectrum 1, respectively.

the rubber wheel. Figure 16.4a and e exhibits the wear surface of the LM25 alloy at 72 m abrading distance, 1.2 m s^{-1} wheel speed and 5 N applied load. The wear surface features microcracking, scratches and pits. Figure 16.4c and f exhibits the wear surface of the base material at 288 m abrading distance, 2.4 m s^{-1} wheel speed and 20 N applied load. The wear surface features more depth and width of grooves, deeply damaging the sample surface and larger and deeper pits. Figure 16.4b and d shows the EDS results of spectrum 30 and 1, respectively, which confirm the presence of oxygen peaks in the sample surface. This recommends that the oxidation occurs in the abrasion process, consequently oxidation wear in the sample.

Figure 16.5a exhibits the wear surface of the LM25–SiCp composite at 72 m abrading distance, 1.2 m s^{-1} wheel speed and 5 N applied load. It reveals flakes, microcracking and pits; these are examined for the base alloy, as in Figure 16.4a and e. The craters formed in the composite (see Figure 16.5a) are finer than the base alloy (see Figure 16.4a). The SiCps of the composite are observed on the worn surfaces, and these SiCps prevent the sample surface from the movement of the sand particle, as exhibited in Figure 16.5e. Figure 16.5c and f exhibits the wear surfaces of the composite at 288 m abrading distance, 2.4 m s^{-1} wheel speed and 20 N applied load. The wear surface represents the coarser pits, microcracks and flakes like debris and matrix materials removal from the dispersoid region. The intensity of protrusion of the silicon carbide phase at the lower applied load (see Figure 16.5e) was more than the higher applied load (see Figure 16.5f). This is because of the complete removal of reinforcements from the base materials. At higher speed and applied load, removing material is more due to ploughing and cutting action, leading to low energy expense on plastic deformation. Figure 16.5b shows the EDS results for spectrum 19, representing the oxidation effect on the worn surface; however, almost no oxidation effect was observed on the worn surface for spectrum 20 of Figure 16.5d.

16.4 CONCLUSIONS

The following conclusions of the present research work can be obtained:

1. The LM25–SiCp composite has excellent abrasion resistance compared to the LM25 alloy in as-received and T6 heat treatment state because of the addition of SiCp (hard dispersoid) in the matrix.
2. The abrasion rate of the materials is reduced with increasing abrading distances due to the increasing abrasion-induced plastic deformation and deterioration of the cutting ability of sand particles.
3. The steady-state abrasive abrasion rate of the materials is improved with applied load and wheel speed. This is because the penetration depth increases with a rise in wheel speed and applied load.
4. The material removal mechanism varies with the applied load and wheel speed. At lower applied load and wheel speed, the removal of the materials occurs due to the rolling actions of the abrasive particle, whereas at higher wheel speed and applied load, material removal occurs due to scratching/ploughing/cutting action of the abrasive medium.

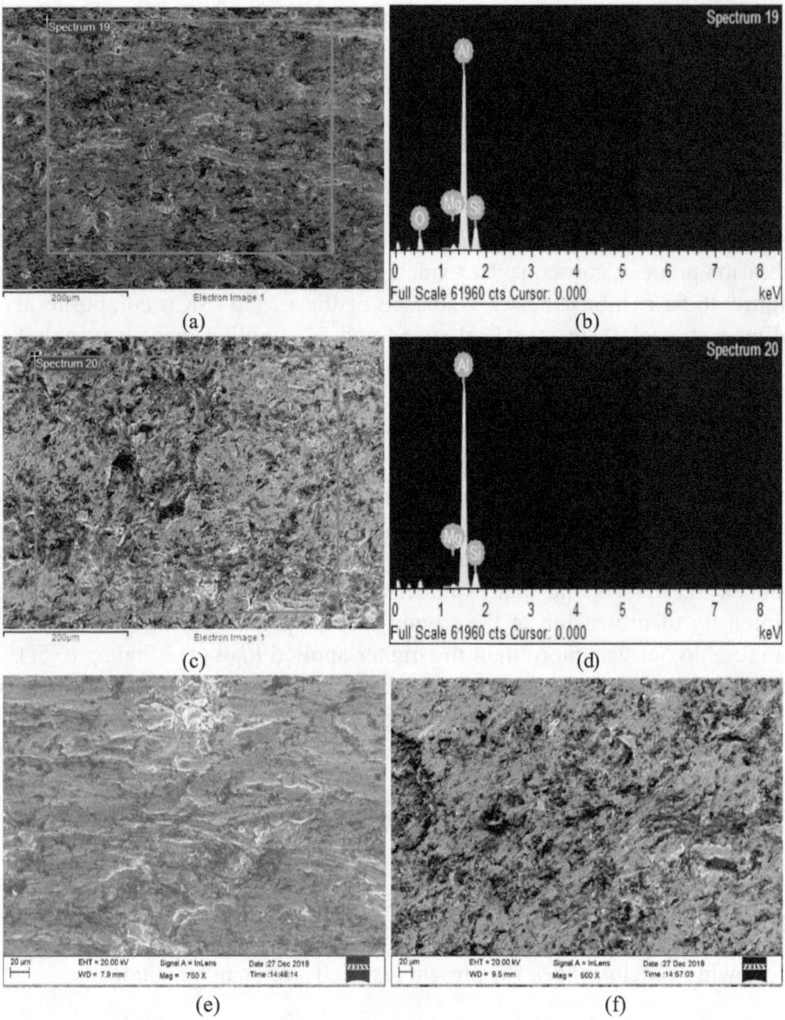

FIGURE 16.5 Worn surfaces of the as-cast composite (a, e) for 72 m abrading distance, 1.2 m s^{-1} wheel speed and 5 N applied load and (b) EDS analysis 2 result of spectrum 19, respectively, and (c, f) for 288 m abrading distance, 2.4 m s^{-1} wheel speed and 20 N applied load and (d) EDS analysis result of spectrum 20 observed on the worn surface for spectrum 20.

The effect of different particle sizes of reinforcement, the influence of different dispersoids and abrasive medium, cryogenic treatment on wear behaviour, and other types of wear such as fretting, adhesive, corrosion, etc. can be studied in future.

REFERENCES

1. Singh, J., & Chauhan, A. (2016). Characterisation of hybrid aluminium matrix composites for advanced applications–a review. *Journal of Materials Research and Technology*, *5*, 159–69. Doi: 10.1016/j.jmrt.2015.05.004.

2. Wang, A. G., & Hutchings, I. M. (1989). Wear of alumina fibre–aluminium metal matrix composites by two-body abrasion. *Journal of Materials Science and Technology, 5*, 71–76. Doi: 10.1179/mst.1989.5.1.71.
3. Natarajan, N., Vijayarangan, S., & Rajendran, I. (2006). Wear behaviour of A356/25SiCp aluminium matrix composites sliding against automobile friction material. *Wear, 261*, 812–822. Doi: 10.1016/j.wear.2006.01.011.
4. Surappa, M. K. (2008). Dry sliding wear of fly ash particle reinforced A356 Al composites. *Wear, 265*, 349–360. Doi: 10.1016/j.wear.2007.11.009.
5. Kucukomeroglu, T. (2010). Effect of equal-channel angular extrusion on mechanical and wear properties of eutectic Al–12Si alloy. *Materials & Design, 31*, 782–789. Doi: 10.1016/j.matdes.2009.08.004.
6. Shivamurthy, R. C., & Surappa, M. K. (2011). Tribological characteristics of A356 Al alloy–SiCp composite discs. *Wear, 271*, 1946–1950. Doi: 10.1016/j.wear.2011.01.075.
7. Feyzullahoglu, E., Ertürk, A. T., & Güven, E. A. (2013). Influence of forging and heat treatment on wear properties of Al-Si and Al-Pb bearing alloys in oil lubricated conditions. *Transactions of Nonferrous Metals Society of China, 23*, 3575–3583. Doi: 10.1016/S1003–6326(13)62903-9.
8. Singh, R. K., Telang, A., & Das, S. (2021). Abrasive wear response of Al-Si-SiCp composite: Effect of friction heat and friction coefficient. *International Journal of Materials Research, 112*, 366–372. Doi: 10.1515/ijmr-2020-7869.
9. Khosravi, H., & Akhlaghi, F. (2015). Comparison of microstructure and wear resistance of A356–SiCp composites processed via compocasting and vibrating cooling slope. *Transactions of Nonferrous Metals Society of China, 25*, 2490–2498. Doi: 10.1016/S1003-6326(15)63867-5.
10. Toptan, F., Kerti, I., & Rocha, L. A. (2012). Reciprocal dry sliding wear behaviour of B_4Cp reinforced aluminium alloy matrix composites. *Wear, 290–291*, 74–85. Doi: 10.1016/j.wear.2012.05.007.
11. Kumar, H., Prasad, R., Kumar, P., Tewari, S. P., & Singh, J. K. (2020). Mechanical and tribological characterisation of industrial wastes reinforced aluminium alloy composites fabricated via friction stir processing. *Journal of Alloys and Compounds, 831*, 154832. Doi: 10.1016/j.jallcom.2020.154832.
12. Shah, K. B., Kumar, S., & Dwivedi, D. K. (2007). Aging temperature and abrasive wear behaviour of cast Al–(4%,12%,20%) Si–0.3% Mg alloys. *Materials & Design, 28*, 1968–1974. Doi: 10.1016/j.matdes.2006.04.012.
13. Singh, R. K., Telang, A., & Das, S. (2020). Microstructure, mechanical properties and two-body abrasive wear behaviour of hypereutectic Al-Si-SiC composite. *Transactions of Nonferrous Metals Society of China, 30*, 65–75. Doi: 10.1016/S1003-6326(19)65180-0.
14. Singh, R. K., Telang, A., & Das, S. (2019). Microstructure, mechanical, and high-stress abrasive wear behaviour of as-cast and heat-treated Al-Si-SiCp composite. *International Journal of Materials Research, 110*, 121–129. Doi: 10.3139/146.111727.
15. Mondal, D. P., & Das, S. (2006). High-stress abrasive wear behaviour of aluminium hard particle composites: Effect of experimental parameters, particle size and volume fraction. *Tribology International, 39*, 470–478. Doi: 10.1016/j.triboint.2005.03.003.
16. Das, S., Mondal, D. P., Sawla, S., & Ramakrishnan, N. (2008). Synergic effect of reinforcement and heat treatment on the two-body abrasive wear of an Al-Si alloy under varying loads and abrasive sizes. *Wear, 264*, 47–59. Doi: 10.1016/j.wear.2007.01.039.
17. Singh, M., Modi, O. P., Dasgupta, R., and Jha, A. K. (1999). High stress abrasive wear behaviour of aluminium alloy–granite particle composite. *Wear, 233–235*, 455–461. Doi: 10.1016/S0043-1648(99)00186-6.
18. Singh, M., Jha, A. K., Das, S., & Yegneswaran, A. H. (2000). Preparation and properties of cast aluminium alloy-granite particle composites. *Journal of Materials Science, 35*, 4421–4426. Doi: 10.1023/A:1004821426862.

19. Singh, M., Mondal, D. P., Modi, O. P., & Jha, A. K. (2002). Two-body abrasive wear behaviour of aluminium alloy–sillimanite particle reinforced composite. *Wear, 253*, 357–368. Doi: 10.1016/S0043-1648(02)00153-9.
20. Prasad, K. N. P., & Ramachandra, M. (2018). Determination of abrasive wear behaviour of Al-Fly ash metal matrix composites produced by squeeze casting. *Materials Today: Proceedings, 5*, 2844–2853. Doi: 10.1016/j.matpr.2018.01.075.
21. Chandra, B. T., Sanjeevamurthy, & Shankar, H. S. S. (2018). Effect of heat treatment on dry sand abrasive wear behavior of Al7075-albite particulate composites. *Materials Today: Proceedings, 5*, 5968–5975. Doi: 10.1016/j.matpr.2017.12.198.
22. Zheng, B., Li, W., Tu, X., Xu, F., Liu, K., & Song, S. (2020). Effect of titanium binder addition on the interface structure and three-body abrasive wear behavior of ZTA ceramic particles-reinforced high chromium cast iron. *Ceramics International, 46*, 13798–13806. Doi: 10.1016/j.ceramint.2020.02.169.
23. Radhika, N., & Raghu, R. (2017). Characterisation of mechanical properties and three-body abrasive wear of functionally graded aluminum LM25/titanium carbide metal matrix composite. *Materials Science and Engineering Technology (MAWE), 48*, 882–892. Doi: 10.1002/mawe.201700559.
24. Canakci, A. (2011). Microstructure and abrasive wear behaviour of B_4C particle reinforced 2014 Al matrix composites, *Journal of Materials Science, 46*, 2805–2813. Doi: 10.1007/s10853-010-5156-2.
25. Modi, O. P., Yadav, R. P., Prasad, B. K., Jha, A. K., Das, S., & Yegneswaran, A. H. (2001). Three-body abrasion of a cast zinc–aluminium alloy: Influence of Al_2O_3 dispersoid and abrasive medium, *Wear, 249*, 792–799. Doi: 10.1016/S0043-1648(01)00813-4.
26. Shankar, B. L., Anil, K. C., & Patil, R. (2016). A study on 3-body abrasive wear behaviour of aluminium 8011/graphite metal matrix composite, *IOP Conf. Series: Materials Science and Engineering, 149*, 012099. Doi: 10.1088/1757-899X/149/1/012099.
27. Sridhar, H. S., Sanman, S., Prasad, T. B., & Chandra, B. T. (2020). Effect of reinforcement and applied load on three-body dry sand abrasive wear behavior of A356 bottom ash metal matrix composites. *Materials Today: Proceedings, 26*, 2814–2816. Doi: 10.1016/j.matpr.2020.02.586.
28. ASTM E 92-82. 2003. *Standard Test Method for Vickers Hardness of Metallic Materials*. ASTM International, West Conshohocken, PA.
29. ASTM Standard G65-15. 1981. *Standard Practice for Conducting Dry Sand/Rubber Wheel Abrasion Tests*. ASTM International, West Conshohocken, PA.
30. Singh, R. K., Telang, A., & Khan, M. M. (2017). Effect of T6 heat treatment on microstructure, mechanical properties and abrasive wear response of fly ash reinforced Al-Si alloy. *Materials Today: Proceedings, 4*, 10062–10068. Doi: 10.1016/j.matpr.2017.06.321.
31. Prasad, B. K., Prasad, S. V., & Das, A. A. (1992). Abrasion-induced microstructural changes and material removal mechanisms in squeeze-cast aluminium alloy-silicon carbide composites. *Journal of Materials Science, 27*, 4489–4494. Doi: 10.1007/BF00541584.
32. Mutton, P. J., & Watson J. D. (1978). Some effects of microstructure on the abrasion resistance of metals. *Wear, 48*, 385–398. Doi: 10.1016/0043-1648(78)90234-X.

17 Post-Annealing Influence on Structural, Surface and Optical Properties of Cu_3BiS_3 Thin Films for Photovoltaic Solar Cells

Sampat G. Deshmukh and Kailash J. Karande
SKN Sinhgad College of Engineering

Koki Ogura
Kyushu Sangyo University

Vipul Kheraj
S. V. National Institute of Technology

Rohan S. Deshmukh
SKN Sinhgad College of Engineering

CONTENTS

DOI: 10.1201/9781003220176-17

17.1 INTRODUCTION

Amongst the numerous semiconductors, research on non-toxic, environmental friendly and economical materials having abundance in earth's crust is demanded due to the environmental consideration (Cd), growing cost (Ge, In) and earth abundance problem (Te, In) in CdTe (CT) and Cu(In, Ga)Se_2 (CIGS)–based solar cells [1]. This has been becoming a motto for the investigation of alternative semiconductor materials based on sulphur, since they are abundant and non-toxic in nature. Investigators reveal that the class I-V-VI compound has technological uses within the area of thin-film devices and photovoltaic energy conversions [2]. The most favourable material from I-V-VI family is Cu_3BiS_3. It occurs naturally in the wittichenite mineral with an orthorhombic structure ($a = 0.7723$ nm, $b = 1.0395$ nm and $c = 0.6715$ nm) [3]. Cu_3BiS_3 shows suitable *p*-type electrical conductivity (10^{-4} to $10^{-2}/\Omega$.cm) [4], optical bandgap about 1.45 eV [5] and strong absorption coefficient 10^{-5} cm^{-1} [6], fetching one of the most suitable absorbers in solar cells with a very thin layer. Due to these promising characteristics, different methods have been used to prepare Cu_3BiS_3 thin films by means of many researchers. In Table 17.1, some studies introduce the invention of Cu_3BiS_3 thin films.

In addition to this facts present in Table 17.1, the literature incorporates a few other studies about Cu_3BiS_3 [12]. Other grown methods are more costly than chemical bath deposition (CBD); they sometimes require a vacuum system, high temperature, understanding of the subject and take a long time to prepare [12,13]. CBD is economical and easier than other techniques, thus it was utilized in the current work.

Firstly, Nair and Huang in 1997 [14] revealed the development of *p*-type Cu_3BiS_3 absorber material in photovoltaic through annealing chemically deposited Bi_2S_3 on CuS film. Further, Colombara et al. [3] stated that Cu_3BiS_3 thin films synthesized via

TABLE 17.1
Synthesis Methods Used for Deposition of Cu_3BiS_3 Films

Method	Copper Source	Bismuth Source	Sulphide Source	Deposition Time	Deposition Temp.	Substrate	Ref.
CBD	Copper acetate	Bismuth nitrate	Thiourea	50 minutes	55°C	Glass	[6]
CBD	Copper chloride	Bismuth nitrate	Thiourea	40 minutes	50°C	Glass	[7]
SC	Cuprous chloride	Bismuth chloride	Thiourea	2 minutes	240°C–340°C	Mo-glass	[8]
SP	Copper acetate	Bismuth Tris	–	–	150°C–200°C	Glass, ZnO/ITO	[4]
ED	Copper Nitrate	Bismuth nitrate	Sulphur	10 minutes	RT	Mo-glass	[9]
DC	Cuprous Chloride	Bismuth chloride	Thiourea	5 seconds	280°C	Glass	[10]
e-beam evaporation	Copper chloride	Bismuth chloride	Thiourea	–	200°C–500°C	Glass	[11]

CBD, chemical bath deposition; DC, dip coating; ED, electrodeposition; SC, spin coating; SP, spray pyrolysis.

TABLE 17.2
Photovoltaic Parameters of Cu_3BiS_3-Based Solar Cells under AM1.5 Illumination of 100 mW/cm²

Sample	Voc (mV)	Jsc (mA cm^{-2})	FF (%)	η (%)	Ref.
CdS/Cu_3BiS_3	320	18.2	30.3	1.7	[1]
CdS/Cu_3BiS_3	190	2.37	37.56	0.17	[8]
TiO_2/Cu_3BiS_3	448	4.449	60.8	1.281	[16]
CdS/Cu_3BiS_3	89	3.7	32.8	0.11	[17]
TiO_2/Cu_3BiS_3	90	6.09	25.44	0.139	[18]

sulfurization of Bi-Cu metal precursors can achieve maximum external quantum efficiency (EQE) of about 10%. Later on, Hussain et al. [5] and Chakraborty et al. [15] examined the photo-response characteristics of Cu_3BiS_3 thin films prepared by thermal evaporation and facile hot injection methods respectively. Recently, Kamimura et al. [9] and Li et al. [10] achieved to produce Cu_3BiS_3, which has photoelectrochemical characteristics via electrodeposition and dip coating synthesis techniques. Great importantly, photovoltaic Cu_3BiS_3-based solar cell device with specific conversion efficiency (η) was validated by various investigators, and its photovoltaic parameters are shortened in Table 17.2.

Table 17.2 exposes the photovoltaic device performance exhibited by Cu_3BiS_3. This helps to improve the efficiency of Cu_3BiS_3-based solar cell devices towards more competitive conversion efficiency.

Within this work, we report the preparation of Cu_3BiS_3 films using a CBD approach. The post-annealing influence on the structural, Raman, surface morphological, wettability and optical properties of the films has been investigated.

17.2 EXPERIMENTAL SECTION

17.2.1 Resources

In the present work, resource reagents were used as Finar make copper chloride dihydrate ($CuCl_2.2H_2O$, 99.9%); bismuth (III) nitrate pentahydrate ($Bi(NO_3)_3{\cdot}5H_2O$, > 99%), triethanolamine (TEA) (>99%) from Merk and thiourea (CH_4N_2S, > 99.9%) from S d Fine Chem Ltd. All chemical substances have been of analytical grade (AR) and utilized without extra refinement. Besides, throughout the experiment distilled water was used.

17.2.2 Preparation of Cu_3BiS_3 Precursor Solution

Copper chloride dihydrate ($CuCl_2{\cdot}2H_2O$), bismuth (III) nitrate pentahydrate ($Bi(NO_3)_3{\cdot}5H_2O$) and thiourea (H_2NCSNH_2) had been utilized as the basic ingredients for the precursor solutions. Here, the CH_4N_2S was employed as the source of sulphur as well as the stabilizer of Cu_3BiS_3 precursors [8]. Further, aqueous stock solutions of 30 and 45 mM of $CuCl_2{\cdot}2H_2O$ and CH_4N_2S were prepared respectively. Then, bismuth (III) nitrate pentahydrate was dissolved in TEA and prepared its 10 mM aqueous stock solutions. The final precursor solution was set up by adding the Cu^{2+}, Bi^{3+} and S^{2-} stock solutions, yielding a perfect blue colour solution. Here, excess CH_4N_2S was utilized to ensure both the sufficient supply of S^{2-} ions during the reaction and the complete coordination with Cu^{2+}, Bi^{3+} ions by avoiding the precipitation of hydroxides and metallic sulphides [8]. All these precursor solutions were ultrasonicated before being utilized for film depositions.

17.2.3 Cu_3BiS_3 Thin-Film Deposition

For the deposition of Cu_3BiS_3 thin film by CBD, equal volume of each precursor solution was mixed in 50 mL glass beaker, causing blue colour of the solution. The published procedure [19] was used to scrub the glass substrate. The scrubbed glass slides were set aside vertically in 50 mL precursor solution at room temperature. This precursor bath was then put into a hot water bath maintained at 50°C (± 1°C). As the temperature of precursor bath was risen gradually, the precipitation began in the precursor bath. During the precipitation, a heterogeneous reaction happened and deposition of Cu_3BiS_3 took place on the surface of the glass substrate in the bath, and finally, the precursor turns brown. After 2,400 seconds, Cu_3BiS_3 thin film deposited substrate was taken out from the chemical bath. These thin films were rinsed in H_2O and dried in air at 300 K, as described earlier [20]. This as-deposited Cu_3BiS_3 thin film on glass substrate was additionally annealed at 573, 673 and 773 K in sulphur environment for 1,800 seconds to investigate the post-annealing influence on its properties. The as-deposited Cu_3BiS_3 film was tagged as P1 and annealed at 573, 673 and 773 K were tagged as P2, P3 and P4, respectively.

17.2.4 Cu_3BiS_3 Thin-Film Characterization

The crystalline phases and structural properties of P1, P2, P3 and P4 Cu_3BiS_3 thin films were examined by X-ray diffraction (XRD) pattern taken on Ultima IV Rigaku with CuKα (λ = 1.5406 Å). All samples were scanned in the range of 2θ values from 15° to 60. Raman spectra were taken out in the range of 100–500 cm^{-1} on Bruker RFS-27 for all Cu_3BiS_3 films. Scanning electron microscope (SEM) images of P1, P2, P3 and P4 Cu_3BiS_3 films were collected from JEOL, JBM-6360A operating at an accelerating potential of 20 kV to study its surface morphology. A contact angle meter Rame-Hart, Inc. was utilized to measure the WCA of P1, P2, P3 and P4 Cu_3BiS_3 films. The study of optical properties of Cu_3BiS_3 thin films was done by means of transmission and reflection measurement set up established at our research centre.

17.3 RESULTS AND DISCUSSION

17.3.1 Structural Analysis

Figure 17.1 displays XRD profile of P2, P3 and P4 Cu_3BiS_3 films synthesized by non-vacuum chemical route. The XRD pattern of P1 Cu_3BiS_3 thin film (not shown here) revealed no XRD peaks, indicating the amorphous nature as described in the past report [20]. As an annealing temperature increases from 573 to 773 K, the improvement of crystallinity of P1 takes place as shown in Figure 17.1. For P2 and

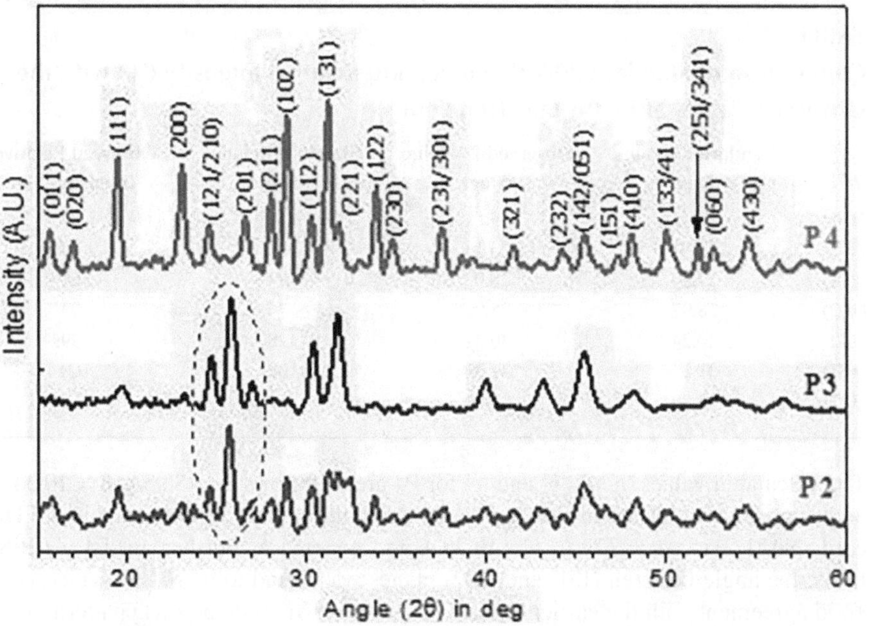

FIGURE 17.1 X-ray diffraction (XRD) profile of P2, P3 and P4 Cu_3BiS_3 films.

P3 Cu_3BiS_3 thin films, peak at 25.98° (shown by dotted circle in Figure 17.1) of (002) indicates that grains are developed along z-direction, i.e. vertical to the substrate surface, which would be shown in the SEM images [Figure 17.3]. The intensity of (002) peak was increased as the annealing temperature increased. This outcome is due to the combined effect of growth rate and reorientation of Cu_3BiS_3 particles. In the film crystallization, annealing temperature plays a significant role. The Cu_3BiS_3 phase is a prominent one in the P4 thin film annealed at 773K. The main diffraction peaks of P4 cantered at $2\theta = 15.68°$, 17.01°, 19.51°, 23.14°, 24.59°, 26.68°, 28.02°, 28.99°, 30.31°, 31.21°, 31.83°, 34.00°, 34.75°, 37.43°, 41.49°, 44.17°, 45.44°, 47.34°, 48.10°, 50.00°, 51.79°, 52.69° and 54.47° are indexed to (011), (020), (111), (200), (121/210), (201), (211), (102), (112), (131), (221), (122), (230), (231/301), (321), (232), (142/051), (151), (410), (133/411), (251/341), (060) and (430) planes of wittichenite orthorhombic Cu_3BiS_3 phase [JCPDF#71-2115]. The predominant sharp peak observed at $2\theta = 31.21°$ indicates the (131) plane for the P4 which is well-matched with the data of JCPDF #71-2115. For P4, the observed 2θ values and relative intensity (%) were consistent with those of JCPDF #71-2115, as presented in Table 17.3. This approves the establishment of wittichenite Cu_3BiS_3.

The lattice parameters '*a*', 'b' and '*c*' of P4 Cu_3BiS_3 film were determined from the analysis of the XRD pattern and were estimated from the formula of orthorhombic system:

$$\frac{1}{d^2} = \frac{h^2}{a^2} + \frac{k^2}{b^2} + \frac{l^2}{c^2} \tag{17.1}$$

TABLE 17.3
Comparison of Standard 2θ-Value (deg) and Relative Intensity (%) with the Observed Values of P4 Cu_3BiS_3 Thin Film

(hkl)	Standard 2θ-Value (Deg)	Observed 2θ-Value (Deg)	Standard Relative Intensity (%)	Observed Relative Intensity (%)
(111)	19.47	19.51	32	74
(200)	23.01	23.14	27	64
(211)	27.97	28.02	43	51
(102)	28.97	28.99	76	93
(112)	30.24	30.31	28	39
(131)	31.24	31.21	100	100
(122)	33.80	34.00	54	57

The calculated values of '*a*', '*b*' and '*c*' for P4 are as follows: $a = 8.03$Å, $b = 10.30$ Å and $c = 6.66$ Å which are in close agreement with the standards mentioned in JCPDF card (No.71-2115), $a = 7.72$ Å, $b = 10.39$ Å and $c = 6.71$ Å. Further, for P4 Cu_3BiS_3 films, the angle between (131) and (102) planes was found to be 57.88°, which is in good agreement with the earlier report [7]. A deeper analysis of XRD pattern of P4 confirms that there was no evidence of any other possible secondary phases of Bi_2S_3, CuS, $CuBiS_2$, Cu_2S and Bi_2O_3, which indicates that pure orthorhombic Cu_3BiS_3 phase, can be obtained under the current development conditions. The crystallite size (D) of P4 Cu_3BiS_3 film was determined via the Debye-Scherrer equation [21]:

$$D = \frac{k \cdot \lambda}{\beta \cdot \cos\theta} \tag{17.2}$$

where λ is the wavelength of X-rays ($\lambda = 1.5406$ Å) used, k is the dimensionless constant (0.9), θ and β are diffraction angle in radian and full width at half maximum (FWHM), respectively.

The crystallite size was estimated as 23.31 nm for the major (131) peak of P4. The existence of broad XRD peaks is a sign of lesser crystallite size in the range of nanoscale, confirming the nanocrystalline nature of the films. The estimated crystallite size equals earlier reports [15].

Moreover, the number of defects, i.e. the dislocation density, $\delta = 1/D^2$ and the microstrain, $\varepsilon = \beta.\cos\,\theta/4$ [21] of P4 was determined. For P4 Cu_3BiS_3 thin film, FWHM, crystallite size, dislocation density and microstrain are concise in Table 17.4. A lesser value of dislocation density and microstrain directs a greater crystallinity of P4 Cu_3BiS_3 thin film.

17.3.2 Raman Spectroscopy

However, as argued earlier, the Cu_3BiS_3 is a ternary compound; the presence of secondary phases is always predictable in the compound synthesized by the chemical route. Hence, for the investigation of the presence of Cu_2S, Bi_2S_3 and $CuBiS_2$ phases

TABLE 17.4
FWHM (β), Crystallite Size (D), Dislocation Density (δ) and Micro Strain (ε) of P4 Cu_3BiS_3 Film

(hkl)	FWHM (β)	D (nm)	δ (nm^{-2}) ($\times 10^{-3}$)	ε($\times 10^{-3}$)
(111)	0.3746	21.53	2.158	1.610
(200)	0.3335	24.32	1.690	1.425
(211)	0.3143	26.07	1.472	1.330
(102)	0.3987	20.59	2.359	1.683
(112)	0.3430	24.01	1.735	1.444
(131)	0.3540	23.31	1.840	1.487
(122)	0.3238	25.66	1.519	1.351

in the P1, P2, P3 and P4 samples, Raman Scattering was performed using Bruker RFS27 with Nd:YAG laser source. Figure 17.2 displays the Raman spectra in the range of 100–500 per cm of all Cu_3BiS_3 thin films.

The Raman spectra of the as-deposited P1 thin film have two intense Raman peaks at 279 and 468 cm^{-1}, confirming the formation of the wittichenite phase of Cu_3BiS_3 in the film as described in the previous report [20]. However, for the films P2 and P3, major peak was observed at 282 cm^{-1}, and for P4, it is at 290 cm^{-1} [22]. As the annealing temperature increased, the major peak of 279 cm^{-1} shifted towards 282 cm^{-1} for P2, P3 and 290 cm^{-1} for P4 sample, which may be due to the smaller size effect [23]. Furthermore, an increase in intensity and shifting of strong major peak of P2, P3 and P4 Cu_3BiS_3 films with a rise in annealing temperature, indicates the transformation of phase into the crystalline [07], similar to XRD. Thus, the presence of three peaks at 169, 290 and 470 cm^{-1} [22] in the Raman spectra of P4 confirms the formation of orthorhombic Cu_3BiS_3 phase in the compound, which may be favourable for its use as an absorber layer in the solar PV cells. More importantly, the absence of a strong peak at 472, 474 cm^{-1} [24] and 237 cm^{-1} or 259 cm^{-1} [25] denies the presence of CuS, Cu_2S and Bi_2S_3 phase in P4 Cu_3BiS_3 sample. Overall, the Raman spectrum eliminates the probability of the existence of CuS, Cu_2S and Bi_2S_3 phase and confirms the formation of the dominant orthorhombic Cu_3BiS_3 phase in P4, annealed at 773K. These outcomes of XRD and Raman investigation confirm single-phase Cu_3BiS_3 can be obtained after annealing.

17.3.3 Scanning Electron Microscopy

Figure 17.3a–d shows the scanning electron microscopy (SEM) images of (P1) as-deposited and (P2, P3, P4) annealed Cu_3BiS_3 films in sulphur ambient at 573, 673 and 773 K, correspondingly. For as-deposited (P1) Cu_3BiS_3 film, the vertically united nanoplates having a thickness 40–50 nm with the formation of small clusters on the glass substrate are detected as shown in Figure 17.3a [20]. The mean diameter of nano-clusters was estimated to be 500 nm.

At 573 K annealing temperature, agglomeration of vertically aligned nanoplates takes place which results in the formation of nano-sheets along c-axis to the surface of

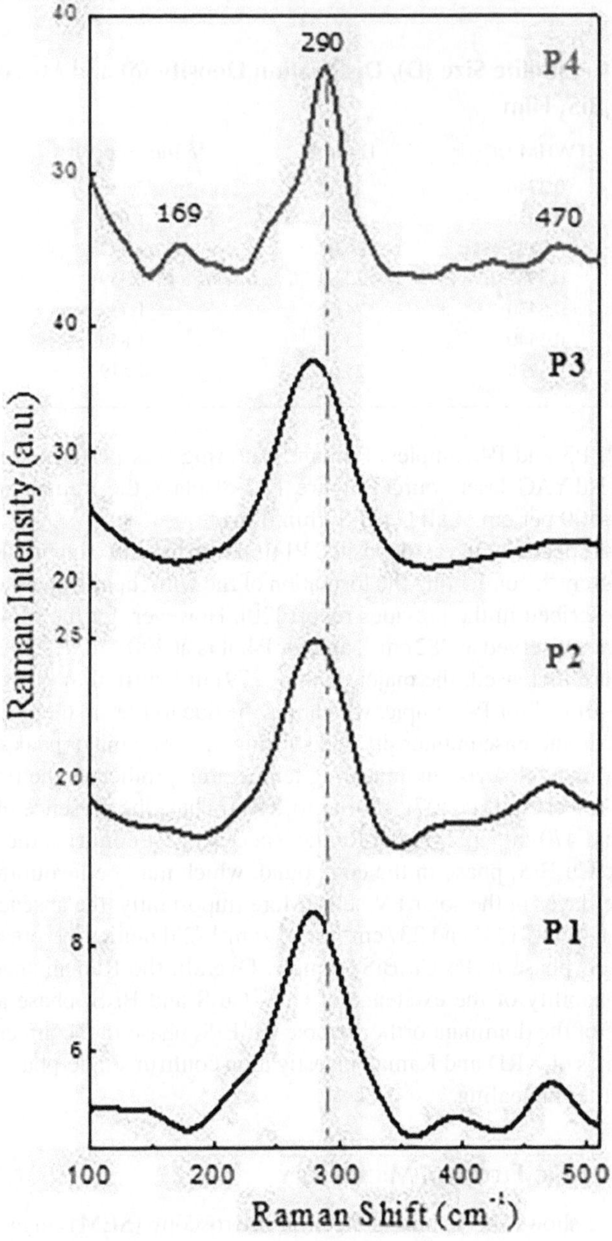

FIGURE 17.2 Raman spectrum of P1 [20], P2, P3 and P4 Cu_3BiS_3 films.

substrate (Figure 17.3b), similar to XRD results. The average thickness of nano-sheets was found to be 60 nm, indicating an increase in crystallinity. These nano-sheets have an additional surface area, which is greatly favoured for high-performance Cu_3BiS_3 thin-film solar cells. As annealing temperature continues to rise at 673 K, these

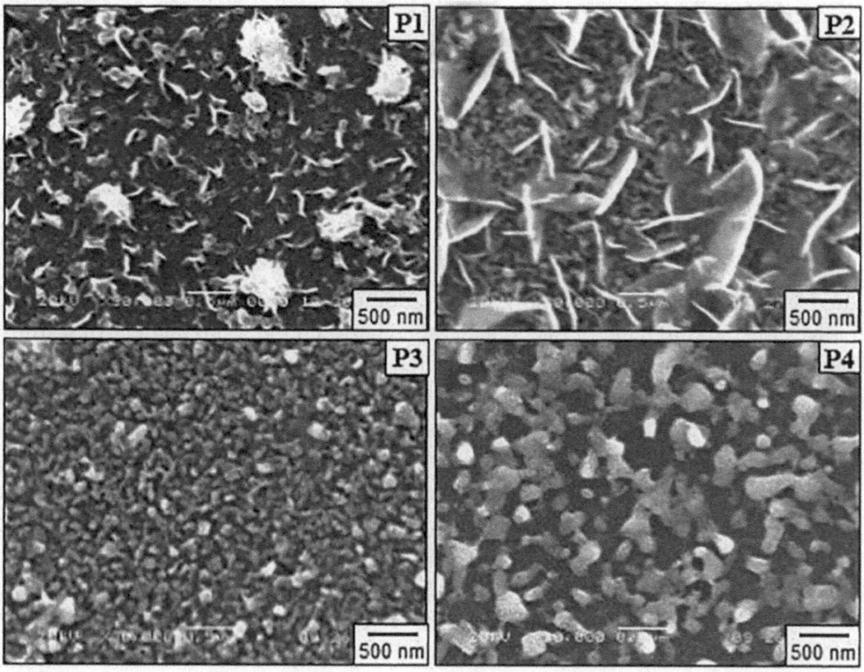

FIGURE 17.3 SEM images of P1 [20], P2, P3 and P4 Cu_3BiS_3 films.

vertically aligned nano-sheets are broken into nano-curled fibres improving the crystallinity of Cu_3BiS_3 (Figure 17.3c). Finally, these nano-curled fibres are fragmented into nano-spherical grains at 773 K with the establishment of a pure orthorhombic Cu_3BiS_3 phase as shown in Figure 17.3d. A dwell time of 30 minutes and an annealing temperature of 773 K, the film shows well-defined nano-spherical grains of an average size of 150 nm, which is superior to the previous report [3]. The morphological result of P2, P3 and P4 agrees well with the XRD results (Figure 17.1).

Both XRD and SEM characterization affirms the formation of large grain, high crystallinity Cu_3BiS_3 thin film after annealing at 773 K. The annealing temperature is lower than those utilized in common metal chalcogenide, viz. CZTS [26] and CIGS [27]. Such a low-temperature development promises the noteworthy potential of the current technique in the formation of Cu_3BiS_3.

17.3.4 Water Contact Angle Studies

The angle amongst the boundary of solid/liquid and the liquid/vapour is referred to as a contact angle. The wettability performance is described by the value of the contact angle, microscopic parameter. However, for a small value of contact angle, good wettability occurs and the surface has hydrophilic nature. On the other hand, if the wettability is poor, the contact angle will be large with hydrophobic surface nature.

The contact angle is a significant factor in surface science and its extent offers a simple and honest routine for the analysis of surface energy. The WCA of P1, P2, P3 and P4 Cu_3BiS_3 films can be determined by Young's relation [28]:

$$\theta = \cos^{-1}\left(\frac{\gamma^{sv} - \gamma^{sl}}{\gamma^{lv}}\right) \tag{17.3}$$

where θ = contact angle, γ^{sv} is the solid surface free energy, γ^{sl} is the solid–liquid interfacial free energy and γ^{lv} is the liquid surface free energy. This method involves the measurement of the contact angle between the water and Cu_3BiS_3 films. The WCA of Cu_3BiS_3 thin films is expected to be governed by chemical composition, homogeneity and surface morphology of the semiconductor materials [29]. Figure 17.4 displays the WCA images of as-deposited (P1) and annealed (P2, P3, P4) at 573, 673 and 773 K Cu_3BiS_3 thin films, correspondingly. For as-deposited (P1) thin film the contact angle was 94°, as per earlier report [20] resulting hydrophobic nature [Figure 17.4a]. This may be due to air trapped in the crevices of vertically aligned nanoplates with clusters prevent the water from adhering to the surface of P1 Cu_3BiS_3 thin film. The WCAs of annealed P2, P3 and P4 Cu_3BiS_3 films were found to be 84°, 79° and 64° [Figure 17.4b–d], respectively, confirming hydrophilic behaviour. This reduction in contact angle of P2, P3 and P4 with rise in annealing temperature attributed to (i) topographical transformation in surface morphology and (ii) nano-curled fibres and/or nano-spherical grains like architecture of Cu_3BiS_3 [Figure 17.3]. Similar behaviour of decrease in contact angle with annealing temperature has been described for TiO_2 [30], Cu_3BiS_3 [21] and ZnO [31] films deposited using chemical and sol-gel routes.

17.3.5 Optical Studies

The optical absorption of P1, P2, P3 and P4 Cu_3BiS_3 thin films has been studied, and the absorption coefficient is confirmed to be greater than 10^5/cm. The nature of the transition and the optical bandgap of all films were determined with the help of the following equation [32]:

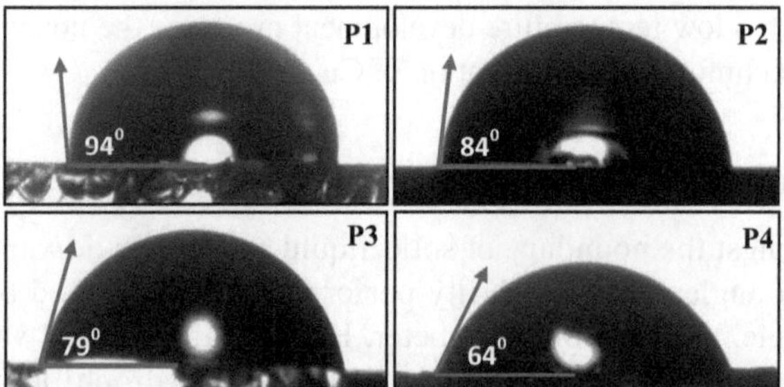

FIGURE 17.4 Contact angle images of P1 [20], P2, P3 and P4 Cu_3BiS_3 thin films.

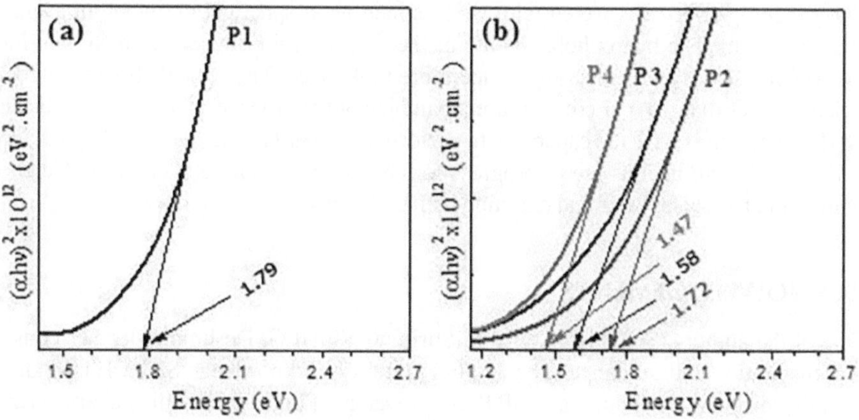

FIGURE 17.5 Plot of $(\alpha h\nu)^2$ vs energy ($h\nu$) of (a) P1 [20] and (b) P2, P3, P4 Cu_3BiS_3 films.

$$\alpha = \frac{a_0(h\vartheta - Eg)^n}{h\vartheta} \quad (17.4)$$

where α is the absorption coefficient, $h\vartheta$ is the incident photon energy, Eg is the optical bandgap of Cu_3BiS_3 material, and a_0 is the constant and is a function of the density of states near valence and conduction band edges, with $n = ½$, 2, 3/2 or 3 depending on whether the transition is direct allowed, indirect allowed, direct forbidden or indirect forbidden correspondingly.

Figure 17.5 shows the variation of $(\alpha h\vartheta)^2$ versus photon energy ($h\vartheta$) for P1, P2, P3 and P4 Cu_3BiS_3 films. However, the nature of the plot proposes a direct allowed energy bandgap nature for all thin films of Cu_3BiS_3. The optical bandgap of P1, P2, P3 and P4 Cu_3BiS_3 thin films was determined by extrapolating a straight portion of the energy axis at $\alpha = 0$. The bandgap values were established to be 1.79, 1.72, 1.58 and 1.47 eV for P1 [20], P2, P3 and P4 Cu_3BiS_3 films, correspondingly. The obtained results of bandgap values are agreeable with those reported by others for Cu_3BiS_3 [13,11,15]. It is noticed that the bandgap values decreased with increase in annealing temperature. This decrease in bandgap may be because of (i) phase transformation from amorphous to crystalline (Figure 17.1) and (ii) change in surface morphology (Figure 17.3) of the films [7]. Finally, high absorption coefficient ($\geq 10^5$/cm) and suitable bandgap (~1.47 eV) make Cu_3BiS_3 material system a great potential candidate for PV solar cell as absorber layer, similar to CIGS thin film [33].

17.4 CONCLUSIONS

In this work, the synthesis and characterization of Cu_3BiS_3 thin films deposited via non-vacuum chemical route have been investigated. The structural analysis shows that the Cu_3BiS_3 thin films are polycrystalline with an orthorhombic structure. The crystallinity of Cu_3BiS_3 films was increased as the annealing temperature increased from 573 to 773 K. The presence of Raman peaks at 169, 290 and 470 cm^{-1} in the Raman spectra of Cu_3BiS_3 confirms the formation of orthorhombic Cu_3BiS_3 phase. Micrographs show that the film surface is covered with well-defined nano-spherical grains of an average size of 150 nm. The hydrophobic nature of the Cu_3BiS_3 thin film was transformed into hydrophilic as a rise in annealing temperature took place. The optical absorption coefficient greater than 10^5 per cm and direct bandgap of the order of 1.79–1.47 eV indicates Cu_3BiS_3 thin films have the capability to perform absorber layer in the PV solar cell. Our collective determinations are in progress to fabricate PV solar cells based on Cu_3BiS_3 with a suitable buffer layer and examine its I-V characteristic.

ACKNOWLEDGEMENTS

One of the authors (SGD) is grateful to Shrimati Vimal G. Deshmukh for her consistent inspiration and guidance. The authors gratefully acknowledge SAIF, IIT Madras; Shivaji University, Kolhapur; and SP Pune University, Pune, for providing characterization facilities. Also, the authors (KJK and SGD) are thankful to PAH Solapur University, Solapur for financial support through PAHSUS/SM-2/2019-20/7461 dated 26.09.2019.

REFERENCES

1. Y. Yang, X. Xiong, H. Yin, M. Zhao, J. Han, Study of copper bismuth sulfide thin films for the photovoltaic application, *J. Mater. Sci.: Mater. Electron.* 30 (2019) 1832–1837.
2. A. Hussain, J. T. Luo, P. Fan, G. Liang, Z. Su, R. Ahmed, et al., p-type Cu_3BiS_3 thin films for solar cell absorber layer via one stage thermal evaporation, *Appl. Surf. Sci.* 505 (2020) 144597.
3 D. Colombara, L. M. Peter, K. Hutchings, K. D. Rogers, S. Schafer, J. T. R. Dufton, M. S. Islam, Formation of Cu_3BiS_3 thin films via sulfurization of Bi-Cu metal precursors, *Thin Solid Films* 520 (2012) 5165–5171.
4. N. Pai, J. Lu, D. C. Senevirathna, A. S. R. Chesman, et al., Spray deposition of $AgBiS_2$ and Cu_3BiS_3 thin films for photovoltaic applications, *J. Mater. Chem. C.* 6 (2018) 2483–2494.
5. A. Hussain, R. Ahmed, N. Ali, N. M. Salam, K. B. Derman, Y. Q. Fu, Synthesis and characterization of thermally evaporated copper bismuth sulphide thin films, *Surf. Coat. Tech.* 320 (2017) 404–408.
6. S. G. Deshmukh, A. K. Panchal, V Kheraj, Chemical bath deposition of Cu_3BiS_3 thin films, *AIP Conf. Proceed.* 1728 (2016) 020033.
7. S. G. Deshmukh, A. K. Panchal, V. Kheraj, Development of Cu_3BiS_3 thin films by chemical bath deposition route, *J. Mater. Sci: Mater. Electron.* 28 (2017) 11926–11933.
8. J. Li, X. Han, Y. Zhao, J. Li, M. Wang, C. Dong, One–step synthesis of Cu_3BiS_3 thin films by a dimethyl sulfoxide (DMSO)–based solution coating process for solar cell application, *Solar Energy Mater. Solar Cells* 174 (2018) 593–598.
9. S. Kamimura, N. Beppu, Y. Sasaki, T. Tsubota, T. Ohno, Platinum and indium sulfide-modified Cu_3BiS_3 photocathode for photoelectrochemical hydrogen evolution, *J. Mater. Chem. A* 5 (2017) 10450–10456.
10. J. Li, X. Han, M. Wang, Y. Zhao, C. Dong, Fabrication and enhanced hydrogen evolution reaction performance of a Cu_3BiS_3 nanorods/TiO_2 heterojunction film, *New J. Chem.* 42 (2018) 4114–4120.
11. P. V. Bhuvaneswari, K. Ramamurthi, R. Ramesh Babu, Effect of substrate temperature on the structural, morphological and optical properties of copper bismuth sulfide thin films deposited by electron beam evaporation method, *J. Materials Sci.: Mater. Electron.* 29 (2018) 17201–17208.

12. S. G. Deshmukh, V. Kheraj, A comprehensive review on synthesis and characterizations of Cu_3BiS_3 thin films for solar photovoltaics, *Nanotechnol. Environ. Eng.* 2 (2017) 15.
13. D. Lee, H. Ahn, S. Park, H. Shin, Y. Um, Post-annealing effects on Cu_3BiS_3 thin films grown by chemical bath deposition technique, *Nanosci. Nanotech. Lett.* 10 (2018) 1–4.
14. P. K. Nair, L. Huang, M. T. S. Nair, H. Hu, E. A. Meyers, R. A. Zingaro, Formation of p-type Cu_3BiS_3 absorber thin films by annealing chemically deposited Bi_2S_3-CuS thin films, *J. Mater. Res.* 12 (1997) 651–656.
15. M. Chakraborty, R. Thangavel, P. Komninou, Z. Zhou, A. Gupta, Nanospheres and nanoflowers of copper bismuth sulphide (Cu_3BiS_3):Colloidal synthesis, structural, optical and electrical characterization, *J. Alloys Compd.* 776 (2019) 142–148.
16. J. Yin, J. Jia, Synthesis of Cu_3BiS_3 nanosheet films on TiO_2 nanorod arrays by a solvothermal route and their photoelectrochemical characteristics, *CrystEngComm* 16 (2014) 2795–2801.
17. J. Hernadez Mota, M. Espindola Rodriguez, Y. Sanchez, I. Lopez, Y. Pena, E. Saucedo, Thin film photovoltaic devices prepared with Cu_3BiS_3 ternary compound, *Mater. Sci. Semicond. Proces.* 87 (2018) 37–43.
18. S.U. Rahayu and M.-W. Lee, The investigation of chemically deposited Cu_3BiS_3 into mesoporous TiO_2 films for the application of semiconductor-sensitized solar cells, *AIP Conf. Proceed.* 2221, 030005 (2020).
19. S. G. Deshmukh, A. Jariwala, A. Agarwal, C. Patel, A. K. Panchal, V. Kheraj, ZnS nanostructured thin-films deposited by successive ionic layer adsorption and reaction, *AIP Conf. Proc.* 1724 (2016) 020033.
20. S. G. Deshmukh, V. Kheraj, K. J. Karande, A. K. Panchal, R. S. Deshmukh, Hierarchical flower-like Cu_3BiS_3 thin film synthesis with non-vacuum chemical bath deposition technique, *Mater. Res. Express* 6(8), (2019) 084013.
21. S. G. Deshmukh, R. S. Deshmukh, V. Kheraj, Investigation of structural, optical and wettability properties of cadmium sulphide thin films synthesized by environment friendly SILAR technique, *Electric. Electron. De., Circuits, Mater.: Technol. Challen. Solutions*, John Wiley & Sons, Inc. (2021) 285–298, Doi: 10.1002/9781119755104.ch15.
22. M. V. Yakushev, P. Maiello, T. Raadik, M. J. Shaw, P. R. Edwards, J. Krustok, A. V. Mudryi, I. Forbes, R. W. Martin, Electronic and structural characterisation of Cu_3BiS_3 thin films for the absorber layer of sustainable photovoltaics, *Thin Solid Films* 562 (2014) 195–199.
23. C. Wang, C. Cheng, Y. Cao, W. Fang, L. Zhao, X. Xu, Synthesis of Cu_2ZnSnS_4 nanocrystallines by a hydrothermal route, *Jpn. J. Appl. Phys.* 50 (2011) 065003.
24. X. Shuai, W. Shen, Z. Hou, S. Ke, C. Xu, C. Jiang, A versatile chemical conversion synthesis of Cu_2S nanotubes and the photovoltaic cativities for dye-sensitized solar cell, *Nanoscale Res. Lett.* 9 (2014) 513.
25. R. Li, Q. Yue, Z. Wei, Abnormal low-temperature behavior of a continuous photocurrent in Bi_2S_3 nanowires, *Mater. Chem. C* 1 (2013) 5866–5871.
26. K. Sun, C. Yan, F. Liu, J. Huang, F. Zhou, J.A. Stride, M. Green, X. Hao, Over 9% efficient kesterite Cu_2ZnSnS_4 solar cell fabricated by using $Zn_{1-x}Cd_xS$ buffer layer, *Adv. Energy Mater.* 6 (2016) 1600046.
27. T. B. Harvey, I. Mori, C. J. Stolle, T. D. Bogart, D. P. Ostrowski, M. S. Glaz, J. Du, D. R. Pernik, V. A. Akhavan, H. Kesrouani, D. A. Vanden Bout, B. A. Korgel, Copper Indium Gallium Selenide (CIGS) photovoltaic devices made using multistep selenization of nanocrystal films, *ACS Appl. Mater. Interf.* 5 (2013) 9134–9140.
28. A. M. More, T. P. Gujar, J. L. Gunjakar, C. D. Lokhande, Oh-Shim Joo, Growth of TiO_2 nanorods by chemical bath deposition method, *Appl. Surf. Sci.* 255 (2008) 2682–2687.
29. A. M. More, J. L. Gunjakar, C. D. Lokhande, O. S. Joo, Fabrication of hydrophobic surface of titanium dioxide films by successive ionic layer adsorption and reaction (SILAR) method, *Appl. Surf. Sci.* 255 (2009) 6067–6072.

30. D. S. Dhawale, D. P. Dubal, R. R. Salunkhe, T. P. Gujar, M. C. Rath, C. D. Lokhande, Effect of electron irradiation on properties of chemically deposited TiO_2 nanorods, *Jr. Alloys Compd.* 499 (2010) 63–67.
31. J. Lü, K. Huang, X. Chen, J. Zhu, F. Meng, X. Song, Z. Sun, Reversible wettability of nanostructured ZnO thin films by sol–gel method, *Appl. Surf. Sci.* 256 (2010) 4720–4723.
32. D. S. Dhawale, A. M. More, S. S. Latthe, K. Y. Rajpure, C. D. Lokhande, Room temperature synthesis and characterization of CdO nanowires by chemical bath deposition (CBD) method, *Appl. Surf. Sci.* 254 (2008) 3269–3273.
33. C.-H. Wu, P.-W. Wu, R.-C. Hsiao, C.-Y. Hsu, Growth and characterization of high quality CIGS films using novel precursors stacked and surface sulfurization process, *J. Mater. Sci. Mater Electron.* 29 (2018) 11429–11438.

18 Self-Cleaning Antireflection Coatings on Glass for Solar Energy Applications

J. Shanthi, R. Swathi and O. Seifunnisha
Avinashilingam Institute for Home Science and Higher Education for Women

CONTENTS

18.1 INTRODUCTION

To meet present energy demand, due to sustainable availability, solar energy is widely used. Many methods and infrastructures are available to harvest solar energy, in which photovoltaic system receives considerable attention because of their widespread ability to generate energy on large scale. A number of solar panels are spread over a large area and produce clean green power. An increase in energy production demands highly efficient solar panels which not only lead towards the progress of

DOI: 10.1201/9781003220176-18

better performance and consistent solar cells but also towards the means to attenuate the outdoor factors, which reduces the efficiency of the solar panel [1].

The design of the solar cell is complex, which mainly consists of three layers. The topmost layer of the solar cell is made up of glass having antireflection property, which protects the inner components of the cell and reduces the reflection and transmits more sunlight. The middle layer of the cell consist of photoactive materials mainly made up of silicon or a thin film of CdTe, which are responsible for the production of electric charge, and a number of electrons and hole transport layers. The bottom layer, consisting of a metallic electrode with a metallic grid, creates electric current.

In the case of solar module, a huge amount of optical loss takes place due to the reflection of light, instead of being absorbed and converted to electricity. The optical loss of solar modules due to the reflection reduces the efficiency badly. High-efficiency solar panel requires high absorption of solar energy. Antireflection coating on solar cover glasses improves the efficiency of the panels by reducing the reflection of light. Reflection is usually an unfavourable phenomenon in many applications such as flat-panel displays, solar cells, photodetectors, lenses, etc. Hence, antireflection coating plays the main role in reducing the reflection. There are several ways for the introduction of antireflectance, in which coating with a modifiable refractive index is the major one. An ideal antireflecting coating has approximately zero reflectance at a particular λ, where its refractive index n_c is equal to $(n_1 n_2)^{1/2}$ and n_1 and n_2 are the refractive indices of air and substrate, respectively.

Cleaning of the dusty panel is cumbersome and time-consuming and causes scratches on the coating. So it is necessary to improve the performance of glass coatings for efficient yield all the days. Non-wettable coating will act as anti-dust coating and reduce the accumulation of dust particles on it. This makes the surface highly water repellent by rolling down with dust particles, and the panel surface remains clean for longer time duration.

18.1.1 Theoretical Aspects of Antireflection and Non-Wettability

18.1.1.1 Antireflection

The concept of antireflection coating was initially introduced by Lord Rayleigh when he observed a rise in the optical transparency due to tarnishing instead of reduction and was led to the concept of antireflection coating. It was then carried out by Fraunhofer in 1817 by producing actual antireflection coating by engraving a surface in an atmosphere of sulphur and nitric acid vapour. The antireflection coating credibly increases the transmission of light by reducing reflection. The transition of light depends on the medium travelled, and the medium is optically characterized by the factor 'n'. The Fresnel provides the mathematical model of the reflection of the coating. According to Fresnel, the two conditions for obtaining an ideal antireflection coating is

1. $n_c = (n_{air} n_s)^{1/2}$, where n_c is the refractive index of the coating and n_{air} and n_s are the refractive index of air and substrate, respectively.
2. The film thickness depends on the quarter of the wavelength.

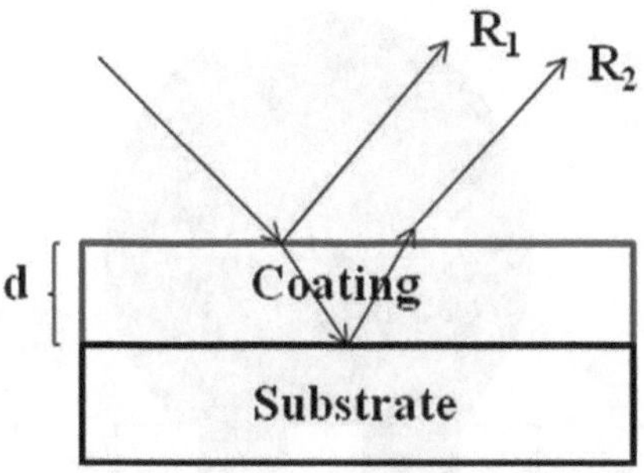

FIGURE 18.1 Light propagation through ARC.

Figure 18.1 shows the illustration of an idyllic single-layer antireflection coating. In the case of ideal antireflection coating, the reflection would be absent, because of the destructive interference between coating-substrate or air-coating interfaces.

Fresnel theory also suggests that the antireflection property also depends on the s- and p-polarization of the light. By tuning the refractive index and thickness of the coating, a minimum reflection can be obtained. When we consider a single-layer coating, to achieve antireflection property, the n of the coating should be 1.22 as the glass has a value of 1.5. But the natural materials of low '*n*' are rare and costly to obtain. Hence, one of the best ways to obtain the antireflection property is by inducing porosity to the coating so as to reduce the '*n*' by controlling the quantity and volume of introduced nano-pores. In the case of multilayer coating, the antireflection property is achieved by applying gradient to reduce the refractive index gradually from substrate to the air.

18.1.1.2 Non-Wettability

Wettability is an imperative asset of solid surface. It entrusts both the surface chemistry and surface topography of the material. It describes the ability of a fluid phase to wet a solid surface. The non-wetting character was derived from the lotus leaf effect, where the surface resists the spreading of water by rolling off to spherical droplets to run away from the surface. The non-wettable behaviour of the surface can be achieved by inducing surface roughness. The surface with a contact angle greater than 90° is known as hydrophobic (hydrophobic) and those having contact angle (CA) less than 90° is known as hydrophilic (hydrophilic). The wetting on rough surfaces is explained by two theories:

1. Wenzel Theory
2. Cassie–Baxter Theory

The Wenzel state (Figure 18.2) describes the surface that is completely wet by the liquid, and the degree of wetness varies with respect to the roughness of the surface. The state of wettability can be stated by

$$\text{Cos}\,\theta_w = r\,\text{Cos}\,\theta_r \qquad (18.1)$$

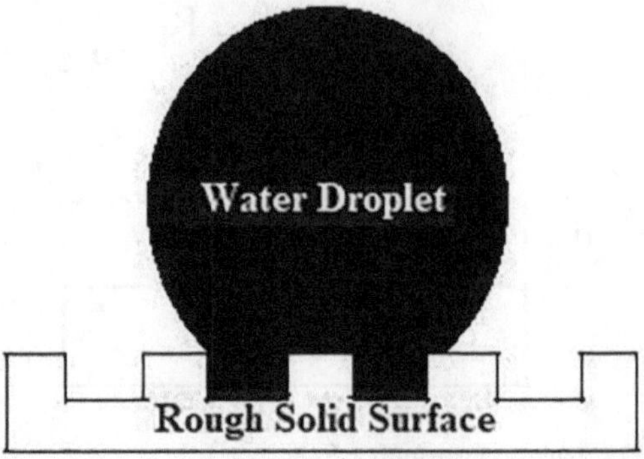

FIGURE 18.2 Wenzel state.

where θ_w is the Wenzel CA, r is the surface roughness factor, and θ_r is Young's CA.

However, the surface's fully wetting texture is less favourable for hydrophobicity. This was then modified by Cassie and Baxter. They explain hydrophobic behaviour of the surface clearly. According to Cassie–Baxter, more stable state is achieved when air cavities are formed in between the surface (Figure 18.3). The effect of trapped air cavities results in a composite solid–liquid–air interface with heterogenic behaviour. The partial wetting of the surface is represented by

$$\mathrm{Cos}\,\theta_{\mathrm{CB}} = r_f f\,\mathrm{Cos}\,\theta_r + f - 1 \tag{18.2}$$

where

θ_{CB} is the Cassie–Baxter CA, r_f is the actual wetted area divided by the projected wetted area of the surface, and f is the fraction of the projected area of the surface that is wetted by the liquid.

Both the theories explain the relationship between surface roughness water contact angle (WCA) and also the role of air cavities. The CA of the surface increases with an increase in surface roughness. The water-repelling surfaces have a wide variety of applications such as self-cleaning, waterproof textiles, oil–water separation, etc. [2,3].

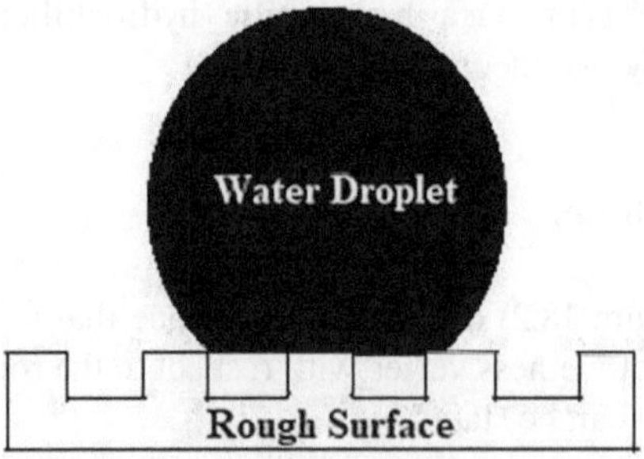

FIGURE 18.3 Cassie–Baxter state.

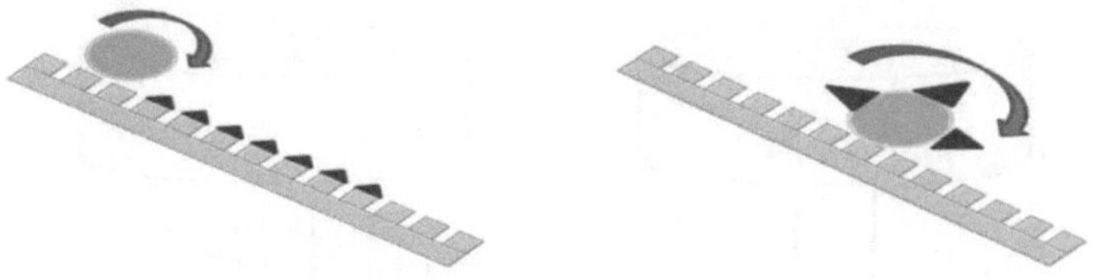

FIGURE 18.4 Self-cleaning performance of a rough surface.

Figure 18.4 represents the self-cleaning performance of a non-wettable surface. During the self-cleaning performance, the water droplet rolls over the contaminated surface and adsorbs the dirt particle and moves away from the surface, which results in a cleaned surface.

18.1.2 Fabrication Technique of Hydrophobic Antireflection Coatings

Over the years, various materials have been synthesized in the form of thin films due to their future technological significance and scientific concern in their properties. There is a rapid evolution of deposition technology because of the high demand of improved quality thin films. From the various deposition techniques, sol–gel processing is a good relevant method for the fabrication of hydrophobic antireflection coating. Sol–gel process is the best way of synthesizing materials from sol by transforming them into gel during a sufficient aging period. Most of the sol–gel synthesis has mainly evolved from the controlled hydrolysis and polycondensation of complex in the present suitable solvent. During sol–gel process, sol is transformed to gel by condensation. It has a variety of merits such as tunable microstructure, capability to control the stoichiometry of precursor solution, simplicity in compositional adjustment, and simple and inexpensive equipment. A number of techniques are available to deposit the sol gels to the substrate, among them spin-coating and dip-coating techniques are preferred due to low cost.

18.1.2.1 Spin-Coating Technique

Spin-coating is one of the major techniques for film deposition from liquid precursors used in the microelectronics industry. During spin-coating, a thin layer of film is formed by the evaporation and spinning due to the centrifugal force. There are four stages for spin-coating such as deposition, evaporation, spin-off and spin-up which are shown in Figure 18.5. At the stage of deposition, an excess amount of fluid is removed from the surface of the substrate. An outward radial flow of liquid starts in the spin-up stage due to the centrifugal force. Further in the spin-off stage, surplus liquid flows to the edge and leaves as droplets. At the ending stage, the primary mechanism of thinning is taken over by evaporation [4].

18.1.2.2 Dip-Coating Technique

Dip-coating is the simple and easiest method for the synthesis of the thin from chemical solutions at a fast rate with the maximum degree of control. It is the one of the popular ways to deposit the thin film for large area. In this technique, the substrate

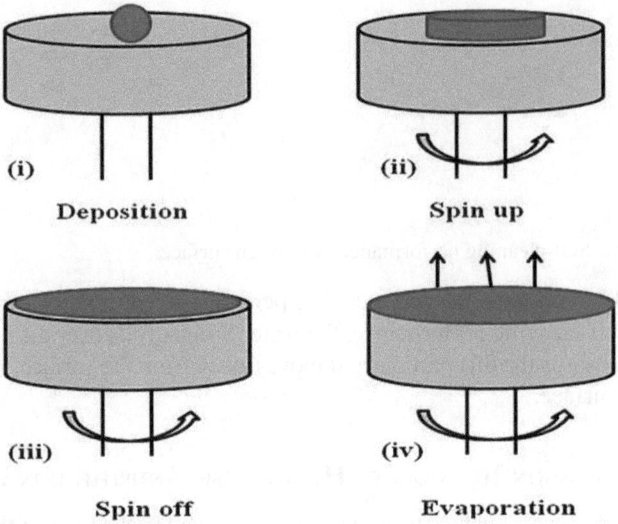

FIGURE 18.5 Stages of spin-coating.

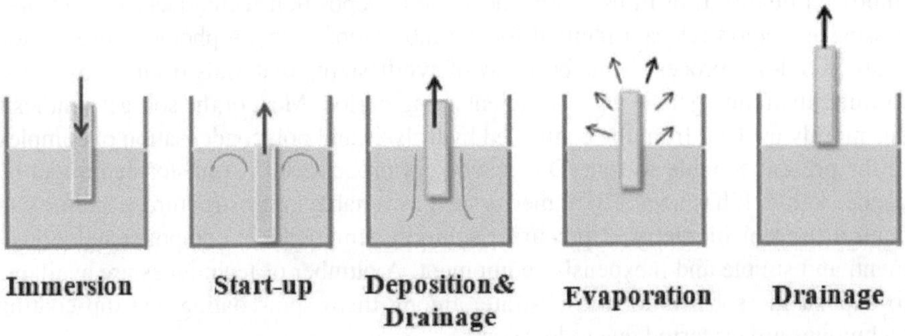

FIGURE 18.6 Stages of dip-coating.

is dipped in the corresponding sol, with a controlled withdrawal speed. The dip-coating involves various stages such as immersion, start-up, deposition, evaporation and drainage, which are shown in Figure 18.6 [5–7].

18.1.3 Recent Progress towards the Self-Cleaning Antireflection Coatings

The researchers showed great interest towards the development of non-wettable anti-reflection coating because of its wide variety of applications in solar and optical fields. In 2020, Zhongfeng Ji and Ji Liu & Du [8] developed a transparent self-cleaning coating (GPMD–DTDA/PFTS) based on the sol–gel method. The coating shows good optical performance of 95% with a contact angle of 107°. Besides, the coating has good mechanical stability with a hardness of 8H.

In 2020, Attila Abraham [9] synthesized a microporous hybrid silica coating using sol–gel method. The hybrid coating prepared using TEOS/MTES/CTAB shows an excellent transmittance of ~99% with a minimum hydrophobic contact angle of 90°. A fumed silica/fluorinated polyacrylate–based coating was reported by Fuchu Zhao [10]. Zhao et al showed more than 95% transmittance with less than 4% haze. Moreover, the composite latex has good hydrophobic behaviour.

Li et al. [11] have studied the synthesis and characterization of the hydrophobic antireflection coating for solar cell cover glasses. The authors developed the coating via sol–gel dip-coating method using SiO_2/PTFE sol. The coating showed a transmittance of 97.8% in the visible region with a maximum contact angle of 104°. They observed that the WCA of the coating increases with the increase of PTFE.

Several works were reported during the past few years related to transparent hydrophobic coating. But as we go through the works, it can be clear that most of the works are based on fluorinated materials and is very hazardous to the environment and ecosystem.

Here, we report less toxic transparent hydrophobic coating based on spin-coating method. The acid-catalysed silica-based coating exhibited maximum optical transmittance of 92.6% in the visible region with hydrophobic property.

18.2 FABRICATION OF HYDROPHOBIC ANTIREFLECTION COATING

18.2.1 Materials

Methyltrimethoxysilane (MTMS, Sigma Aldrich), Pluronic F-127 (Sigma Aldrich), Ethanol (Hayman) and HCl (Merck) were used for the preparation of sol. All the chemicals were used as received. Distilled water was preferred for the preparation of hydraulic sol.

18.2.2 Preparation of Sol and Deposition of Coating

Initially, the ethanol was mixed with an appropriate amount of acidic catalyst HCl and distilled water by using a magnetic stirrer for a while. Then a measured amount of Pluronic F-127 was added to the hydraulic solution and stirred. After proper stirring sufficient amount of silica precursor MTMS was poured into the pluronic sol and stirred for 2 hours continuously in order to obtain a well-mixed sol. After that, the sol was kept at room temperature for aging. The sol after sufficient aging period was dropped on the impurity-free glass substrate using spin-coating unit. The evenly deposited coated substrate was annealed in muffle furnace for an hour.

The coated substrates were annealed at 250°C to eliminate the volatile components. The optical transparencies of the substrates for different molarity of PF-127 were analysed by UV–Visible spectrophotometer and were compared with the transmittance value obtained from a mathematical calculation using pyranometer reading. The presence of functional groups in the coating was observed by Fourier-transform infrared spectroscopy (FTIR) spectrometer. The non-wetting behaviour of the coated substrate was determined by water contact angle.

18.3 RESULTS AND DISCUSSION

The concentration of the polymeric porogen PF-127 was optimized by measuring the transmittance of MTMS/PF-127 coating with different concentrations of PF-127.

18.3.1 Optical Performance of the Coating

Figure 18.7 shows the transmittance of the MTMS/PF-127-based coating for different molarity of PF-127. From Figure 18.7, the pristine MTMS coating exhibits the greatest transparency of 87.2% in the visible region and was less than that of a bare glass of transparency of 90.5%. The transparency of the pristine MTMS increased by ~5% in the visible region with the addition of polymeric porogen PF-127. The concentration of the PF-127 for the MTMS/PF-127-based coating was optimized as 0.1 g, where the coating shows maximum transparency of 92.6% at 482 nm, which is greater than that of uncoated substrate.

The mathematical calculations were followed in accordance with Duffie and Beckman's model [12]. All the optical parameters used for the calculations were simply assumed to be independent of the wavelength of the solar radiation. The solar insolation reading for the coated and uncoated glass slides was measured using a pyranometer on a sunny day. The transmittance of the substrate for different angles of incidence and refractions is tabulated in Table 18.1.

The transmittance values obtained from UV–Visible spectra and mathematical calculations are similar, as shown in Table 18.2.

18.3.2 Structural Determination Using FTIR Spectroscopy

The presence of functional groups in the prepared MTMS/PF-127 coating was identified by the FTIR analysis. Figure 18.8 shows the FTIR spectra of the coating. A long sharp absorption band appeared at 1,086.62 cm^{-1} specifying an asymmetric stretching vibration of Si–O–Si bond [13,14], where the band emerges at 837.99 cm^{-1}

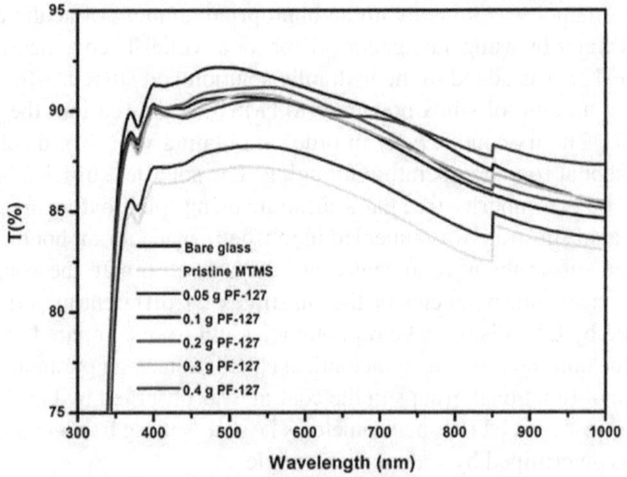

FIGURE 18.7 UV–Visible transmittance spectra of MTMS/PF-127-coated substrate.

TABLE 18.1
Transmittance Value of Coated and Uncoated Substrates Corresponding to Different θ_1 and θ_2

Angle of Incidence (θ_1)	Angle of Refraction (θ_2)		Transmittance (T_{OUT}) (%)	
	Uncoated Glass	MTMS/PF-127 Coated Glass	Uncoated Glass	MTMS/PF-127 Coated Glass
45°.0′	28°.12′	31°.01′	88.39	92.51
40°.26′	25°.61′	28°.20′	87.22	93.35
36°.70′	23°.74′	26°.12′	90.29	91.59
34°.11′	21°.99′	24°.17′	92.62	93.87
33°.20′	21°.48′	23°.61′	90.46	93.23
33°.71′	21°.99′	24°.17′	90.83	92.86

TABLE 18.2
A Comparison on the Optical Transparency of the Substrate using Experimental and Mathematical Method

Sample	UV–Visible Transmittance (%)	Calculated Transmittance (%)
Uncoated glass	90.5	90.3
MTMS/PF-127 (0.2:0.1)	92.6	92.9

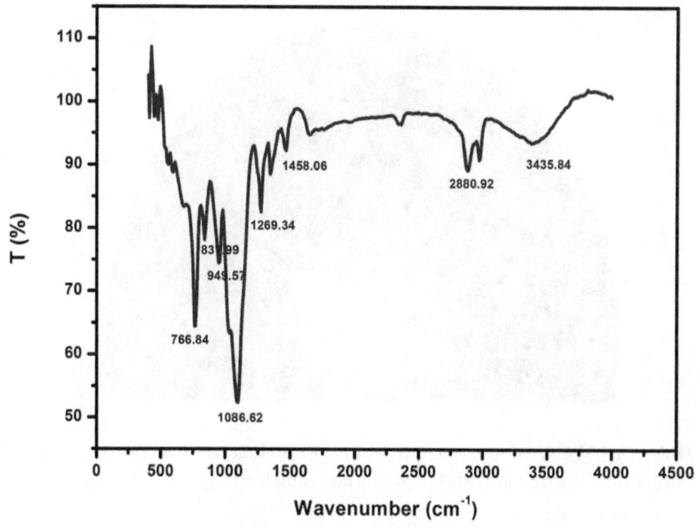

FIGURE 18.8 Fourier transform infrared spectra of MTMS/PF-127-based coating.

indicating the Si-C bond [15,16]. These two bands verify the presence of silica group in the coating. The bands formed at 949.57 and 1458.06 cm^{-1} remark the occurrence of C–H bending vibration [17]. The PF-127 was characterized by the absorption peak 2880.92 cm^{-1}, which represents C-H stretching vibration [18]. In addition, the band at 3,435.84 cm^{-1} was accountable for the stretching vibration of –OH in PF-127 [19].

18.3.3 Wetting Property of the Coating

The wetting property of the prepared MTMS/PF-127–based coating was analysed by measuring the water contact angle. The surface roughness of the coating (Figure 18.9b) plays a crucial role in the wettable behaviour of the surface [20,21]. Figure 18.9a shows the WCA of MTMS/PF-127 coating measured at room temperature.

Surface roughness plays a major role in the non-wettable behaviour of the coating. With an increasing roughness, water droplets cling to the surface and tend to roll off. Here the surface roughness of the MTMS/PF-127 coating was determined by using 3D laser profilometry and was observed as 28 μm. The high surface roughness of the prepared coating ensures the hydrophobic behaviour with CA of 94°. Figure 18.9c shows the water droplet on the dusty-coated and -uncoated glass substrate. The MTMS/PF-127-coated glass substrate with dust particles clings together with the water droplet, ensuring the self-cleaning ability of the coating.

18.4 CONCLUSION

Transparent antireflection coating has a wide variety of applications in optical and optoelectronic devices. In the present chapter, the theory and the preparation of the non-wettable transparent coating were described in detail. In the solar field, the antireflection coating enhances the efficiency of the solar panel, and the self-cleaning ability which encourages the high performance of the coating. The characteristic

(a)

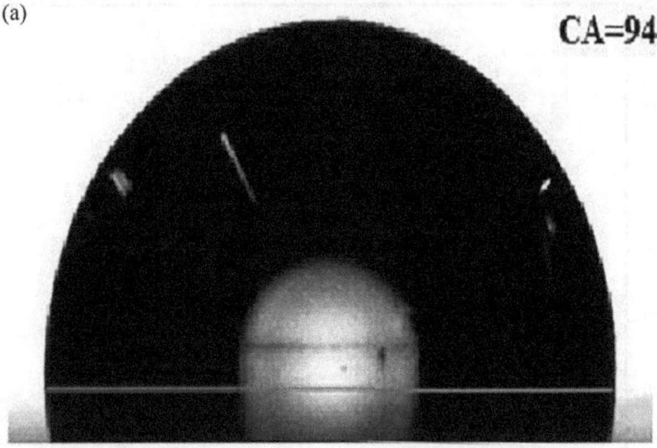

FIGURE 18.9 (a) WCA of the coated film, (b) surface profile of the coated film, and (c) water droplet on the uncoated and coated glass substrate.

(Continued)

(b)

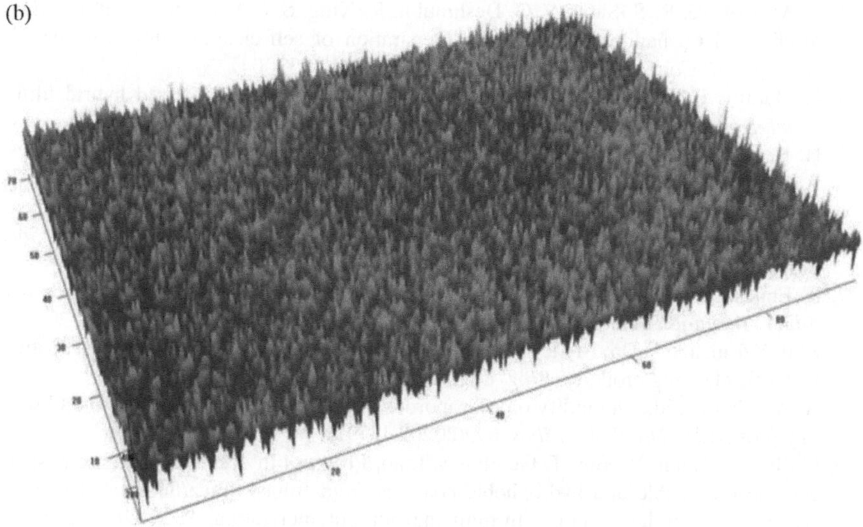

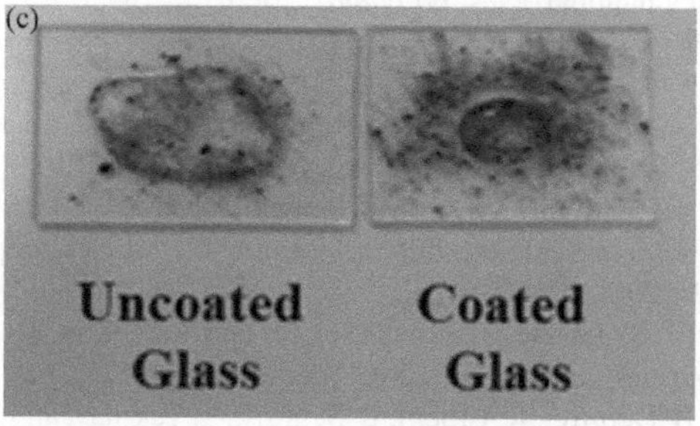

FIGURE 18.9 (*CONTINUED*) (a) WCA of the coated film, (b) surface profile of the coated film, and (c) water droplet on the uncoated and coated glass substrate.

bands present in the coating were confirmed by FTIR analysis. The prepared MTMS/PF-127-based coating exhibited good optical transparency of 92.6% in the visible region with 94° of water contact angle. In conclusion, the prepared antireflective coating can be applied on solar panels and also in optical devices for better performance.

REFERENCES

1. M. Augusto, M. Lopes, D. Jesus, G. Timò, and C. Agustín-sáenz, Solar energy materials and solar cells anti-soiling coatings for solar cell cover glass: Climate and surface properties in fluence, *Sol. Energy Mater. Sol. Cells*, 185(2018), 517–523.
2. C. Agustín-Sáenz, M. Machado, O. Zubillaga, and A. Tercjak, Hydrophobic and spectrally broadband antireflective methyl-silylated silica coatings with high performance stability for concentrated solar applications, *Sol. Energy Mater. Sol. Cells*, 200(2018), 109962.

3. A. M. Kokare, R. S. Sutar, S. G. Deshmukh, R. Xing, S. Liu, and S. S. Latthe, ODS-Modified TiO_2 nanoparticles for the preparation of self-cleaning superhydrophobic coating, *AIP Conf. Proc.*, 1953(2018) Doi: 10.1063/1.5033004.
4. F. Mammeri, Nanostructured flexible PVDF and fluoropolymer-based hybrid films, *Front. Nanosci.*, 14 (2019), 67–101. DOI: 10.1016/B978-0-08-102572-7.00003-9.
5. H. K. Raut, V. A. Ganesh, A. S. Nair, and S. Ramakrishna, Anti-reflective coatings: A critical, in-depth review, *Energy Environ. Sci.*, 4(2011), 3779–3804.
6. M. Alhashemi, O. Albadwawi, and I. Almansouri, Optical simulations and analysis for single and double layer antireflection coatings on Si solar cells, *Sustain. Dev. Soc. Responsib.*, 2 (2020), 149–155.
7. S. Ebnesajjad and A. H. Landrock, Adhesive applications and bonding processes, *Adhes. Technol. Handb.*, 424(2015), 206–234.
8. Z. Ji, Y. Liu, and F. Du, Facile synthesis of solvent-free and mechanically robust coating with self-cleaning property, *Prog. Org. Coatings*, 149(2020), 105923.
9. A. Ábrahám et al., Durability of microporous hybrid silica coatings: Optical and wetting properties, *Thin Solid Films*, 699(2020), 137914.
10. F. Zhao, J. Guan, W. Bai, T. Gu, and S. Liao, Progress in organic coatings transparent, thermal stable and hydrophobic coatings from fumed silica/fluorinated polyacrylate composite latex via in situ miniemulsion polymerization, *Prog. Org. Coatings*, 131(2018), 357–363.
11. X. Li, X. Du, and J. He, Self-cleaning antireflective coatings assembled from peculiar mesoporous silica nanoparticles, *Langmuir*, 26(2010), 13528–13534.
12. J. A. Duffie and W. A. Beckman, *Solar Engineering of Thermal Process*, Wiley, 4th Edition, (2013) 3–10.
13. A. B. D. Nandiyanto, R. Oktiani, and R. Ragadhita, How to read and interpret ftir spectroscope of organic material, *Indones. J. Sci. Technol.*, 4(2019), 97–118.
14. F. Chi, D. Liu, H. Wu, and J. Lei, Mechanically robust and self-cleaning antireflection coatings from nanoscale binding of hydrophobic silica nanoparticles, *Sol. Energy Mater. Sol. Cells*, 200(2019), 109939.
15. S. A. Mahadik, M. S. Kavale, S. K. Mukherjee, and A. V. Rao, Transparent superhydrophobic silica coatings on glass by sol-gel method, *Appl. Surf. Sci.*, 257(2010), 333–339.
16. A. Yildirim, G. B. Demirel, R. Erdem, B. Senturk, T. Tekinay, and M. Bayindir, Pluronic polymer capped biocompatible mesoporous silica nanocarriers, *Chem. Commun.*, 49(2013), 9782–9784.
17. B. Mali, S. N. Moharil, V. Mhasal, and M. B. Narkhede, Drug-excipient interaction study of tramadol HCl with polymers, *World J. Pharm. Res.*, 6(2017), 848–861.
18. T. Al Kayal et al., Evaluation of the effect of a gamma irradiated DBM-Pluronic F127 composite on bone regeneration in Wistar rat, *PLoS One*, 10(2015), 1–19.
19. S. Fu and H. Yu, Redox-sensitive Pluronic F127-tocopherol micelles : Synthesis, characterization, and cytotoxicity evaluation, *Int. J. nanomed.*, 12(2017), 2635–2644,.
20. P. Ragesh, V. Anand Ganesh, S. V. Nair, and A. S. Nair, A review on self-cleaning and multifunctional materials, *J. Mater. Chem. A.*, 2(2014), 14773–14797.
21. S. G. Deshmukh, V. Kheraj, K. J. Karande, A. K. Panchal, and R. S. Deshmukh, Hierarchical flower-like Cu_3BiS_3 thin film synthesis with non-vacuum chemical bath deposition technique, *Mater. Res. Express*, 6(8) (2019), 084013. Doi: 10.1088/2053-1591/ab283f.

Index